D1273065

STATE OF MICHIGAN
JUNG S.
PARK
ENGINEER
NO.
30616
LICENSED PROFESSIONAL ENGINEER

Elements of

COMPUTER PROCESS CONTROL

with Advanced Control Applications

Elements of Computer Process Control With Advanced Control Applications

PRADEEP B. DESHPANDE
Professor of Chemical Engineering
University of Louisville

RAYMOND H. ASH
Manager, Control Systems Technology
Procter and Gamble Company

Instrument Society of America

Elements of COMPUTER PROCESS CONTROL with Advanced Control Applications

INSTRUMENT SOCIETY of AMERICA
67 Alexander Drive
P.O. Box 12277
Research Triangle Park
North Carolina 27709

ISBN: 87664-449-3

Library of Congress Catalog Card No.: 80-82117

Deshpande, P. B. and Raymond H. Ash

Elements of Computer Process Control with Advanced Control Applications.

Research Triangle Park, N.C.: Instrument Society of America

8007 800515

Design and Production by Publishers Creative Services Inc.

Second Printing 1982

This book is dedicated to my parents,
who gave so much and asked for so little (PBD),
and to our wives, Meena and Joanne

Preface

This book is intended for those who are interested in learning the basics of computer process control. To be of sufficient value, the reader should have had a first course in linear control theory. The person should have some familiarity with Laplace transform methods for solving differential equations, block diagram representations of process control systems, the determination of transient open-loop/closed-loop responses and stability of process control systems, and the frequency-response method (based on Bode plots) for finding the tuning constants of conventional controllers.

There are several ways in which computers are used in control applications. In one approach, the digital computer replaces the analog controllers of the conventional control system. The controller equation is executed by means of a digital computer program, and thus, in theory at least, many analog controllers can be replaced by a suitable computer program. This approach to control is known as direct digital control. In this text, we are primarily concerned with the design and implementation of direct digital control systems which we refer to as, simply, computer-control systems.

A second type of application of digital computers in control applications is known as supervisory control. In this approach, a master computer program constantly calculates and updates the set points of the analog controllers based on some predetermined operating strategy. The computer hardware adjusts the set points of the analog controllers so as to maintain the operation of a plant at some optimal level. In another type of application, the computer performs the information function. In this approach, the computer gathers the plant data, processes it, stores it, and communicates it to the operators and other personnel; no direct changes in the process operation are made by the computer in this instance.

The pictorial representation of a typical computer control loop is shown in Figure 1. Of course this loop may be only one of hundreds of such loops in a plant. The computer samples the measured variable by means of a device called an analog-to-digital converter and compares it with the set point so as to generate a measure of error. A digital computer program acts on this error, the result of which is the control command. This command is converted into a continuous signal by means of a device called a digital-to-analog converter and fed to the control valve. This strategy is repeated over and over so as to reduce the error to an acceptable level. As control engineers, we must be able to design, analyze, and implement such computer-control systems.

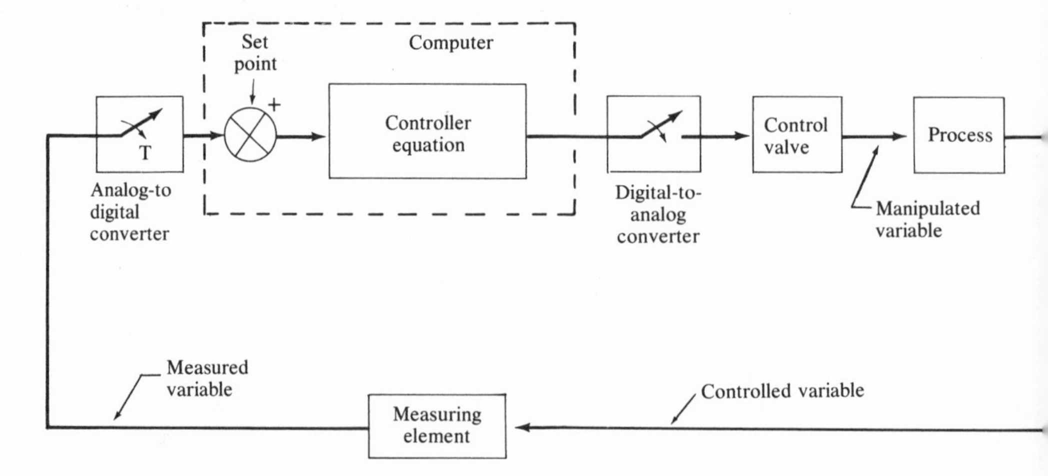

Figure 1
Typical Computer Control System

The book begins with a chapter on an overview of linear control theory and its application to conventional process-control problems. Each subsequent chapter deals with a distinct element of a computer-control loop. The analysis of computer control loops requires a background in Z-transforms. Early in the book, the reader is introduced to the Z-transform and its usefulness in the study of computer-control systems. An understanding of the first twelve chapters should enable the reader to design and analyze computer-control systems. The remaining chapters in the book are devoted to the design and applications of advanced control concepts. These concepts are particularly valuable and fairly straightforward to implement when a control computer is available.

If used as a text, the material in the book should be adequate for a one-semester course in computer process control. Instructors in electrical and mechanical engineering may also find the text suitable for a course in sampled-data control theory. At the University of Louisville, this material has been used in a class consisting of chemical, mechanical, and electrical engineering students, and the response of the students has been quite favorable.

We wish to acknowledge the support of the Procter and Gamble Company and that of Professor Charles A. Plank, Chairman of the Department of Chemical and Environmental Engineering at the University of Louisville, in the preparation of this book. Special thanks are also due to Mrs. Mary Gerstle, senior secretary of the same department, who so patiently typed page after page full of mathematical material. Professor Darrel Chenoweth provided some of the problems on Z-transforms, and some of the computer programs were developed by graduate students R. Alan Schaefer, F. Joseph Schork, and P. C. Gopalratnam. We thank Professor P. M. Christopher for his editorial comments on some of the material in the book.

Pradeep B. Deshpande
Raymond H. Ash

Louisville, Kentucky
Cincinnati, Ohio
August 1980

ADDENDUM TO PREFACE, SECOND PRINTING

Since we turned the final manuscript in to the publisher for first printing some useful advances in advanced control methodology have taken place. To keep the reader abreast of these recent developments we have, with the consent of the publisher, cited the relevant papers at the end of the appropriate chapters most closely related to the topics in which recent developments have taken place.

Louisville, Kentucky
Cincinnati, Ohio
February 1982

Pradeep B. Deshpande
Raymond H. Ash

Contents

Part II
DESIGN AND APPLICATION OF ADVANCED CONTROL CONCEPTS

PART I

Elements of Computer Process Control

CHAPTER 1

Review of Conventional Process Control

In this chapter we will review and highlight some of the important aspects of linear control theory and its application to conventional process control. Linear control theory forms the basis for the development of computer control concepts which is the subject of this book. A thorough review of these concepts will greatly help the reader in understanding the subsequent material on computer control.

1.1. Introduction to Process Control

Within the context of chemical engineering applications, we may define the term *process* as a collection of interconnected hardware (e.g., tanks, pipes, fittings, motors, shafts, couplings, and gauges), each doing its part toward an overall objective of producing some product or a small group of related products. This definition also fits the viewpoint of the *process engineer* whose job is to design this hardware. From the standpoint of the control engineer, the focus is on *physical variables* (e.g., temperature, pressures, levels, flows, voltages, speeds, positions, and compositions) rather than on hardware. The control engineer is interested in knowing how these variables affect one another and how they change with time. The fundamental objective in process control is to maintain certain key process variables as near their desired values, called *set points*, as much of the time as possible. From among all the process variables, certain variables are chosen as the key process variables because maintaining them at specified values means the production objectives will be satisfied. The production objectives are

—To achieve desired production, i.e. throughput.

—To produce at acceptable cost.

—To produce material of acceptable quality.

—To do all of the above in a safe manner with minimum harm to the environment.

In dealing with process control concepts, it is useful to categorize the physical variables according to the following classification:

—*Outputs*—these are the key process variables to be maintained at desired values, that is, controlled.

—*Inputs*—these are the variables that, when changed, cause one or more outputs to change. These inputs may be further subclassified as control inputs and disturbance inputs.

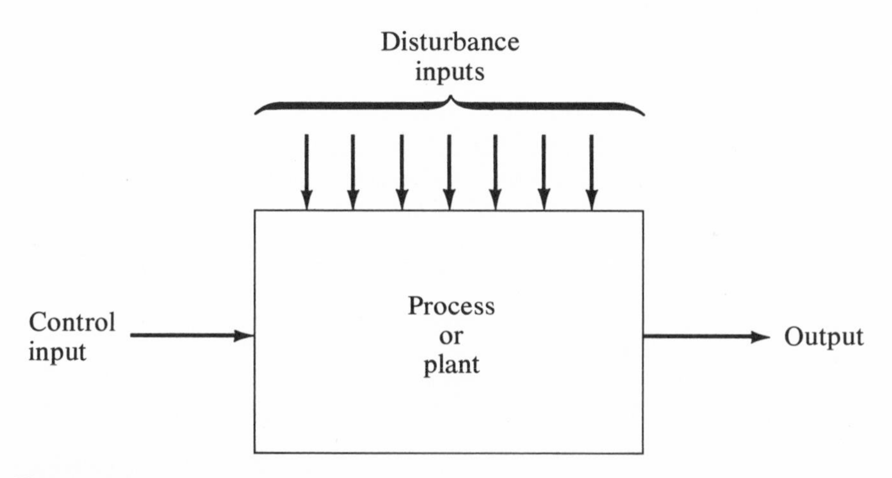

Figure 1.1
Typical Block Diagram

—a) *Control Inputs.* These are also called *manipulated variables.* These variables are changed by the controller to bring and maintain the outputs at set points. For example, the flow of process streams is often changed by actuators, such as control valves or variable speed pumps, so as to control certain outputs.

—b) *Disturbance Inputs.* All other process variables that affect the outputs in any way are called *disturbance inputs* or *process loads.* These disturbance inputs cause unwanted changes in process outputs.

1.2. Process Dynamics and Mathematical Models

The objective of control strategy development is to determine how to change a control input so as to correct a deviation between the desired value and the actual value of the process output. This development is complicated by the inertia or lag inherent in most processes. The presence of lag means that if a sudden change is made in a control input, the output will not follow immediately. There will be a time lag before the output reaches its new value.

To do a good job of controlling a process we need to know how control inputs affect outputs, quantitatively. If a control input is changed by a known amount, we need to know

—How much the output will ultimately change and in which direction.

—How long it will take for the output to change.

—What trajectory the output will follow; that is, what the pattern of output variation with time is.

An answer to these questions is provided by a dynamic mathematical model of the process. This model gives a functional relationship between an input and an output of a process and may be defined as any mathematical expression or formula that when values of the independent variables (i.e., inputs) are given, enables prediction of the value of the dependent variable (i.e., output) by plugging into the formula and calculating the result. Thus a mathematical model is an expression from which physical behavior can be predicted.

A schematic representation of the cause-and-effect relationship between input and output is referred to as a *block diagram*, shown in Figure 1.1. Generally, there is one block for each piece of the entire process. A single block usually has one output, one control input, and several identifiable and/or measurable disturbance inputs. There may also exist several unidentifiable or unknown disturbance inputs that affect the output. The block diagram also represents the quantitative cause-effect relationship (i.e., mathematical model) between an input and an output.

Any given physical variable in a process can be any one of the types shown in Figure 1.1; that is, an output or control input of one process block can be a disturbance input to another process block.

Mathematical models are derived from the laws of physics and chemistry. Some examples are

—Conservation of mass and energy (i.e., unsteady-state material and energy balances).

—Newton's laws of motion.

—Kirchoff's laws of electrical circuits.

—Equations of state for gases.

These models usually take the form of differential equations involving time rates of change of variables as well as the variables themselves. To predict the output behavior resulting from a known change in input, these differential equations must be solved, but they are more difficult to solve than algebraic equations.

Some processes are linear, that is, they are described by linear differential or algebraic equations, and the principle of superposition applies. Many others are nonlinear, and these are described by complex nonlinear differential equations. However, the behavior of the latter around some normal operating level can be approximated by linear differential equations. For processes that are under control the interest is in their behavior around some steady-state operating level, and this approximation is usually applicable.

Linear differential equations can be converted, by Laplace transforms, into algebraic equations from which transfer functions can be

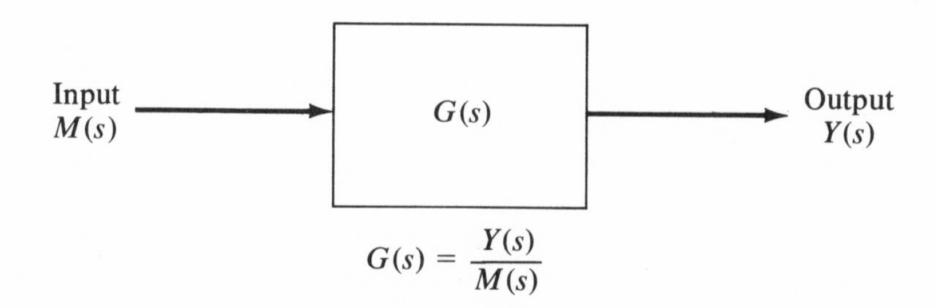

$$G(s) = \frac{Y(s)}{M(s)}$$

Figure 1.2
Transfer Function Shown on a Block Diagram

derived. Since Laplace transforms convert a differential equation into an algebraic equation, the solution is considerably simplified.

A *transfer function* $G(s)$ is a mathematical expression that represents the ratio of the Laplace transforms of a process output $Y(s)$ to that of an input $M(s)$ as shown in Figure 1.2. Recall that the output $Y(t)$ and the input $M(t)$ are expressed as deviation variables, that is,

$$Y(t) = y(t) - y_{\text{steady state}} \tag{1.1}$$

and

$$M(t) = m(t) - m_{\text{steady state}}$$

Therefore, at the steady-state operating level Y and M will be zero. From the transfer function, the response of $Y(t)$ to a specified $M(t)$ can be obtained by inverting the equation

$$Y(s) = G(s)\,M(s) \tag{1.2}$$

1.3. Types of Dynamic Processes

Processes may be categorized as being one of the following types:

—Instantaneous or steady state.
—First-order lag.
—Second-order lag.
—Dead-time or transport lag.

Of course, some processes may be of very high order, but as we shall see shortly, their behavior can often be approximated as a first- or second-order lag plus dead time.

Instantaneous Process

The dynamics of this class of processes is negligible. That is, the output follows changes in input so quickly that the process always remains at steady state, for all practical purposes. The representation of the instantaneous process is shown in Figure 1.3*a*. As an example, consider the operation of a control valve with a linear flow characteristic shown in Figure 1.3*b*. In this case

input is valve stem position X_v 0 to 100%
output is flow Q 0 to Q_{max}

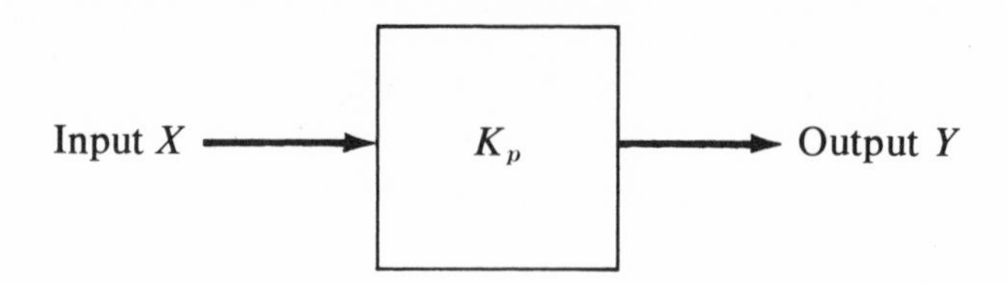

Model: $Y(t) = K_p\, X(t)$

Transfer function: $G(s) = \dfrac{Y(s)}{X(s)}$

where s = Laplace transform variable

Process gain $K_p = \dfrac{\text{change in output}}{\text{change in input}}$

Figure 1.3*a*
Instantaneous Process

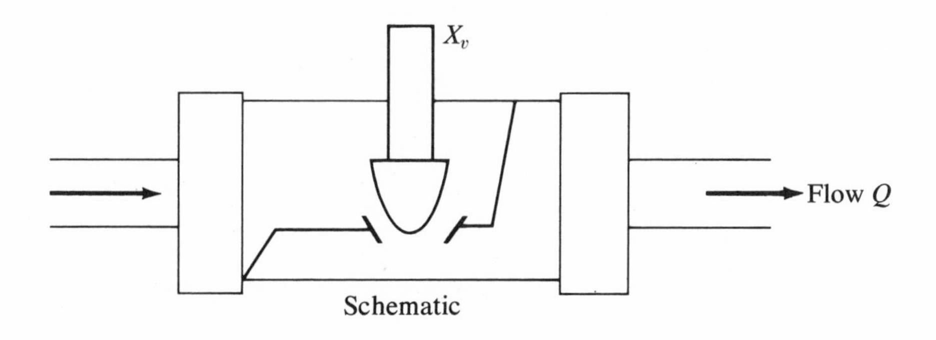

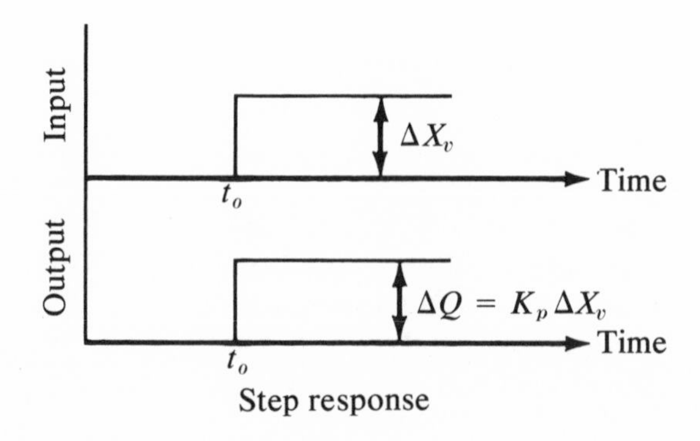

Figure 1.3*b*
Example of Instantaneous Process

$$K_p = \frac{\Delta Q}{\Delta X_v} = \frac{\text{gpm}}{\%} \qquad (1.3)$$

First-order Lag

The output of a first-order process follows the step change in input with classical exponential rise. The model, parameters, and transfer function for this class of processes are

$$\text{Process model:} \quad \tau \frac{dy}{dt} + y = K_p x \qquad (1.4)$$

where

x = process input
y = process output
K_p = process steady-state gain
τ = time constant

$$\text{Transfer function:} \quad G_p(s) = \frac{Y(s)}{X(s)} = \frac{K_p}{\tau s + 1} \qquad (1.5)$$

Block diagram: $X(s) \longrightarrow \boxed{\dfrac{K_p}{\tau s + 1}} \longrightarrow Y(s)$

Dynamic parameters:

$$\text{Steady-state gain } K_p = \frac{\text{final steady-state change in output}}{\text{change in input}}$$

Time constant, τ: time to reach 63.2% of final value in response to a fixed (i.e., step) change in input

The step response of a first-order lag is shown in Figure 1.4.

Examples of first-order lags include many sensor/transmitters. final control elements, and numerous processes. One simple example of a first-order process is a stirred tank where flow in equals flow out, as shown in Figure 1.5.

Input: temperature of incoming liquid T_i
Output: temperature of the effluent stream (assumed to be equal to that of the liquid in the tank because of good mixing) T_o

$$\text{Transfer function:} \quad \frac{T_o(s)}{T_i(s)} = \frac{1}{\tau s + 1} \qquad (1.6)$$

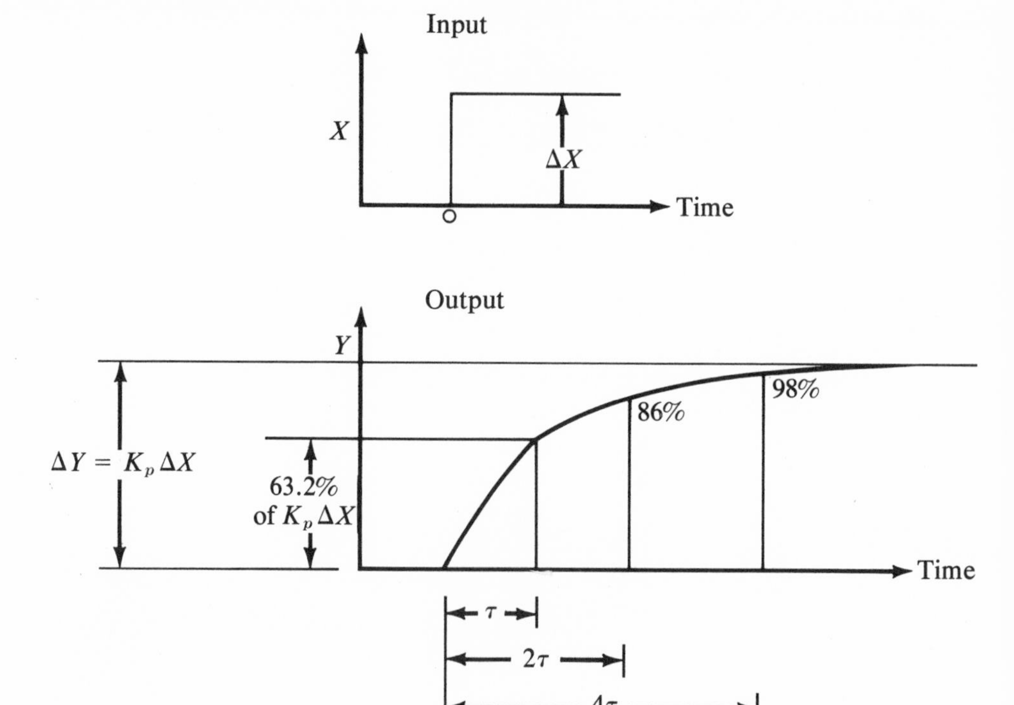

Figure 1.4
Step Response of a First-order Process

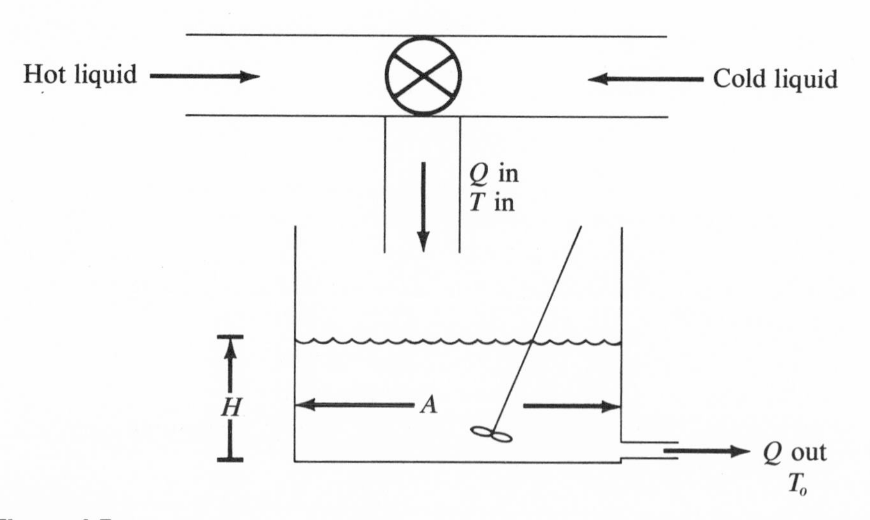

Figure 1.5
Example of a First-Order Process

where

$$\tau = \text{residence time of liquid in the tank}$$

$$= \frac{\text{tank volume}}{\text{volumetric flow}}$$

$$= \frac{H\,A}{Q_{\text{out}}}$$

where

A = cross-sectional area of the tank
H = constant height of liquid in the tank

Second-order processes

Some processes are inherently second order, including a spring/mass system, an inductor/capacitor system, a U tube manometer, and so on. Such processes have the following characteristics

$$\text{Process model:} \quad \tau^2 \frac{d^2y}{dt^2} + 2\,\zeta\tau\,\frac{dy}{dt} + y = K_p\,X \tag{1.7}$$

where

τ = time constant of the second-order process

$\omega_n = \frac{1}{\tau}$ = natural frequency of the process

ζ = damping factor
K_p = steady-state gain

$$\text{Transfer function:} \quad G_p(s) = \frac{Y(s)}{X(s)} = \frac{K_p}{\tau^2 s^2 + 2\zeta\tau s + 1} \tag{1.8}$$

Block diagram: $X(s) \longrightarrow \boxed{\dfrac{K_p}{\tau^2 s^2 + 2\zeta\tau s + 1}} \longrightarrow Y(s)$

The step response of second-order processes is shown in Figure 1.6.

Two first-order lags in series, in which the output of first process is input to second, can be described by a second-order process model which may be represented as

Block diagram: $X(s) \rightarrow \boxed{\dfrac{K_1}{\tau_1 s + 1}} \rightarrow \boxed{\dfrac{K_2}{\tau_2 s + 1}} \rightarrow Y(s)$

Process model: $$\tau_1 \tau_2 \frac{d^2 y}{dt^2} + (\tau_1 + \tau_2)\frac{dy}{dt} + y = K_1 K_2 x \qquad (1.9)$$

A comparison of this equation with that of the inherent second-order model gives

$$\tau = \text{time constant} = \sqrt{\tau_1 \tau_2}$$

$$\zeta = \frac{1}{2}\frac{\tau_1 + \tau_2}{\sqrt{\tau_1 \tau_2}} \geqslant 1.0 \text{ for two first-order processes in series}$$

$$K_p = \text{process gain} = K_1 K_2$$

Dead Time or Transport Delay

For a pure dead-time process whatever happens at the input is repeated at the output Θ_d time units later, where Θ_d is the dead time. For example, consider the flow of a liquid through an insulated pipe having a cross-sectional area A and length L at a volumetric flow rate q. At steady state the temperature of the liquid at the entrance x will be the same as that of the liquid at the exit of the pipe. If the temperature at the entrance changes, the change will not be detected at the

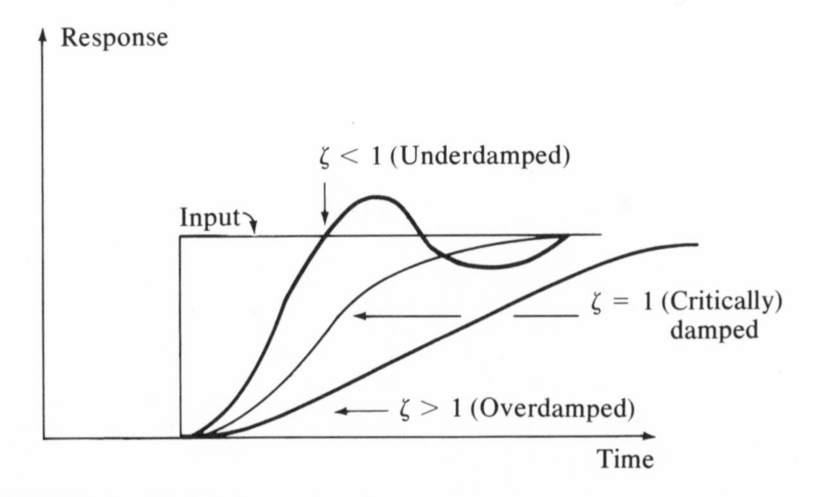

Figure 1.6
Step Response of Second-order Processes

exit until Θ_d seconds later. This dead time is simply the time required for the liquid to flow from the entrance to the exit and is given by

$$\Theta_d = \frac{\text{volume of pipe, m}^3}{\text{volumetric flow rate, m}^3\text{/s}} = \frac{L\,A}{q} \tag{1.10}$$

The process model for pure dead time is given by

$$y(t) = x(t - \Theta_d) \tag{1.11}$$

where y is the exit temperature. At steady state

$$y_s = x_s \tag{1.12}$$

The last two equations give us a model in terms of deviation variables

$$Y(t) = y(t) - y_s = x(t - \Theta_d) - x_s = X(t - \Theta_d) \tag{1.13}$$

Taking Laplace transform gives the transfer function of the pure dead-time element

$$G_p(s) = \frac{Y(s)}{X(s)} = e^{-\Theta_d s} \tag{1.14}$$

The response Y for an arbitrary input X is shown in Figure 1.7a. An

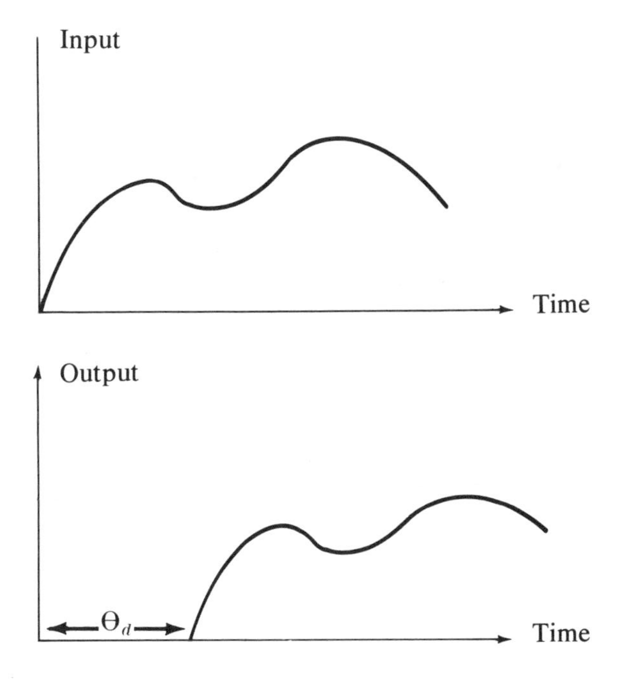

Figure 1.7*a*
Response of Dead-time Element

example of pure dead time in paper manufacture is shown in Figure 1.7*b*.

Higher-order Processes and Approximate Models

Most processes have many dynamic elements, each having a different time constant. For example, each mass or energy storage element in a process can provide a first-order dynamic element in the model. A 50-tray distillation column thus has 50 mass storage elements and 50 energy storage elements. The exact mathematical model relating distillate composition to feed composition would be greater than 100th order, when the dynamics of condenser and reboiler are included. Fortunately, it is possible to approximate the behavior of such high-order processes by a system having one or two time constants and a dead time. We shall demonstrate the justification for this approximation in this section.

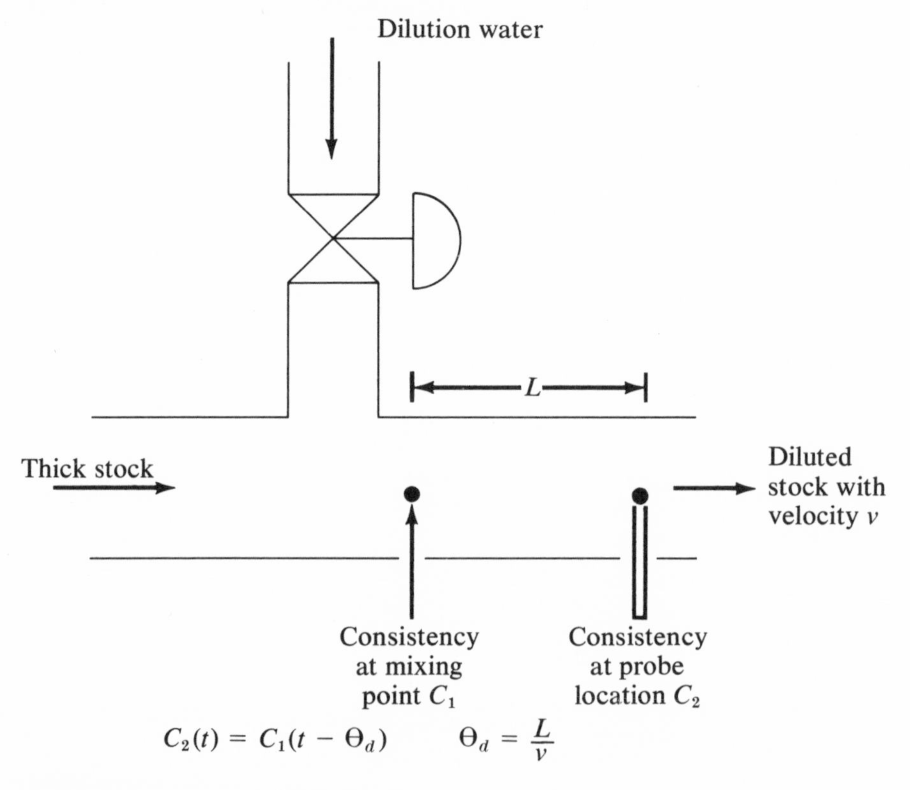

Figure 1.7*b*
Process Having Pure Dead Time

Consider a process having N first-order elements in series as shown in Figure 1.8*a*. In this figure each element has a time constant τ/N. The sum of these time constants for the N elements is τ.

When one or two time constants dominate (i.e., are much larger than the rest), as is common in many processes, all the smaller time constants work together to produce a lag that very much resembles pure dead time. This is clearly seen from Figure 1.8*b* which shows

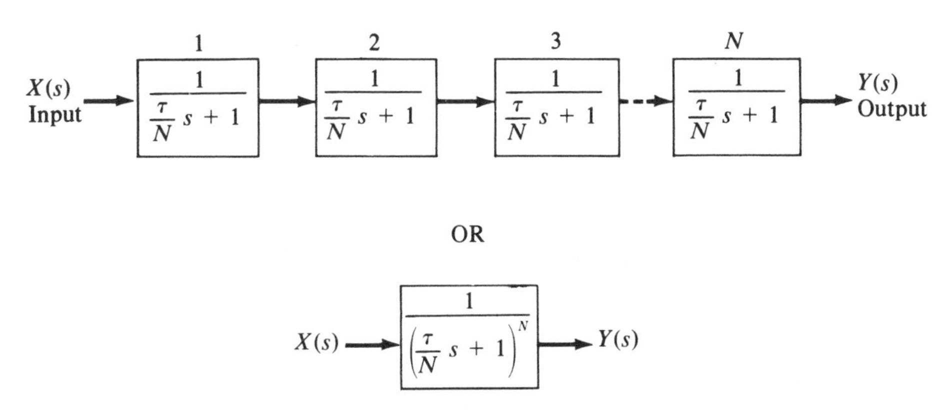

Figure 1.8*a*
***N* First-order Elements in Series**

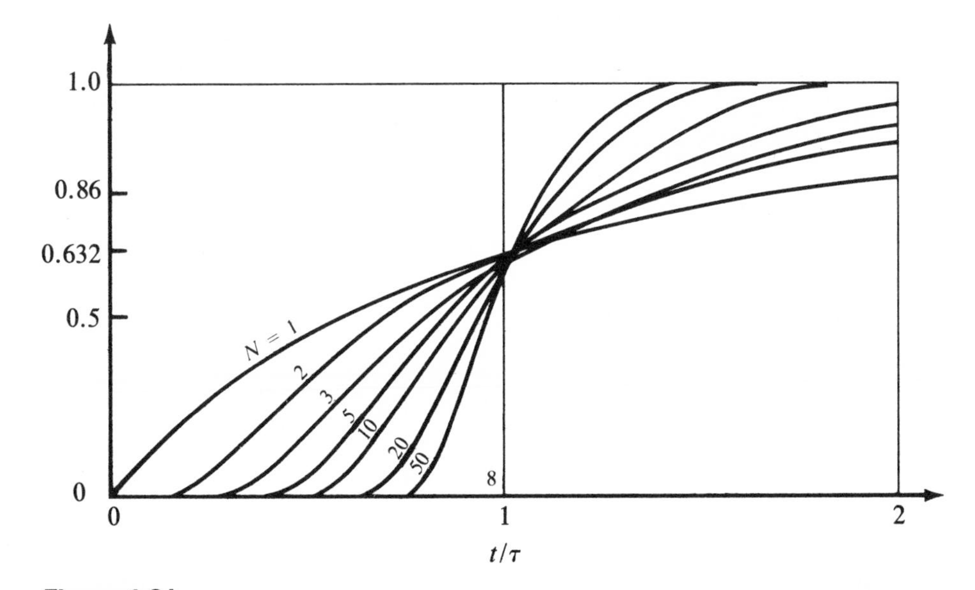

Figure 1.8*b*
Response of Cascaded First-order Lags

the response of the process to a unit step change in X as N is varied from 1 to ∞. This figure shows that as N gets larger, the response shifts from exact first order to pure dead time (equal to τ). It is therefore possible to approximate the actual input-output mathematical model of a very-high-order, complex, dynamic process with a simplified model consisting of a first- or second-order process combined with a dead-time element, as shown in Figure 1.8*c*.

The proper selection of model parameters Θ_d, τ_1, and τ_2 can make the response of the second-order model very close to the response of

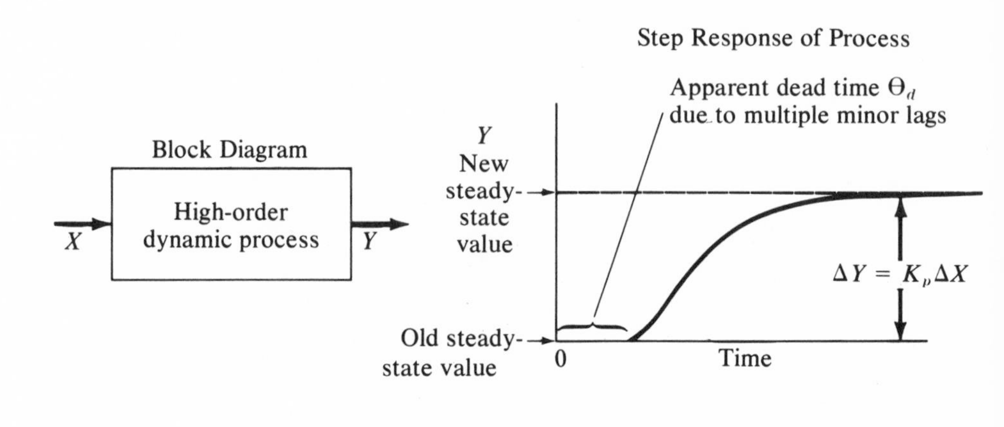

Approximating Models

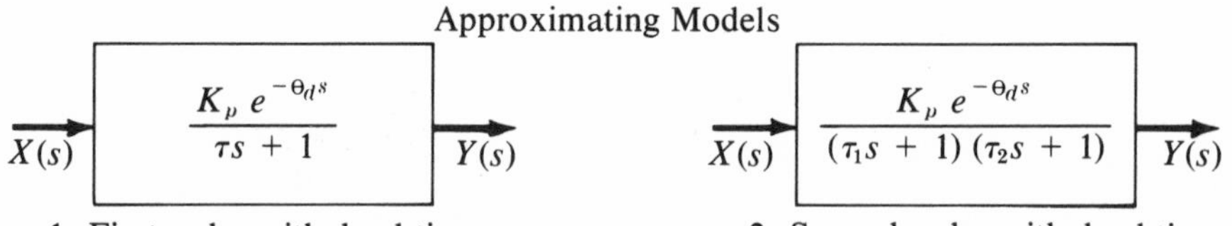

1. First order with dead time

2. Second order with dead time

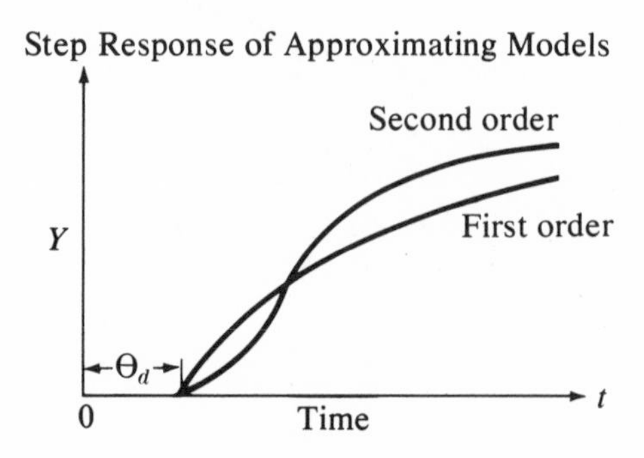

Figure 1.8*c*
High-order Processes and Approximating Models

many processes. The second-order model will reduce to the first-order model if one of the two time constants of the former model is much smaller than the other. Indeed, the response of some processes may be adequately described by a first-order model with dead time.

Complete Process Model

The complete process model includes the model of the process, the measuring element, and the final control element. A block diagram of the complete process model is shown in Figure 1.9.

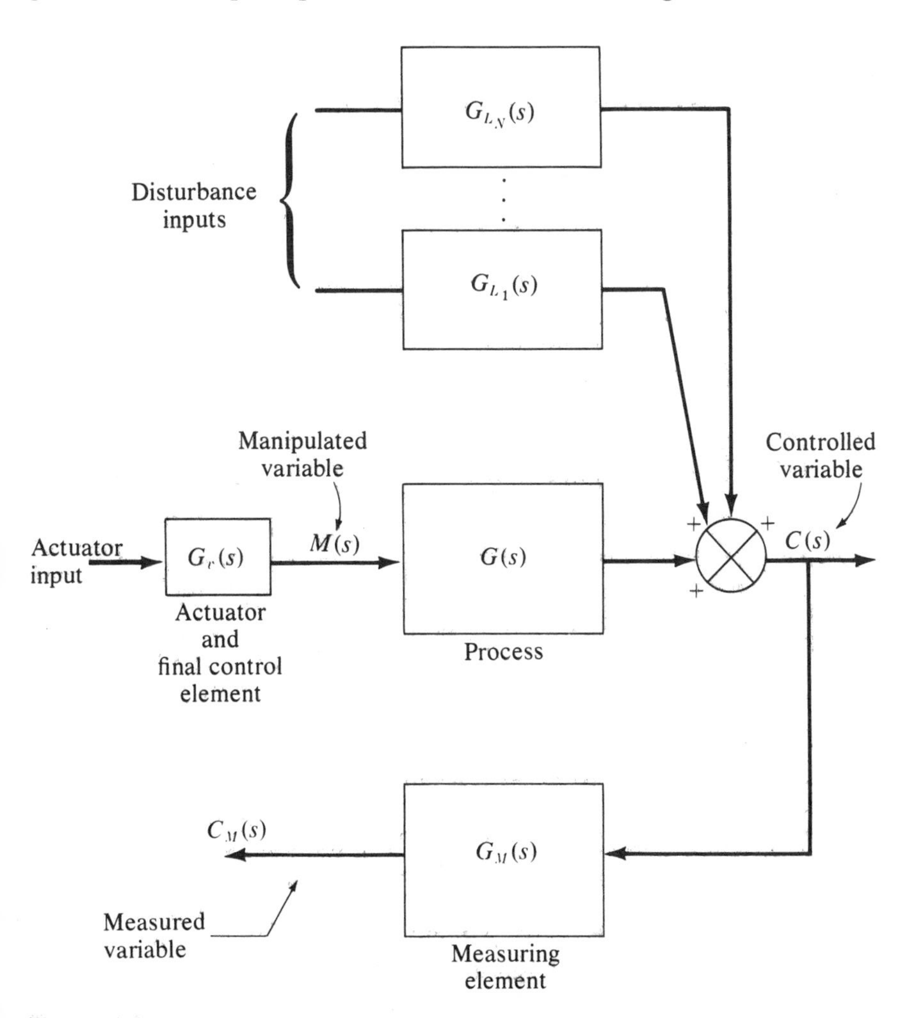

Figure 1.9
Block Diagram of the Complete Process Model

If load dynamics and process dynamics are the same, the block diagram of the process model can be as shown in Figure 1.10. Bear in mind that the notion of a transfer function applies only to linear systems. If the process is nonlinear, as most industrial processes are, the model must be linearized in the vicinity of the desired operating level of the plant. For a different operating level, new parameters will have to be established for the process model. Note also that the arrangement of the block diagram, Figure 1.10, implies that the disturbances affect output, linearly (i.e., superposition principle applies).

1.4. Basic Feedback Control

If there are no disturbances, no control is needed once steady-state operating conditions are achieved. Load upsets cause outputs to deviate from desired values; therefore, one or more control inputs must be changed so as to maintain the outputs at set point. A *control strategy* is required to achieve this objective. In this context, we define control strategy as a set of rules by which control action is determined when the output deviates from set point, that is, which control input(s) should be changed, in which direction, by how much, and when? In other words, control strategy is an algorithm or equation that determines controller output (to the actuator) as a function of the present and past measured errors. The block diagram of the control-

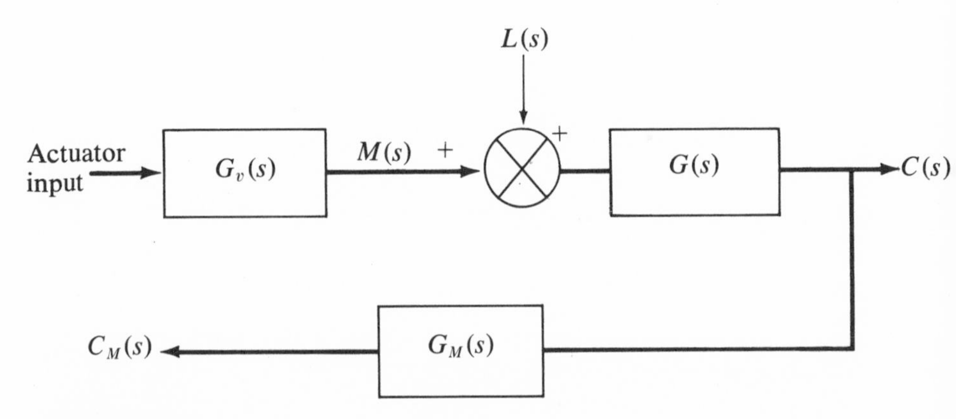

Figure 1.10
Block Diagram of the Process Having the Same Load Dynamics and Process Dynamics

ler portion of an automatic feedback control loop is shown in Figure 1.11.

A block diagram of the basic feedback control loop, which combines the process and the controller, is shown in Figure 1.12. In some applications the actuator and sensor dynamics are negligible compared with the dynamics of the process, in which case the block dia-

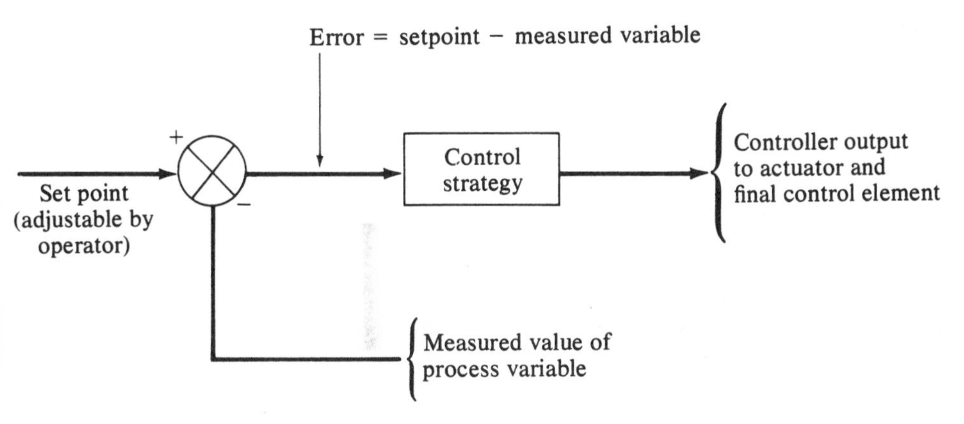

Figure 1.11
Controller Block Diagram

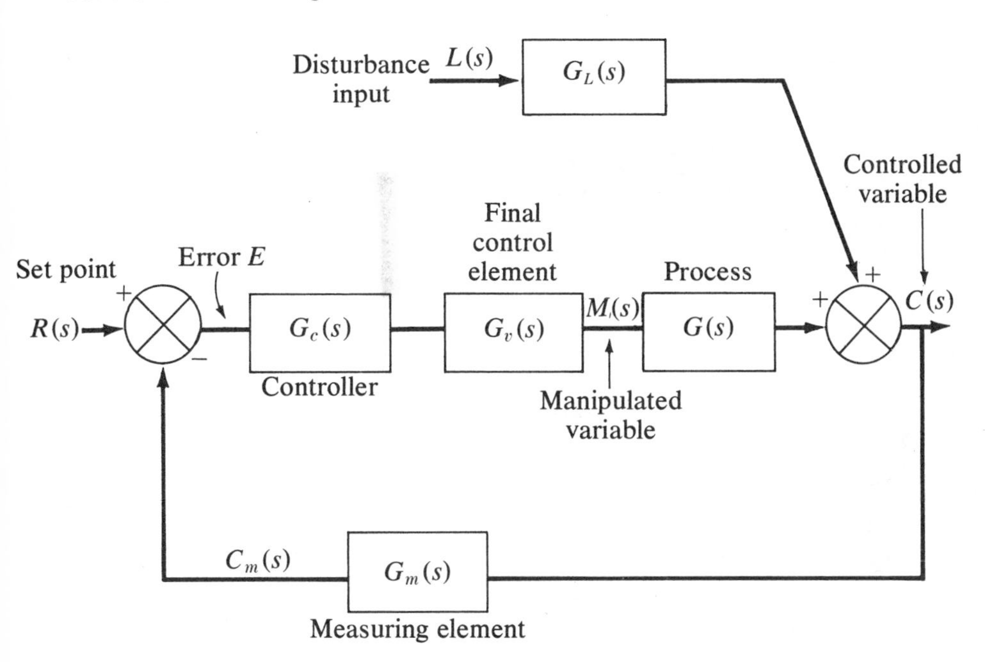

Figure 1.12
Block Diagram of a Feedback Control Loop

gram can be somewhat simplified. In some others the process transfer function includes the actuator and sensor dynamics, as shown in the block diagram of Figure 1.13.

Basic Control Strategy

The basic control strategy in conventional feedback control is to compare the measured variable with the desired value of that variable and if a difference exists, to adjust the controller output to drive the error toward zero. The hardware that performs this function is the automatic controller. The operation of the ideal three-mode controller is described by the following equation:

$$m(t) = K_c \left[e(t) + \frac{1}{\tau_I} \int_0^t e(t)\, dt + \tau_d \frac{de(t)}{dt} \right] + m_s \qquad (1.15)$$

where

K_c = proportional gain
$e(t)$ = error
τ_I = integral time, seconds or minutes
τ_d = derivative time, seconds or minutes
m_s = steady-state controller output that drives error to zero

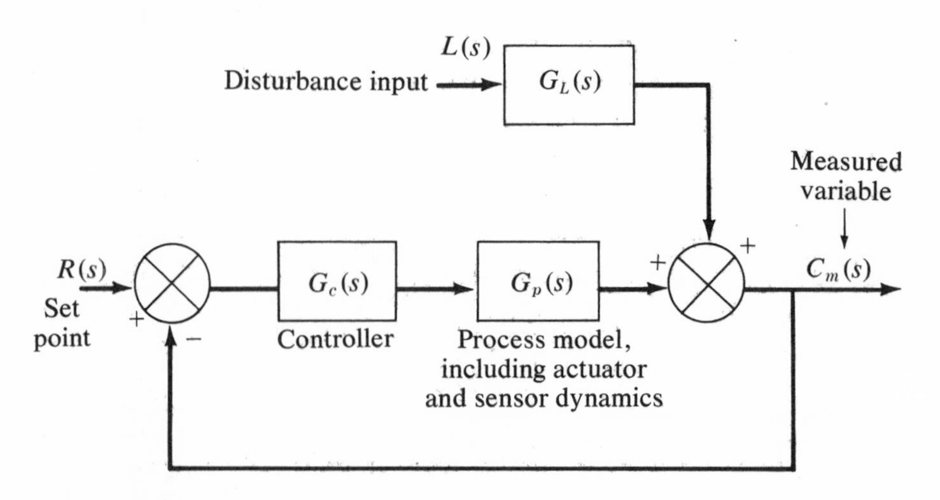

Figure 1.13
Block Diagram Showing a Composite Process Model

The transfer function of the ideal controller equation is

$$\frac{M(s)}{E(s)} = G_c(s) = K_c\left[1 + \frac{1}{\tau_I s} + \tau_d s\right] \tag{1.16}$$

The ideal controller equation has a pure differentiator in it and therefore is not physically realizable. The transfer function of most commercial controllers is

$$G_c(s) = K_c\left[1 + \frac{1}{\tau_I s}\right]\left[\frac{\tau_d s + 1}{\beta\tau_d s + 1}\right] \tag{1.17}$$

where β is a constant $<<1$, typically in the range of 0.01 to 0.1.

Many industrial controllers use the terms *proportional band* and *reset*. They are defined as

$$\text{Proportional band in per cent, } PB = \frac{100}{K_c} \tag{1.18}$$

$$\text{Reset } R_I \text{ in repeats per second or repeats per minute} = \frac{1}{\tau_I}$$

To illustrate the effect of the proportional (P), proportional integral (PI), and proportional + integral + derivative (PID) modes on control, the response of a first-order system with dead time operating under the three types of controllers to a step change of magnitude ΔL in load is sketched in Figure 1.14.

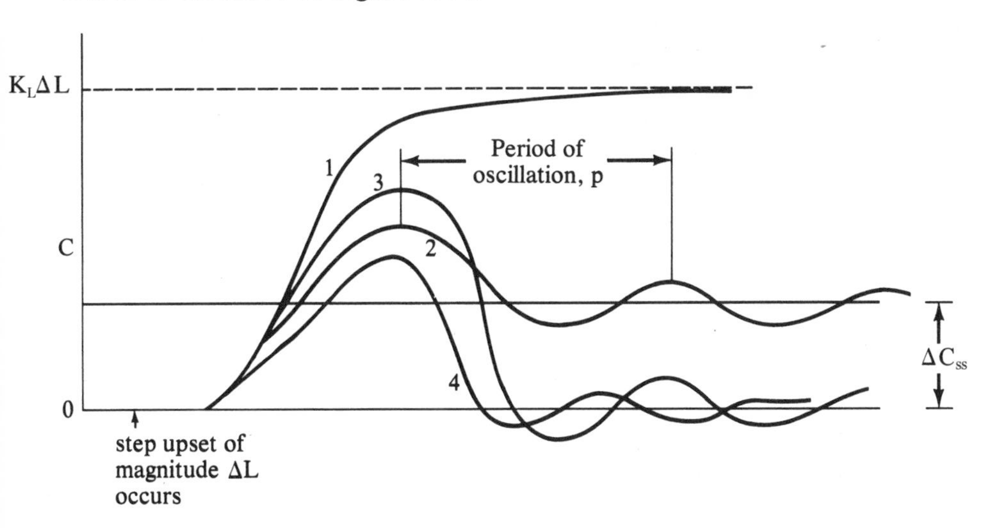

Figure 1.14
Step Response of a First-order + Dead Time System to Load Change

Figure 1.14 shows the following characteristics of the three controller modes.

Curve (1): no control: ultimate deviation in process output, $\Delta C = K_L \Delta L$

Curve (2): proportional only control: gives rise to steady-state offset,

$$\Delta C_{ss} = K_L \, \Delta L \left[\frac{1}{1 + K_p K_c} \right] \tag{1.19}$$

Curve (3): proportional + integral control: eliminates steady-state offset; however, has slightly higher peak offset and slightly longer period of oscillation as compared to proportional only control.

Curve (4): proportional + integral + derivative control: reduces the peak offset as well as the period of oscillation over PI control.

The PID controller is seen to give the best response of the three controller types. However, the derivative mode is sensitive to process or measurement noise, is less forgiving of process parameter changes, and is generally more difficult to tune. For these reasons PI controllers are used in a vast majority of industrial applications.

Controller Tuning

The automatic controllers can be tuned according to one of two methods. The first, called the *ultimate-cycle method*, was originally proposed by Ziegler and Nichols.[1] To use this method, the reset and

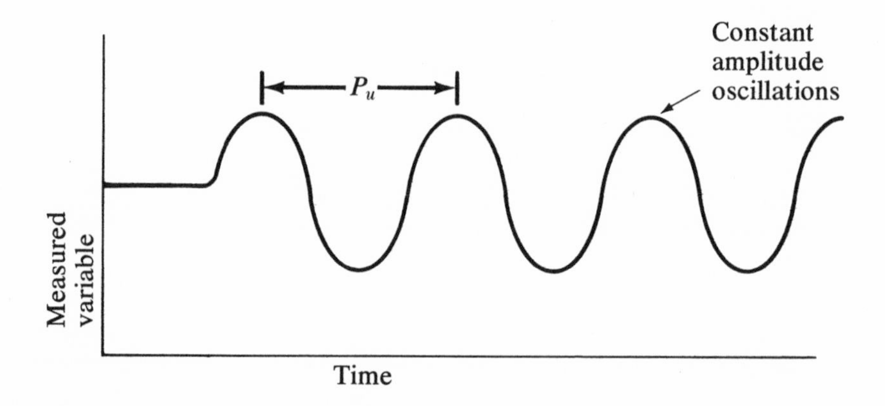

Figure 1.15
Sustained Cycling of Measured Variable under Proportional Control

derivative modes are set at their lowest values, (i.e., lowest values of R_I and τ_d). With the controller in automatic, the gain, K_c, is slowly increased until the measured variable begins to oscillate, as shown in Figure 1.15. The gain setting that results in sustained cycling is called the *ultimate gain*, K_u, and the period of the oscillation is the *ultimate period*, P_u. The Ziegler–Nichols controller settings are determined by the formulas shown in Table 1.1. The modified ultimate cycle method which incorporates the quarter amplitude criteria (see Figure 1.16) yield somewhat different controller settings, which are also shown in Table 1.1.

Recall that the ultimate gain, K_u, and the ultimate period, P_u, can also be determined from the open-loop frequency response diagram of the process, as shown in Figure 1.17.

The second method is called the Cohen–Coon method.[2] It assumes that the process can be represented as a first-order lag plus dead time. To find the tuning constants by this method, the controller is placed in manual, and a step change in its output of magnitude Δm is made. From the step response plot shown in Figure 1.18 the process gain and time constant and the dead time are estimated by the following equations:

$$
\begin{aligned}
K_p &= \frac{\Delta C_{ss}}{\Delta m} \\
\tau_p &= 1.5\,(t_{0.63} - t_{0.28}) \qquad (1.20) \\
\Theta_d &= 1.5\,\left(t_{0.28} - \frac{1}{3}\,t_{0.63}\right) \\
\alpha &= \frac{\Theta_d}{\tau}
\end{aligned}
$$

The controller tuning constants are then determined by the formulas shown in Table 1.1.

Measures of Control Quality

We have mentioned earlier that many processes can be described by a first- or second-order model with dead time. In this section we show how dead time affects the control quality of even properly tuned loops.

Let us define a term, *controllability ratio*, according to the equation

Table 1.1
Controller Tuning Constant Formulas

Type of Controller	*Ziegler–Nichols[1] Original Method*	*Modified Method*	*Cohen–Coon[2] Method*
Proportional	$K_c = 0.5\,K_u$	Adjust the gain to obtain quarter amplitude decay response to a step change in set point	$K_c = \frac{1}{K_p}\left(\frac{1}{\alpha} + 0.333\right)$
Proportional + integral	$K_c = 0.45\,K_u$ $\tau_I = \frac{P_u}{1.2}$ (min)	Adjust the gain to obtain quarter amplitude decay response to a step change in set point $\tau_I = P_u$ (min)	$K_c = \frac{1}{K_p}\left(\frac{0.9}{\alpha} + 0.082\right)$ $\tau_I = \tau_p\left[\frac{3.33\alpha + 0.333\alpha^2}{1 + 2.2\alpha}\right]$
Proportional + integral + derivative	$K_c = 0.6\,K_u$ $\tau_I = \frac{P_u}{2}$ (min) $\tau_d = \frac{P_u}{8}$ (min)	Adjust the gain to obtain quarter amplitude decay response to a step change in set point $\tau_I = \frac{P_u}{1.5}$ (min) $\tau_d = \frac{P_u}{6}$ (min)	$K_c = \frac{1}{K_p}\left[\frac{1.35}{\alpha} + 0.270\right]$ $\tau_I = \tau_p\left[\frac{2.5\alpha + 0.5\,\alpha^2}{1 + 0.6\,\alpha}\right]$ $\tau_d = \tau_p\left[\frac{0.37\alpha}{1 + 0.2\alpha}\right]$

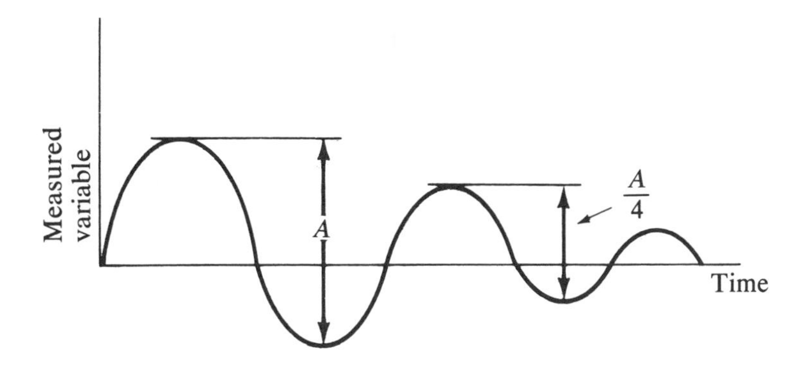

Figure 1.16
Quarter Amplitude-Decay Response

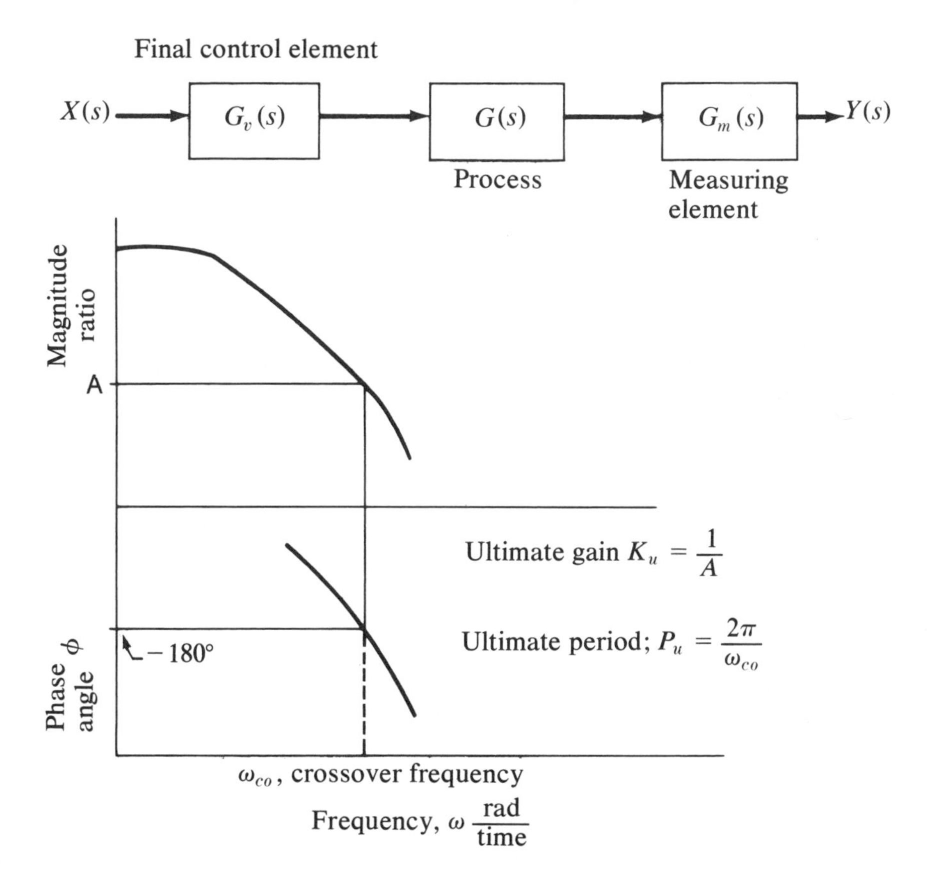

Figure 1.17
Frequency Response Diagram of $[G_v(s)]\ [G(s)]\ [G_m(s)]$

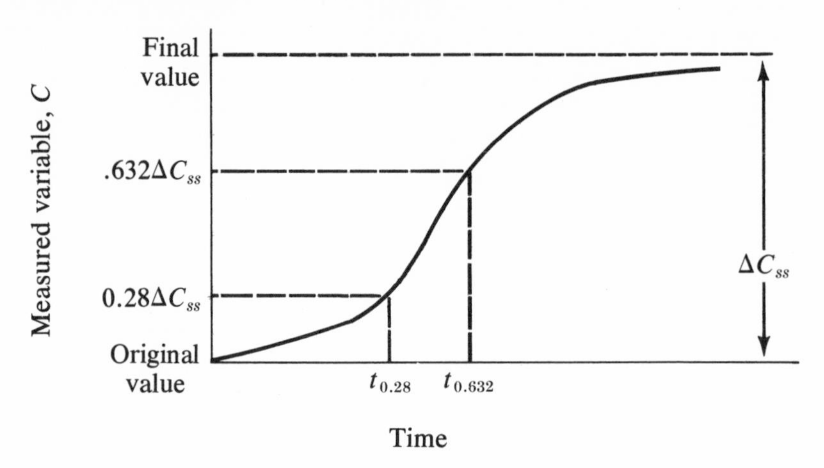

$t_{0.28}$ ($t_{0.632}$) = time at which response reaches 28% (63.2%) of the final value

Figure 1.18
Process Reaction Curve

$$\alpha = \frac{\Theta_d}{\tau_p} \tag{1.21}$$

where

$$\Theta_d = \text{dead time}$$
$$\tau_p = \text{time constant of first-order model}$$
or
$$= (\tau_1 + \tau_2) \text{ for the second-order model}$$

For properly tuned loops the loop gain $K_c\ K_p$ as a function of α is shown in Figure 1.19. As seen from this figure, the loop gain decreases steadily as α increases and for large values of α approaches 0.45 for loops tuned by the Ziegler–Nichols method and 0.08 for loops tuned by Cohen–Coon method. The consequences do not end here. Figure 1.20 shows the peak offset to a step change in load as a function of α. For properly tuned PI control loops this peak offset is given by

$$\frac{\Delta C_{\text{peak}}}{K_L\ \Delta L} = \frac{1.65}{1 + K_c K_p} \tag{1.22}$$

For $\alpha >> 1$ the peak offset approaches 1 which is the open loop value, an unhappy situation indeed.

Figure 1.21 shows the period of oscillation, P_u, as a function of α. For large α

$$\frac{P_u}{\tau_p} = 1 + 3\,\alpha \tag{1.23}$$

or

$$P_u = \tau_p + 3\alpha\,\tau_p \Rightarrow 3\Theta_d$$

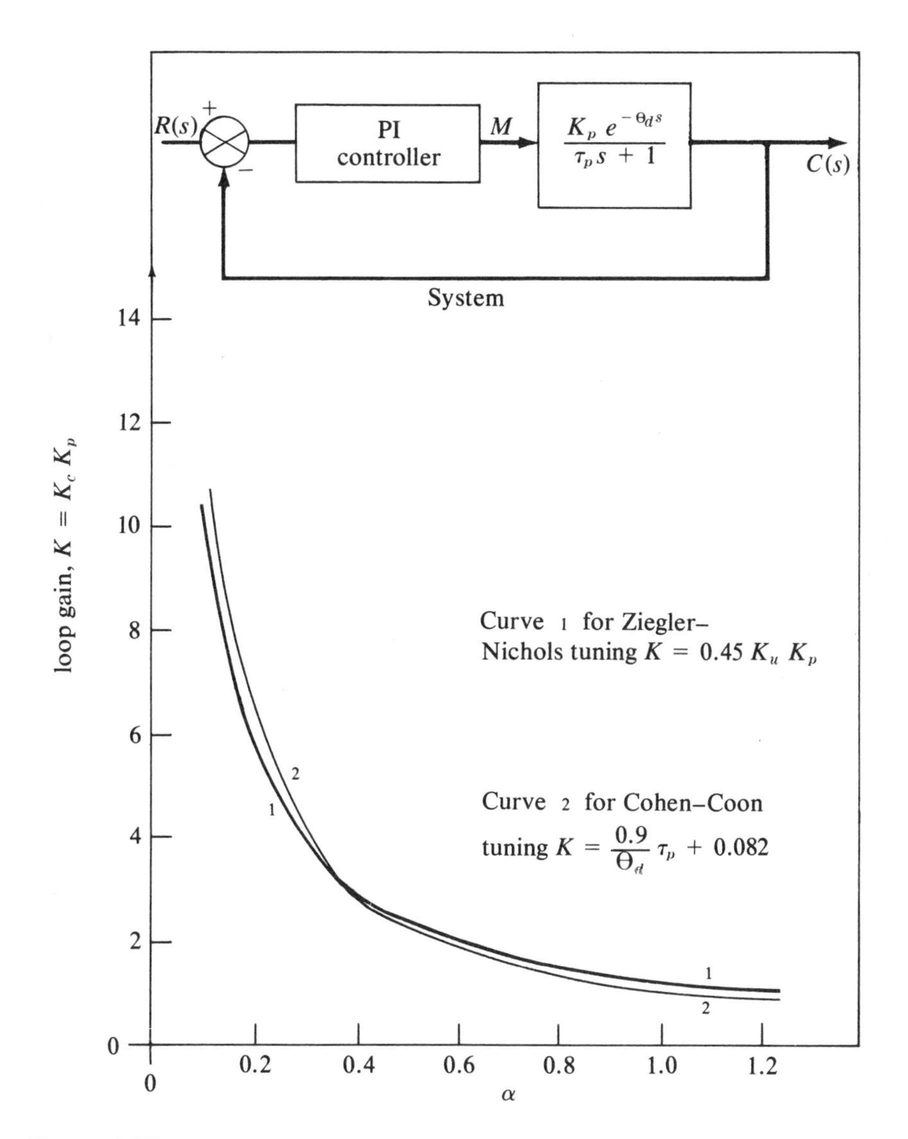

Figure 1.19
Loop Gain versus α for Properly Tuned PI Control Loops

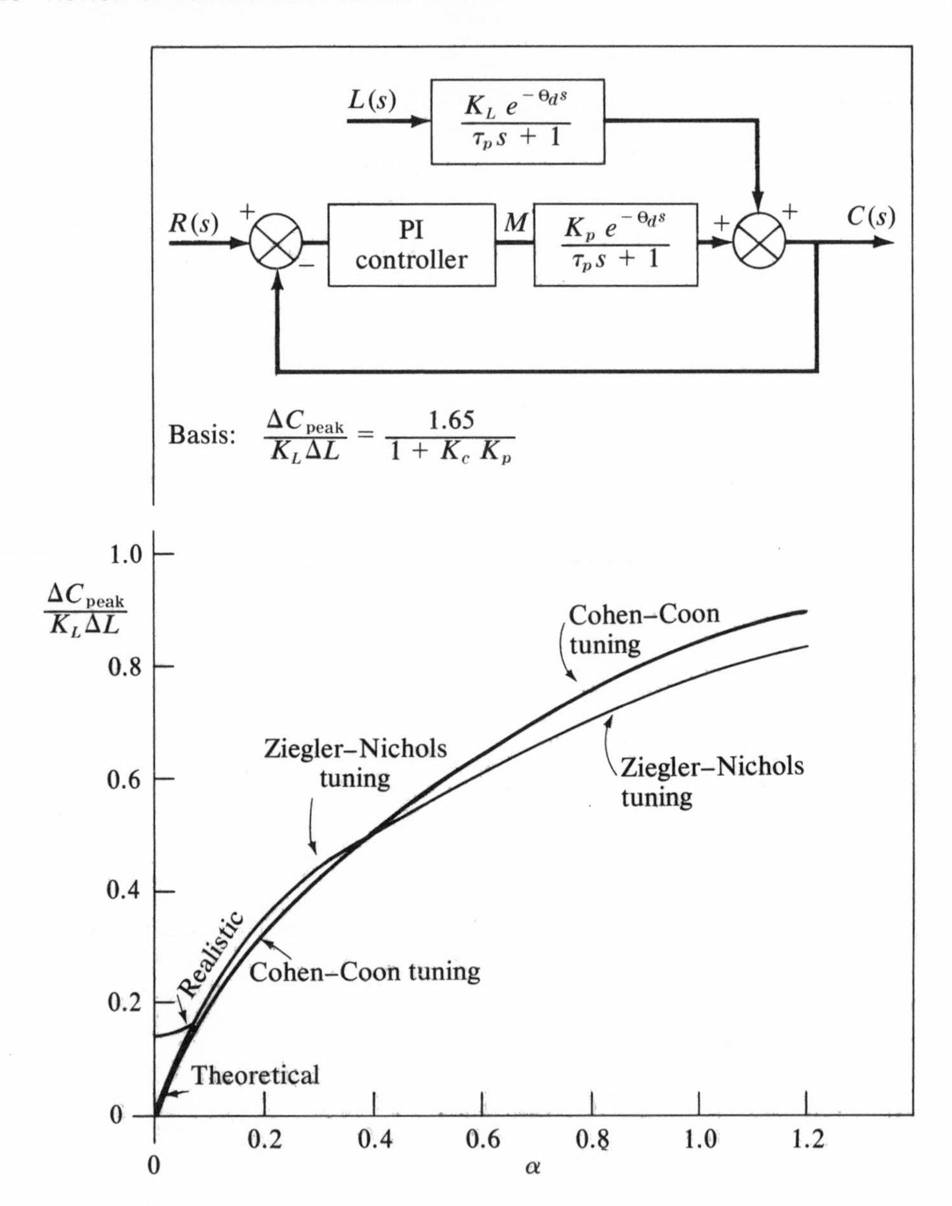

Figure 1.20
Peak offset versus α for Properly Tuned PI Control Loops

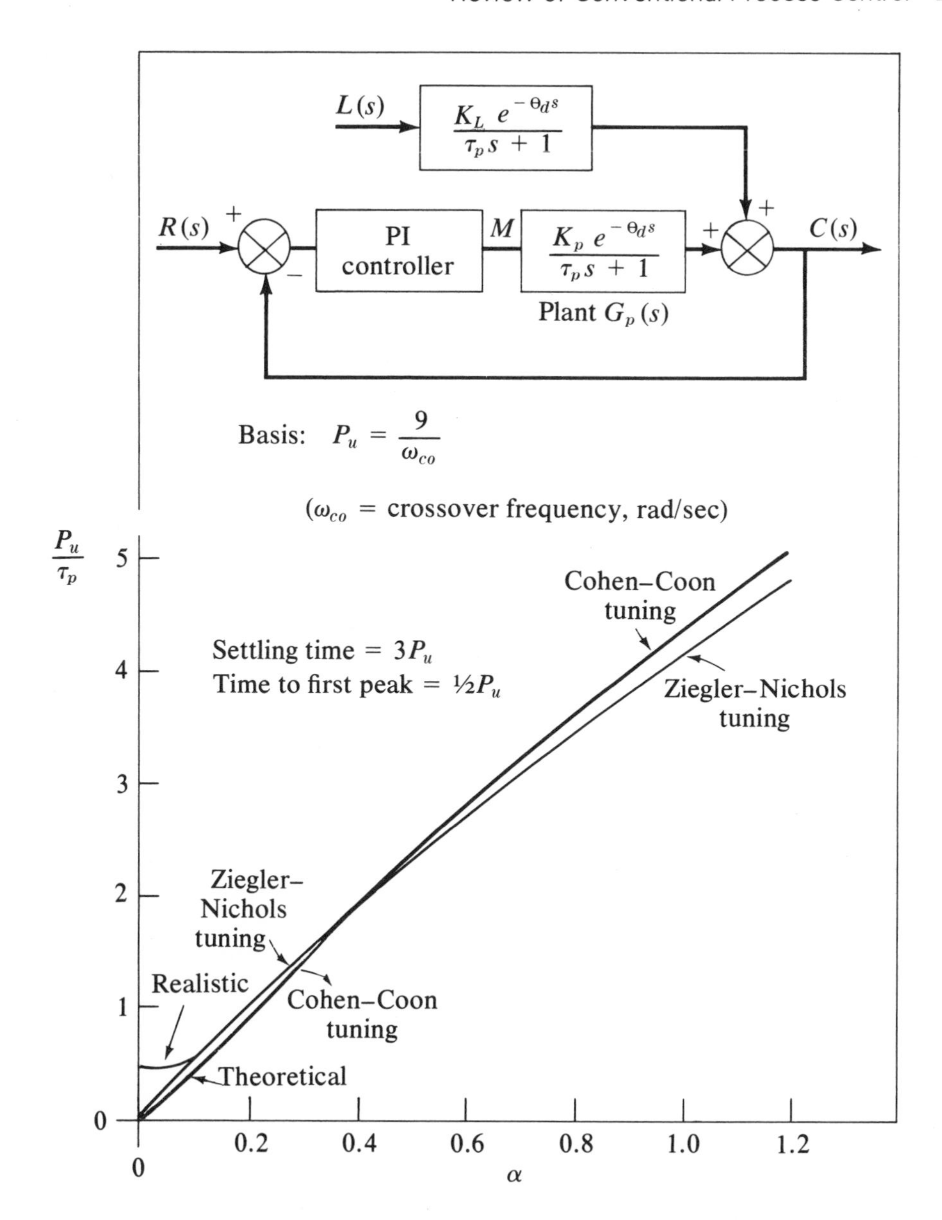

Figure 1.21
Period of Oscillation versus α for Properly Tuned PI Control Loops

Assuming the controlled variable will return to the set point in three periods of oscillation, we can say that for large α, the settling time approaches $3P_u$ or $(3)(3\Theta_d) = 9\,\Theta_d$.

These equations show that large dead times result in a large peak offset, a longer period of oscillation, and a longer settling time. These factors are all indicative of poor control.

1.5. Stability of Conventional Control Systems

Whether a conventional control system is stable or not depends on the nature of the roots of the characteristic equation. To answer the question of stability we start with the closed loop transfer function of the system shown in Figure 1.22. The transfer function relating $C(s)$ to $R(s)$ and $L(s)$ is

$$C(s) = \frac{G_c(s)\,G_1(s)}{1 + G_c(s)\,G_1(s)\,H(s)}\,R(s) + \frac{G_1(s)}{1 + G_c(s)\,G_1(s)\,H(s)}\,L(s) \tag{1.24}$$

In this equation $G_c(s)\,G_1(s)\,H(s)$ is the open loop transfer function which we denote as $G(s)$. The characteristic equation is the denominator of the right side of the equation, that is,

$$1 + G(s) = 0 \tag{1.25}$$

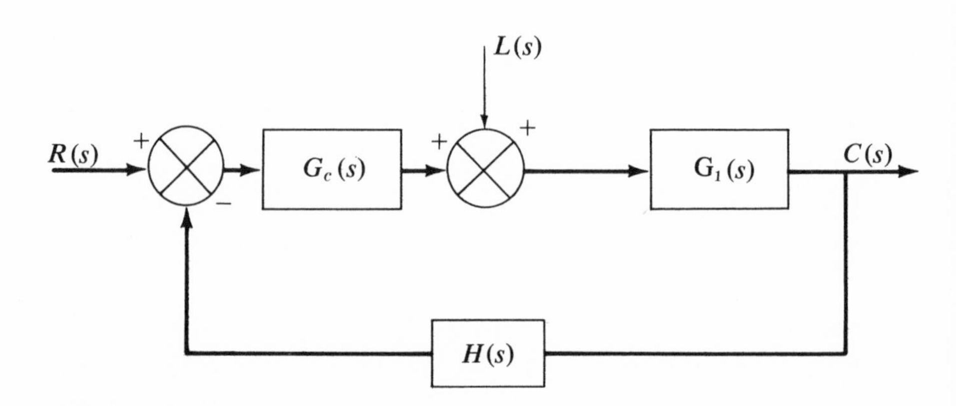

Figure 1.22
Typical Feedback Control System

The roots of the characteristic equation determine stability. For a stable system all the roots must lie to the left of the imaginary axis on the s plane, as shown in Figure 1.23.

One need not physically obtain the roots of the characteristic equation to determine stability. Routh's procedure[3] provides a simple test to determine if the control system is stable. To apply the test we write the characteristic equation $1 + G(s)$ in the form

$$a_0 s^n + a_1 s^{n-1} + a_2 s^{n-2} + a_3 s^{n-3} + \cdots + a_n = 0 \qquad (1.26)$$

If a_0 is negative, multiply both sides of this equation by -1. Then, if any of the coefficients $a_0, \ldots, a_n$ is negative, the system is unstable. If all the coefficients are positive, the system may or may not be stable. Routh's test should then be applied to determine stability. To apply the test, construct the Routh array as shown below:[4]

Row				
1	a_0	a_2	a_4	a_6
2	a_1	a_3	a_5	a_7
3	b_1	b_2	b_3	
4	c_1	c_2	c_3	
5	d_1	d_2		
6	e_1	e_2		
7	f_1			
$n + 1$	g_1			

The elements are filled in for a seventh-order characteristic equation ($n = 7$). For any other n the elements are determined using the same procedure described here.

In general, there will be $n + 1$ rows in all. If n is even, there will be one more element in the first row than the second. The remaining elements in the array are computed according to the following equations:

$$b_1 = \frac{a_1 a_2 - a_0 a_3}{a_1}; \qquad b_2 = \frac{a_1 a_4 - a_0 a_5}{a_1} \cdots$$
$$c_1 = \frac{b_1 a_3 - a_1 b_2}{b_1}; \qquad c_2 = \frac{b_1 a_5 - a_1 b_3}{b_1} \cdots \qquad (1.27)$$

The elements of the other rows are found from the equations that correspond to those above. According to Routh theorems the system is stable if all the elements of the first column are positive and nonzero. The Routh test does not give the roots of the characteristic equation nor the degree of stability (i.e. how far the roots are from the

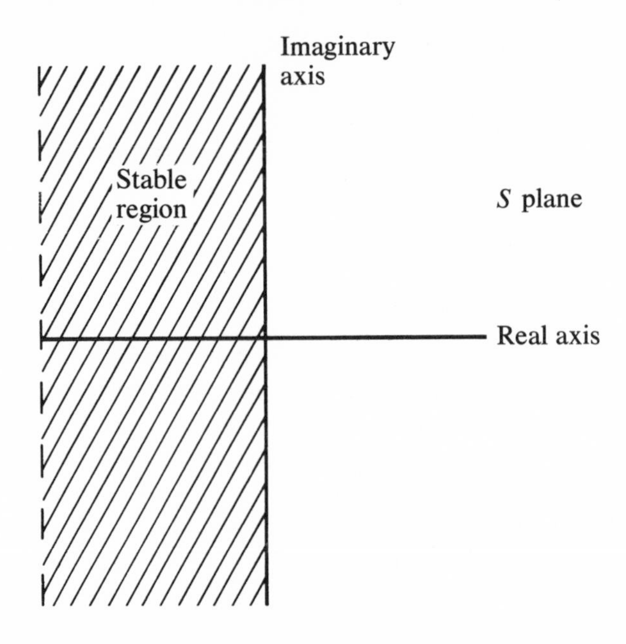

Figure 1.23
Stable Region in the *S* Plane

imaginary axis) of the control system. If this information is needed, we must determine the roots of the characteristic equation. Also, the Routh test cannot be applied to systems containing dead time. For these cases frequency response analysis can answer the questions of stability.

1.6. Problem Control Situations

PI or PID control works "adequately" well for perhaps 90% of industrial control requirements, if properly tuned (unfortunately, all loops are *not* properly tuned). In some situations PID control does not work very well, even if properly tuned.

Poor control is recognizable by a wide band painted on the process variable recorder; that is, the process variable spends a lot of time far away from its setpoint. This is usually caused by one or more of the following problem control situations:

—*Nonlinear Processes*—Gain (K_p) and/or dynamic parameters (e.g., τ_p, Θ_d) change with operating point. This causes the process to exhibit highly variable behavior—sluggish to respond at some times, wholly oscillasory (even unstable) at others; "deviation band" wide and wild at some times, narrow and well-behaved at others. Some form of adaptive control can improve this situation.

—*"Coupled" (Interacting) Processes.* These are processes in which the output is affected by the control input of one or more of the other loops as well as its own. Thus setpoint changes or load upsets in one loop affect other loops. When loops are *cross-coupled,* they "fight" each other—two loops, each stable by itself, may become unstable if strongly enough cross-coupled. Multivariable control decoupling can help this situation.

—*Problem Dynamics*—When process "apparent dead time" (Θ_d) equals or exceeds the "dominant time constant" (τ_p), (i.e., process controllability ratio $\alpha = \Theta_d/\tau_p > 1$), even with "best" PID tuning, peak offsets can approach those of the uncontrolled situation, and the settling time approaches $9\Theta_d$ as we saw earlier. Dead-time compensation can help these situations, often significantly.

—*Problem Disturbances*—These lead to a wide "deviation band," relative to desired range and accuracy of control. Disturbances that have a significant frequency content in the range around the loop's resonant frequency of oscillation are especially troublesome. Disturbances that act quickly on the outputs (i.e., load dynamics faster than process dynamics) cause problems resulting in larger deviations in the controlled variable than normal. Large inherent process noise (e.g., flow or pressure loops) or measurement noise also causes problems.

In many, if not most, of problem disturbance cases, cascade and/or feedforward control can contribute to substantial improvement in control performance.

Advanced Control Strategies

The original development of control theory was in late 1930s and early 1940s. Much theoretical investigation led to several promising approaches to better control. Some (like cascade and, to a lesser extent, feedforward) proved useful right away and have seen extensive application; others (like dead-time compensation and multivariable control decoupling) were essentially shelved because of a lack of practical hardware to implement them.

Now that digital computers are in widespread use for control and have appeared in "packaged" systems, hardware limitations have been removed, and advanced control techniques are seeing wider application in industry.

References

1. Ziegler, J. G., and Nichols, N. B., Optimum Settings for Automatic Controllers, *Trans. ASME*, **64**, 11, (November 1942) 759.
2. Cohen, G. H., and Coon, G. A., Theoretical Considerations of Retarded Control, Taylor Instrument Company's Bulletin, TDS-10A102.
3. Routh, E. J., *Dynamics of a System of Rigid Bodies, Part II, Advanced*, Macmillan, London 1905.
4. Coughanowr, D. R., Koppel, L. B., *Process Systems Analysis and Control*, McGraw-Hill, New York, 1965.

CHAPTER 2

Computer-Control Hardware and Software

In Chapter 1 we reviewed some of the important fundamentals of linear control theory and its application to the solution of conventional process control problems. In the following section we point out the basic differences between conventional control systems and computer-control systems and introduce the reader to the subject of sampled-data control. In the subsequent sections we describe the hardware and software needed in computer-control applications.

2.1 Conventional Control versus Computer Control

In a conventional control loop all the signals are continuous functions of time, as shown in Figure 2.1*a*. The pneumatic control loops employ air pressure for signal transmission, whereas the electronic control loops use voltage or current signals.

In a conventional control system the measuring element senses the value of the controlled variable and transmits it to the controller. This value is compared with the desired value or set point so as to generate a deviation signal, which we call *error*. The controller acts on this error to produce a control signal. The control signal is then fed to the final control element, which in many cases is an automatic positioning valve, in order to reduce the error.

In contrast to conventional control systems, the sampled-data control system involves discrete signals wherein the signal is a train of very narrow pulses of suitable height, as shown in Figure 2.1*b*. A computer-based control system is an example of a sampled-data system.

The schematic of a basic sampled-data control system is shown in Figure 2.2. The controlled variable is measured as before, and the continuous electrical signal, which represents the controlled variable, is fed to a device called an *analog-to-digital converter*, where it is sampled at a predetermined frequency. The sampling period, which is the duration between successive samples, is usually constant in process control applications. The value of the discrete signal thus produced is then compared with the discrete form of the set point in the digital computer to produce an error. An appropriate computer program representing the controller, called a *control algorithm*, is executed which yields a discrete controller output. This discrete signal is

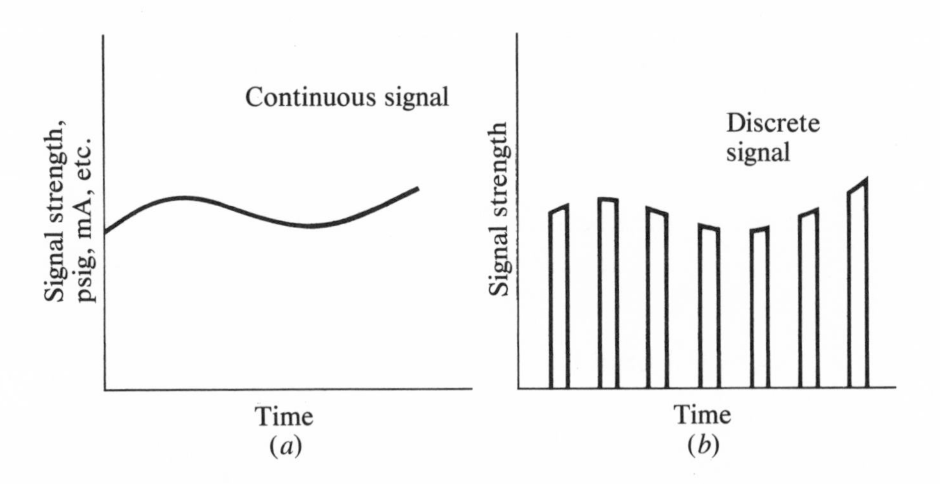

Figure 2.1
Continuous and Discrete Signals. The Discrete Signal Strength, as We Show Later, Is Proportional to the Area of the Corresponding Pulse.

then converted into a continuous electrical signal by means of a device called a *digital-to-analog converter*, and the signal is then fed to the final control element. This control strategy is repeated at some predetermined frequency so as to achieve the closed-loop computer control of the process. As mentioned earlier, this type of sampled-data control technique is referred to as direct-digital computer control, and it is the primary type of computer-control technique with which we concern ourselves in this text.

In many industrial applications the output of the measurement device and the input to the final control element are pneumatic signals. Because the computer works with electrical signals, signal transducers will be required for the conversion of electrical signals into pneumatic signals and vice versa for such applications.

From the foregoing discussion of conventional and sampled-data control systems, we note that the analog controller in the conventional control system is replaced by a digital computer, and the control action that is produced by the controller in the conventional loop is initiated by a computer program in the sampled-data control system. Thus one factor in justifying computer control for a given application is the number of conventional controllers that will be replaced by the digital computer. (It should be understood that this is only one factor in the computation of total cost of a computer control project; such things as the cost of transducers, software develop-

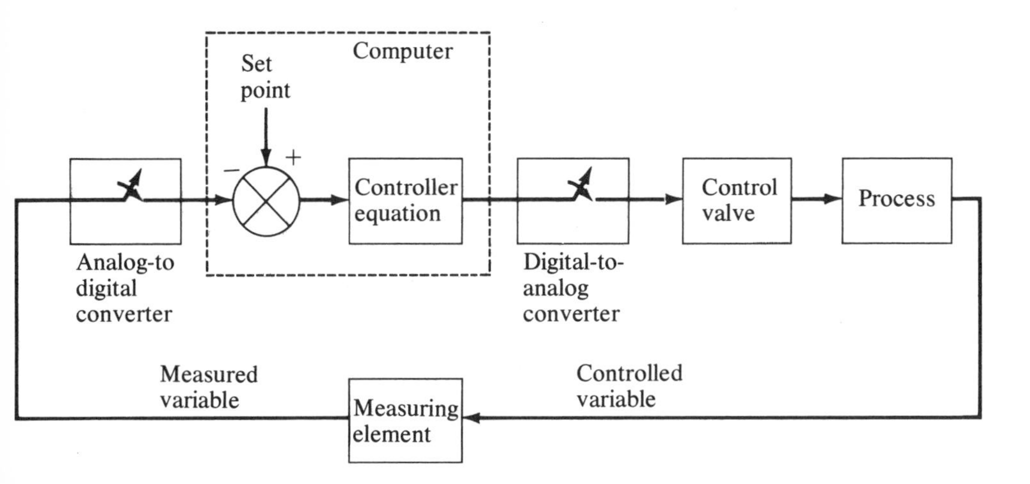

Figure 2.2
Typical Computer Control System

ment, and backup control systems for critical loops must be considered in arriving at the total cost of the project.)

Since digital computers were first used in control applications in the 1950s, their numbers have steadily increased in the last several years. With the development of microprocessors, based on LSI (large-scale integration) technology, the size and cost of digital computers have decreased. Also, with the introduction of microcomputers, the concept of distributed control, wherein small microcomputers control different parts of a large plant, has become attractive. The advantage of this approach is that in the event one of the units fails, only that portion of the plant is affected. As a result of these developments, the number of computer-control applications is expected to increase in coming years.

A possible justification for computer control may come from better performance. Owing to the computational power of digital computers, we will be able to implement control strategies that are otherwise either impractical or impossible with conventional hardware. Examples of such techniques include dead-time compensation, multivariable control decoupling, and optimal control.

The subject matter in this book can be grouped into logical subtopics. Later on in this chapter we describe the hardware and software required for computer-control applications. Chapter 3 presents a detailed procedure for converting a single conventional control loop into a computer control loop using the digital equivalent of the conventional PID (proportional + integral + derivative) controller. With an understanding of the material in that chapter, the reader should be able to operate single or multiloop processes under computer control of P, PI, or PID control algorithms.

The design and analysis of sampled-data control systems requires a knowledge of Z-transforms. At this stage it suffices to say that Z-transforms are to sampled-data systems what Laplace transforms are to conventional-control systems. Therefore, the next several chapters are devoted to the study of Z-transforms and their applications to the design and analysis of computer control systems.

As was pointed out earlier, the availability of control computers makes the implementation of advanced control strategies a relatively straightforward task. The final chapters in this book are concerned with the design and application of advanced control strategies, including feedforward control, dead-time compensation, and multivariable control.

2.2 Basic Concepts of Computers

The remainder of this chapter is directed at introducing the basic ideas of how computer systems are structured and how they work. Although our focus is on computers used in on-line systems to monitor and control processes, the principles developed apply for the most part to all digital computer systems. The scope of this text permits only a brief introduction to the basic concepts of computers and a broad overview of computer hardware and software. References 1 and 2 at the end of the chapter provide the interested reader with a much more detailed, yet still introductory level study of the vast field of digital computers. Both are highly recommended.

From the smallest microcomputer to the largest digital computer in use today, all digital computers are, in their essence, *serial number processors*. They take in data (numbers) from the outside world, process the data (make some calculations or perform some other operations based on the data), and then deliver results (also numbers) to the outside world. By *serial* we mean that the computer performs operations on numbers sequentially, one operation at a time.

Often, relative to the user (human being, process or, other outside-world device), the data are available or required in other than numerical form, for example, characters typed on a typewriter or voltages from sensors of physical variables such as temperature or flow. However, within the computer, data are processed in numerical form only and thus must be converted from the original form to numbers for processing by the computer and then back to whatever form is required for its end use in the outside world. Devices that perform the conversion of information (data) to and from numerical form are called *peripheral* devices and form a part of the overall computer system.

In basic operation a digital computer is similar to a simple electronic calculator. In a calculator the human operator enters numbers (data) and arithmetic operations to be performed on the data (e.g. add, subtract, multiply, divide) using the calculator's keyboard and receives the results of those operations on the calculator's numerical display. A computer is similar in that it performs arithmetic operations on numbers as its primary internal function and communicates with the outside world (in and out) to get data and deliver results of its internal operations. The range of peripheral or outside-world devices from or to which a computer gets data or delivers results is much wider than the simple keyboard and display of the calculator, however. Computers communicate not only with human beings but directly with processes, using a variety of devices to be discussed later.

In addition to the basic arithmetic processing and input/output function of the ordinary calculator, all digital computers have three additional features which greatly increase their power. These features are memory, automatic execution of instructions, and decision-making capability.

1. *Memory*. The memory of a digital computer can store numbers and retrieve them at some later time. Modern computer systems have very large memories capable of storing many thousands of numbers concurrently. Memory is also used to store *instructions*, making possible the next feature.

2. *Automatic Execution of Instructions*. An *instruction* is the vehicle used to tell the computer which operations to perform on what data and what to do with the result. Instructions are numbers in a special format recognizable to the computer and are put into memory by the user. A set of instructions in computer memory is called a *program*. When the computer "runs" a program, it retrieves the instructions from its memory, one-by-one, without operator intervention, and executes them. Programs are task oriented and frequently consist of many thousands of instructions.

3. *Decision-Making Capability*. A digital computer has the capability of changing its normal sequence of instruction execution as a function of the numerical value of a specified internal variable which may represent the result of a calculation or an input variable. This provides the ability of executing different sequences of instructions under varying process or program conditions.

The tremendous power of digital computers does not arise from a vast number of different available operations. A digital computer can execute only a relatively small number of different instruction types (operations), typically fewer than 100. Rather, the power of computers arises from the three features just described. There is virtually no limit on the complexity or sophistication of tasks that can be done with long sequences of basic operations having built-in alternate instruction sequences.

Conceptual Organization of a Computer System

There are three basic parts, or subsystems, in any computer

system, no matter how large or small. They are the *central processing unit* (or CPU), *memory* and the *input/output* (I/O) *interface*. Figure 2.3 shows the conceptual organization of a computer system and is a reference diagram for the discussion which follows.

Central Processing Unit (CPU)

The CPU is the "heart" of a digital computer, the subsystem which performs the actual execution of instructions. The CPU also controls sequencing of operations and controls communication with the memory and I/O interface subsystems.

Within the CPU are a number of hardware devices called *registers* which are used for temporary storage and manipulation of numbers and instructions. Registers are made up of electronic circuits called *flip-flops* which are two-state devices; that is, they operate in one of two stable states, either off or on, and they can be changed from one state to the other at any time on command from control circuits within the CPU.

Within digital computers, numbers are represented and manipulated in binary (i.e., base 2) form, in which each digit (called a *bit*) is

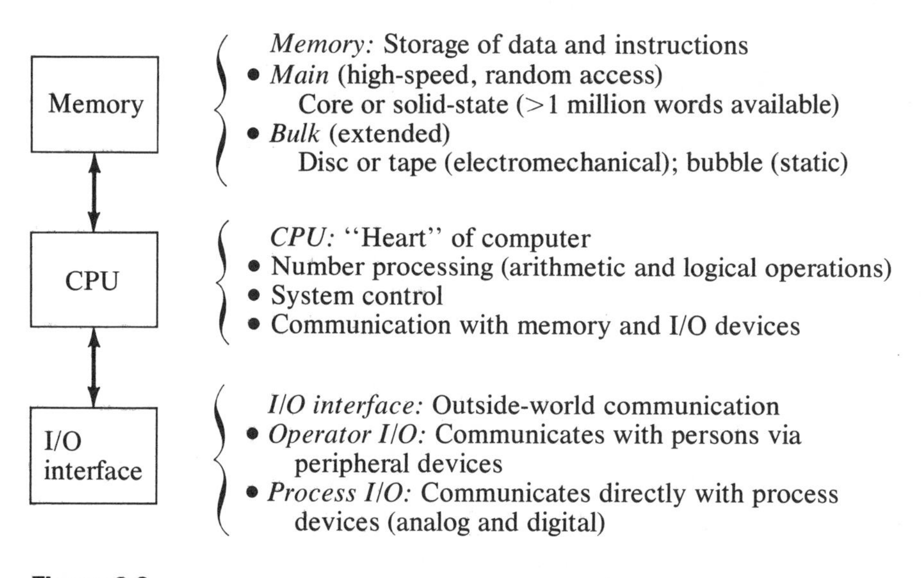

Figure 2.3
Conceptual Organization of a Computer System

either zero or one. Each flip-flop in a register holds one bit of a binary number. The number of flip-flops in a single register corresponds to the number of bits the CPU transfers to or from memory in a single operation. This number is called the *word length* of the computer. Thus a *word* is a group of binary digits used to represent a number or instruction. Sixteen bits is the most common word length for minicomputers.

Arithmetic and logical operations are performed within the CPU on numbers contained in registers according to instructions contained in other registers, and results are stored in the same or other registers.

Memory

The memory subsystem may consist of a number of different physical devices, but conceptually there are only two types of memory: high-speed, random-access memory (often called main memory) and bulk memory (often called extended memory).

High-speed random-access memory is made up of miniature magnetic cores or very-high-density solid-state flip-flop circuits. Each core or flip-flop stores one bit of information. A typical computer has many thousands of words of main memory and can access each and every word on a completely unconstrained (i.e., random) basis. The CPU can read or change the contents of any selected word in main memory in a single operation.

A high-speed random-access memory that permits both read and write operations by the CPU is called RAM (for random access memory). Another kind of memory, which the CPU can read but not change, is called ROM (for read only memory). Variations that can be reprogrammed by off-line devices are called PROM (for programmable ROM) or EPROM (for erasable programmable ROM).

Rapid advances in solid-state technology have drastically lowered the cost and reduced the size of main memory devices in recent years, and the trend will continue. Today (1980), circuit boards only a few square inches in area can contain over 64,000 16-bit words of solid-state RAM, and the entire main memory (up to 1 million or more words in many minicomputers), along with the CPU and several I/O interface boards, can be housed in a single chassis containing a dozen or so small, plug-in circuit boards.

Bulk (or extended) memory is much less expensive per word than main memory and has a much greater storage capacity. It generally resides on peripheral devices and is connected to the CPU through devices called *controllers*. The most common devices in use today for bulk storage are *disc* (rotating magnetic platters) or *magnetic tape* (cassette or reel-to-reel). Not all computer systems have bulk memory, but they all have main memory.

Bulk memory devices are much slower to access than main memory and cannot be accessed for read and write operations in the same random manner which main memory can. Bulk memory is accessed by *blocks*. A *block* is a fixed number of words which varies in size between computer manufacturers—256 words is a commonly used block size. In some systems block size can be specified by the programmer. Reading from or writing to bulk memory requires the transfer of an entire block of words between the bulk device and a predesignated area of main memory, often called a *buffer*. The CPU then reads or writes the desired word(s) in the main memory buffer.

By using removable magnetic discs or tapes, there is virtually no limit on the amount of data that can be stored in bulk memory. As this text is written (1980) most bulk memory devices in use are electromechanical magnetic devices (disc or tape), which tend to be the least reliable components in a computer system. Within the decade, however, bulk memory technology will move rapidly toward static devices built with high-density, high-capacity bubble-memory chips which have recently appeared in the marketplace. A vast improvement in performance (i.e., memory access time) and reliability should result. The equipment is almost certain to change, but the concepts of structure and operation will remain intact.

Input/Output (I/O) Interface

The I/O interface is the subsystem through which the CPU communicates with the outside world. There are two basic kinds of I/O: *operator I/O* and *process I/O*, the latter found only in real-time control and information systems.

Operator I/O communicates with people. Process operators use such devices as pushbuttons, thumbswitches, and keyboards to input data or commands to the computer, and they receive information from the computer via such devices as black-and-white or color CRT (cathode-ray tube) screens, LED (light-emitting diode) numerical displays, and pilot lights. Computer operators and program developers use *computer terminals* (a combination of CRT screen or typewriter with a keyboard) to communicate with the computer. Until recently, punched cards or punched paper tapes were widely used to store programs, and thus card and paper-tape readers and punches were commonly used computer I/O devices. Within the past 2 or 3 years, however, low-cost floppy-disc (a flexible plastic disc with a magnetic surface) and cassette-tape drives have become available and are rapidly replacing punched cards and paper tape.

Process I/O communicates directly between the CPU and all manner of process devices. Typical devices in a real-time process system include sensors, limit switches, tachometers, and shaft encoders for input; control valves, motor starters, stepping motors, and motor-speed controllers for output. Process devices, and some operator devices such as pushbuttons and pilot lights, are connected to the computer through two parts of the I/O interface called the *analog* and *digital I/O subsystems*.

To convert process input signals (usually voltages or currents from transmitters) to the digital (numerical) form needed by the CPU, the *analog I/O subsystem* uses hardware devices called *analog-to-digital* (A/D) *converters*. Because precision A/D converters are still relatively expensive, each A/D converter is generally shared by several (typically 8 to 16) individual analog input lines. A device called a *multiplexer* connects any one of these inputs to the A/D converter to be read on command from the CPU. The analog I/O subsystem also contains hardware used to "condition" the analog signals, for example, circuits to reject noise and smooth the input.

For output, the analog I/O subsystem converts numerical (digital) outputs from the CPU into analog voltages or currents by means of a *digital-to-analog* (D/A) *converter*. Since process analog output signals connected to devices such as control valves must be available to the device continuously, a separate D/A converter is used for each analog output device. A D/A converter sustains its output value until it is changed, which is done in most computer control systems on a periodic basis.

The *digital I/O subsystem* communicates with process devices which have only two possible states: on or off. Digital inputs arise most often from contact closures (e.g., limit switches). Digital outputs are "switches" (reed-relays or solid-state triacs) opened or closed by the I/O subsystem on command from the CPU. Most computer systems read or change more than one digital input or output concurrently; typically, 16 inputs (one word) are read or 16 outputs are updated in a single operation. A temporary location in main memory is used for the CPU to read or change individual digital inputs or outputs. The contents of this temporary memory location are transferred to or from the I/O subsystems by the CPU to actually change or read digital inputs and outputs.

Hardware and Software

A complete computer system, regardless of whether it is being used for process control, data processing, or any other function, consists of two basic kinds of components: Hardware and Software.

Hardware refers to all *physical* components of the system such as the CPU, main memory, bulk memory devices (e.g., disc and tape drives), the I/O interface and peripheral I/O devices such as terminals. In essence, anything which can be touched is classified as hardware. *Software* refers to all the *programs* (sets of instructions that tell the computer which operations to execute, step by step, to perform tasks) which are necessary for the operation of the system.

Tradeoffs are possible between hardware and software in computer systems. For example, the arithmetic unit in the CPU of some very small computers can only add two numbers and change the sign of a number; thus add is the only arithmetical function that can be accomplished with a single instruction. Subtraction requires two instructions (change sign of one number, then add). Multiplication and division are accomplished by repeated sequences of the basic add and change-sign operations which are controlled by programs; that is, multiplication and division are software operations in such computers. Additional hardware can be purchased that performs the multiplication and division operations in a single instruction step, thus making them hardware operations. The advantage of having hardware to perform additional arithmetic operations is simply speed. With multiply and divide hardware, these functions can be executed by the computer in a single instruction step. Without such hardware, multiplying or dividing two numbers typically requires between 10 and 30 instruction cycles, depending on the particular numbers involved.

Program Example

Many of the concepts developed in the foregoing sections can be illustrated with an example program for accomplishing a relatively simple task which is representative of the kinds of tasks performed by on-line computer systems in real-time process-control applications.

The task is as follows:

Read the value of a process sensor into memory, display this value on an LED numerical display, and set an external alarm if the value is outside the limits stored in memory.

Figure 2.4 illustrates this program in flowchart form, with a detailed description of what happens in each step depicted by the flowchart. Flowcharts are graphical descriptions of the sequential steps in a computer program. Each block represents a step in the sequence, and the arrows connecting the blocks define the order in which they are executed. The shape of the block defines the kind of operation: Rectangular blocks are CPU and memory operations, trapezoidal blocks are input/output operations involving the I/O interface, and diamond-shaped blocks are decision points in the program, at which an operation is performed and then one of two or more possible alternative paths in the program is taken, depending on the result of the operation. The oval blocks do not necessarily represent computer operations but mark the beginning and end of a program. The operations described in each block of a program flowchart are typically not single computer instructions but "subtasks" requiring many sequentially executed computer instructions to perform.

The steps required for this example program are:

1. Read the value of the process sensor voltage from the analog input subsystem into a holding register in the CPU. In typical computer systems, this requires connecting the A/D converter to the input line for the desired sensor with the multiplexer, waiting several microseconds for the A/D converter output to stabilize, and then transferring the resulting digital value into the CPU register.

2. Convert the A/D converter reading to physical units. Suppose that the sensor is a temperature-measuring instrument spanning the range 60°F to 300°F, corresponding to a 0 to 10 volt range on the analog input line. For simplicity, assume that the A/D converter output is a number between 0 and 10,000 (0 = 0 volts, 10,000 = 10 volts) and thus represents the number of millivolts present on the input line. (More typically, the A/D output span would be from 0 to 4096 to 0 to 16384, depending on the device precision). To convert the A/D output number to degrees, it is multiplied by a scale factor and then an offset is added; i.e. temperature (°F) = (scale factor) × (A/D output number) + offset. In this example the scale factor is 240/10,000 = 0.024, and the offset is 60. These conversion factors are stored in memory.

3. Store the converted value of the process temperature in a location assigned to it in memory.

4. Display the process temperature on an output LED numerical

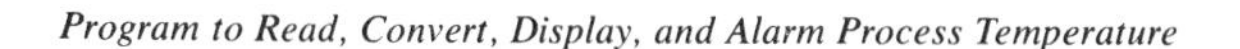

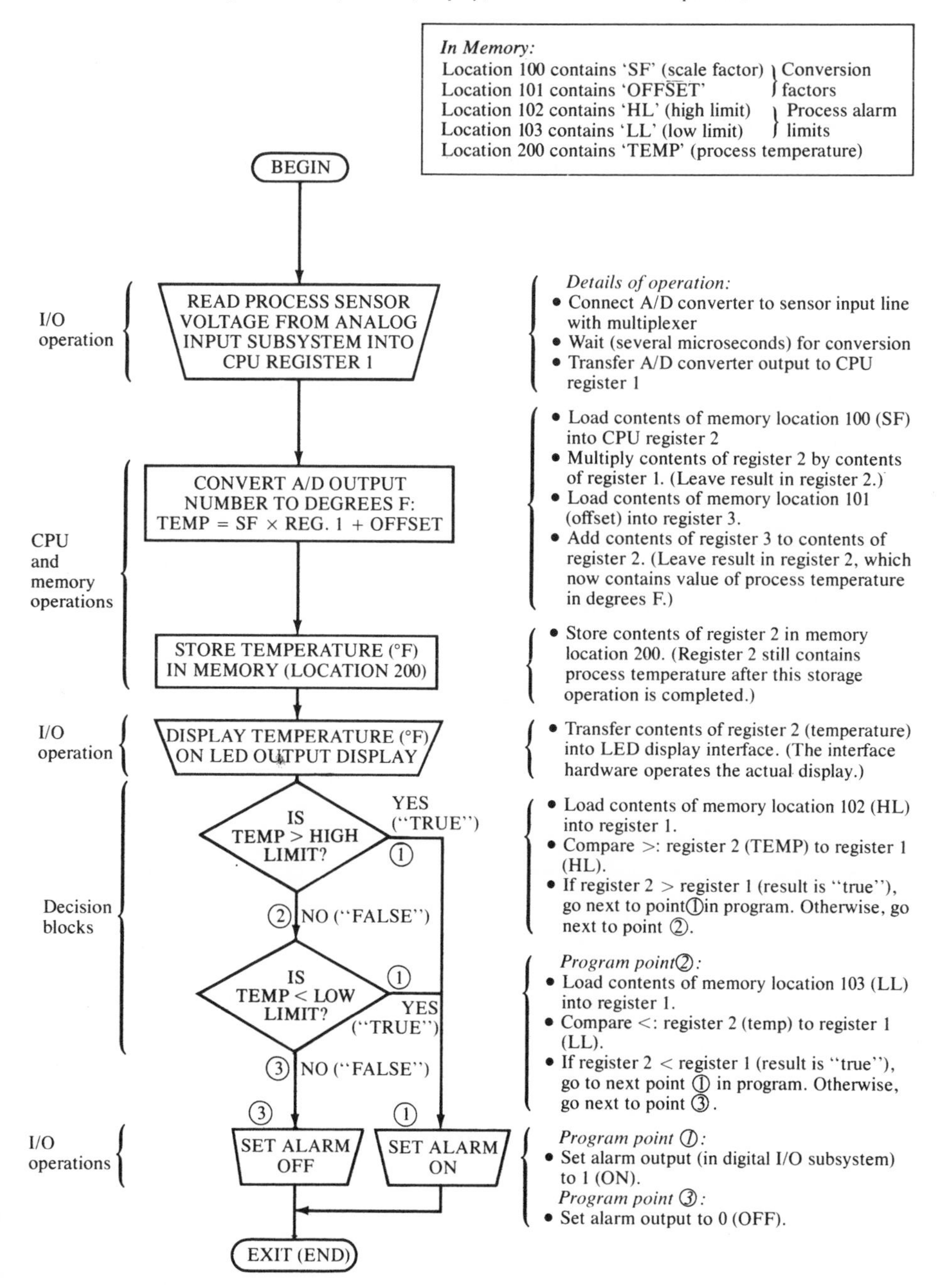

Figure 2.4
Example Program

display. Such displays hold their values until changed by the computer.

5. Compare the process temperature to high and low process limits stored in memory. If out of limits, set alarm (horn and/or blinking light) on. If within limits, set alarm off.

The decision or comparison steps in this program are examples of logical operations in the CPU. The result of a logical operation can have only one of two values: true or false. In a compare $> (A \rightarrow B)$ operation, true results if $A > B$, false otherwise. In a compare $< (A \rightarrow B)$ operation true results if $A < B$, false otherwise. A logical operation causes a flag (called a *condition code*) in the CPU to be set with its result (true or false). After the logical operation is complete, the next program instruction will read the internal flag and set one of two possible program routes for the ensuing calculations, depending on the flag's value.

In a typical real-time on-line computer system, a program such as this one would be executed periodically, typically once per second. Another program, called an *executive*, performs the automatic scheduling and execution of such programs.

2.3 Computer System Hardware Concepts

In this section we expand on the basic principles of computer hardware structure and operation introduced in the foregoing section. Although details of hardware structure and operation (called *architecture*) vary substantially from one computer to another, there are a number of basic principles involved which are essentially the same in any digital computer. Understanding of these basic principles will provide the student with a foundation on which to build further knowledge of specific computer systems.

Binary Numbers and Words

The hardware of a digital computer consists primarily of thousands or even millions of *two-state devices*, as discussed earlier. Each can be viewed, at any given time, as containing a single-digit number with only two possible values: 0 if the device is off, or 1 if the device is on. Use of such devices leads in a natural way to the use of binary (base 2) numbers for internal number representation and calculations.

In the binary number system there are only two digits, 0 and 1, as compared to the 10 digits, 0 through 9, in the familiar decimal (base 10) number system. Multidigit binary numbers are built on powers of two (i.e., 1, 2, 4, 8) for each higher significant digit, rather than powers of 10, as in the decimal system. Each digit in a binary number is called a *bit*. Using three-digit binary numbers, one can count from 0 to 7 as depicted in Table 2.1, which shows all possible combinations of three binary digits.

Table 2.1
Binary Numbers: 0 to 7

2^2: *4's digit*	2^1: *2's digit*	2^0: *1's digit*	
0	0	0	$= 0 + 0 + 0 = 0$
0	0	1	$= 0 + 0 + 1 = 1$
0	1	0	$= 0 + 2 + 0 = 2$
0	1	1	$= 0 + 2 + 1 = 3$
1	0	0	$= 4 + 0 + 0 = 4$
1	0	1	$= 4 + 0 + 1 = 5$
1	1	0	$= 4 + 2 + 0 = 6$
1	1	1	$= 4 + 2 + 1 = 7$

In a similar way, four binary digits enable counting from 0 to 15. In general, n binary digits have 2^n possible combinations ranging from 0 to $2^n - 1$. The largest decimal number which can be represented with 12 bits, for example, is 4095 ($2^{12} - 1$), while with 16 bits it is 65,535 ($2^{16} - 1$).

Computers manipulate and store binary digits in groups called *words*. The number of bits in a single word (called the *word length*) varies between different computers, but 16-bit word length has become the most common for small computers. Larger systems use 32-bit words. A group of 8 bits is called a *byte*, a near-universal standard of reference. Early microcomputers used a word length of one byte, though the trend today is toward two-byte (16-bit) words.

Memory Organization and Addressing

RAM systems are organized in units of either bytes or words, with each unit (byte or word) being assigned a unique address so that it can be individually accessed for reading or writing. Addresses are assigned sequentially beginning with 0 and ending at the total number of bytes or words in main memory. Main memory sizes exceeding 1 million words are available and economically feasible in modern minicomputer systems. Memory sizes are usually expressed in units of K (for kilo) words or bytes, 1K actually being 1024 or 2^{10} units. Thus a 32K word memory segment contains 32 × 1024 or 32,768 words.

Bulk memory devices (disc and tape systems) are organized and addressed in terms of *blocks*, as discussed earlier. Each block, or group of words, has a unique address relative to the beginning of device. In many bulk memories, blocks can be organized into groups called *files* which are identified by unique names assigned by programmers. Such file-structured devices have directories on the device which are tables associating specific device block locations with file names.

Words or bytes in memory can represent one of two things: data or instructions. The following two sections discuss each in further detail.

Data and Number Representation

Stored data can represent either numbers in binary form or coded alphanumeric symbols. Alphanumeric symbols are letters, digits, and punctuation or mathematical symbols printable on a typewriter. Such symbols are used for computer input and output and are stored internally in coded form. The most common code used for internal representation is called ASCII, in which 7 bits are used to represent 96 different symbols and 32 control codes such as tab and carriage return. Each ASCII coded symbol is stored in 1 byte of computer memory.

Numbers are represented internally in one of two forms: fixed point or floating point.

In *fixed-point* (also called *integer*) representation, one word is used to represent a single whole number or integer. Thus with a 16-bit word length, 2^{16} or 65,536 different numbers can be represented. To handle negative as well as positive numbers, the span of the numbers is taken from −32,768 to +32,767. A negative number is represented by the "two's complement" of the corresponding positive number, which is obtained by binary subtraction of the number from 0, or (equivalently) complementing each bit in the number (i.e., all 0s are changed to 1s and vice-versa) and then adding 1 to the result. Two's complement number representation allows very efficient arithmetic operations within the CPU.

In some computers, double-precision numerical storage and

operations are possible. In the double-precision mode, two words are used to represent a single number, which greatly increases the range of integers and precision with which numerical calculations can be carried out.

In *floating-point* (also called *real-number*) representation, numbers are stored and manipulated in the equivalent of a scientific notation format, in which a number is represented by a *characteristic*, or exponent and a *mantissa*, or fraction. For example, in decimal scientific notation, the number 321.3 is represented as 0.3213×10^3, where the fraction is 0.3213 and the exponent is 3. The fraction is ''normalized'' to a value between 0.100 and 0.999, and the exponent can be viewed as the number of digits to the right (left for negative exponents) the decimal point must be shifted to convert the fraction to the actual number. Binary scientific notation follows the same principles, with the exponent and fraction being binary numbers. Floating-point representation in digital computers assigns a certain number of bits to the fraction and a certain number of bits to the exponent. Most 16-bit computers use two words for each floating-point number, but no standard at all applies in bit assignment for exponent and fraction between manufacturers. Typical, but by no means universal, is to use 1 bit for the sign of the number, 8 bits for the exponent, and 23 bits for the fraction. This allows a range of about $\pm 10^{\pm 38}$ with a precision equivalent to about seven significant decimal digits. Double-precision-floating point numbers, where available, add one or two additional words of precision to the fraction. Zero is a special floating-point number, because it falls outside the range of standard representation. Usually, zero is represented by all zeros (which corresponds to zero exponent and zero fraction).

Arithmetic operations on floating-point numbers are accomplished either with special, extra-cost floating-point hardware or with software (special programs that add, subtract, multiply, and divide floating-point numbers using sequences of fixed-point arithmetic and shifting operations). Floating-point software is available from most computer vendors as part of the software library provided with their hardware.

Floating-point representation is almost always used in applications requiring substantial numerical calculations, because it allows fixed precision independent of magnitude and a much wider range of numbers than fixed point. However, unless special arithmetic hardware is purchased, floating point arithmetic operations are much slower to execute than fixed-point operations and double the storage required in most minicomputers.

Instructions and Basic Computer Operations

Instructions stored in the memory of a computer are coded commands with which the programmer specifies what the computer is to do—what operation to perform, where to get the data, and what to do with the result. *Programs* are groups of instructions, usually stored in adjacent locations in memory, that are intended to be executed sequentially or in another order specified by the instructions themselves.

Each instruction defines a single operation to be performed by the CPU. The instruction is coded in a special format, shown in Figure 2.5. Although the specific operations which can be performed by the CPU and the specific instruction layout differ between computers, the structure shown is essentially universal to all computers. The operation code, or OP-code, specifies the operation and occupies from 3 to 10 bits of the instruction, depending on the number of different instruction types available in the computer's instruction set. Sixteen-bit minicomputers typically allocate 8 bits to the OP-code and have 50 to 100 instruction types in their instruction set. The total length of an instruction can be one, two, or even three words depending on the particular instruction. Bits beyond the OP-code are used for additional information defining the operation in more detail, for example, specifying CPU registers to be used or memory addresses of data to be fetched and/or stored as part of the operation.

No two computers, except possibly different models in the same family from the same manufacturer, have identical instruction sets. Nevertheless, the types of basic operations performed by the CPU are the same in all computers. It is conceptually convenient to classify instruction types into three categories: operations on data, program execution control operations, and input/output operations.

Operations on data perform arithmetic or logical operations on one or two numbers (called *operands*) located in memory or CPU registers; for example, an ADD instruction adds the contents of a specified memory location to the contents of a specified CPU register. Both the memory address and the CPU register are specified in the additional-information bits of the instruction. At the conclusion of this operation, the register would contain the sum of the contents of the memory location and the previous number contained in the register.

The following instruction types from this category are representative of basic operations performed by the CPU of a digital computer:

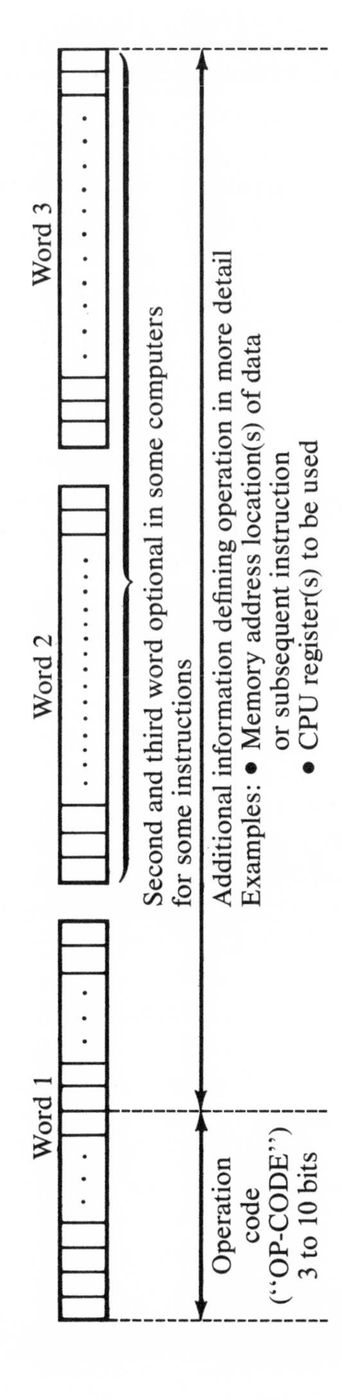

Figure 2.5
Instruction Format

- *Arithmetic Operations* — ADD, SUBTRACT (plus MULTIPLY and DIVIDE with added hardware) on two operands. CHANGE SIGN, INCREMENT (add 1), DECREMENT (subtract 1) on single operands.
- *Logical Operations* — AND, OR, EXCLUSIVE OR on two operands. NOT, SHIFT (left or right) on single operands.
- *Transfer Operations* — LOAD (memory to register), STORE (register to memory) and MOVE (register to register or memory to memory).

Program execution control operations alter the normal sequential execution of instructions and are used to provide alternate execution paths as illustrated in the program example. Instruction in this category are known broadly as BRANCH instructions.

In normal sequential operation the CPU gets the next instruction to be executed from the memory location following that of the instruction just completed. BRANCH instructions specify a different memory address for the next instruction.

There are three categories of BRANCH instructions:

—*Branch unconditional* (JUMP) instructions simply specify the memory address of the next instruction to be executed.

—*Branch on condition* instructions also specify a memory location for the next instruction, but the transfer is made only if a *condition*, specified by the instruction and based on the result of the last CPU operation, is met. For example, a BGT (branch on greater than zero) instruction will execute the transfer if the result of the last instruction executed was positive. Otherwise, normal sequential execution continues. All computers have many conditional branch instructions specifying different transfer conditions.

—*Branch and link* instructions are unconditional branch instructions that also cause the address of the next sequential instruction to be stored in a location (memory or CPU register) called a *link*. This allows one program to initiate the execution of (*call*) another group of instructions, called a *subroutine*. The last instruction in a subroutine is a RETURN instruction which executes an unconditional branch to the instruction located at the address stored in the link [i.e., the instruction following the subroutine call (branch and link) instruction], thus resuming execution of the previous program from the point at which the subroutine was initiated.

Input/Output operations transfer data between peripheral devices and the CPU or main memory. Two basic operations are provided: READ, which transfers data into the CPU or memory from a peripheral device, and WRITE, which moves the data from CPU or memory to a peripheral device.

I/O operations, and the instructions provided for them, are highly dependent on the architecture (hardware design) of the particular computer, and there is a wide variation in I/O instruction subsets between manufacturers. In many computers each peripheral device has a program (usually called a *driver*) to manage the detailed operations and timing needed for an I/O operation on that device. The driver is called (as a subroutine) whenever I/O operations are required by a user-written program.

Bulk memory devices (discs and tapes), as well as all peripheral devices, are connected to the CPU and main memory through hardware known as *device controllers*. Collectively, the device controllers comprise the I/O interface subsystem described earlier. Thus bulk memory devices are often considered to be peripherals, and transfers of data in and out of them are viewed as I/O operations. Most bulk memory devices are capable of transferring large blocks of data into or out of main memory in response to a single CPU command. This capability is known as *direct memory access*, or DMA, and allows the CPU to perform other operations while the block transfer is in progress.

The question frequently arises as to how the CPU knows whether the contents of a specific memory location is data or an instruction. The simple answer is that it doesn't, in any direct way. Once execution has begun, the sequence is inherently specified by the program itself, and the CPU assumes that the specified memory locations contain instructions. If, because of programmer error, there is data rather than an instruction at a specified memory location, the CPU will attempt to execute the data as an instruction. If the data happen to represent a valid instruction, it will be executed with, of course, unpredictable results. If it is not a valid instruction, the CPU generates an error message and stops execution of the program.

Program errors are called *bugs* and can wreak havoc in the orderly operation of a computer system, particularly if it is operating in a multitask mode in which several programs are resident in memory concurrently and execution is under control of a master executive program. An error that inadvertently causes even a single instruction in memory to be changed can lead to quick destruction of all programs in memory as more and more incorrect instructions are executed, some of which further change the instructions in memory.

Such a phenomenon is known as "bombing" the system and can be corrected only by reloading the destroyed ("bombed") programs into memory.

Writing error-free programs is a real challenge to users of computer systems. Many programming aids to the tedious and challenging job of debugging have been developed and are usually provided by vendors as part of the system software available with the hardware.

Principles of CPU Operation

Although all CPUs are different, there are certain principles of structure and operation that are essentially universal to all computers. The computer system block diagram shown in Figure 2.6 expands the conceptual diagram of Figure 2.3 and provides more detail for CPU structure.

Computers are *synchronous* devices, which means that they operate under control of an internal clock that steps the CPU circuitry through the operations necessary to execute instructions. A single instruction may require from 2 to 20 or more CPU clock steps (cycles) to execute, depending on the instruction. Most computers of 1980 vintage run at speeds exceeding one million cycles per second.

Within the CPU are located several registers that are key to its operation:

1. The *program counter* (PC), sometimes called the instruction address register, which contains the memory address of the next instruction to be executed.

2. The *instruction register*, into which instructions are loaded from memory.

3. The *accumulator*, which accumulates the results of repeated arithmetic and logical operations.

4. *General-purpose* (GP) *registers*, which are used to temporarily store numbers or memory addresses needed by the program. In many computers, all GP registers can be used as accumulators in arithmetic and logical operations, greatly increasing programming flexibility and eliminating the need for a separate accumulator.

The following steps comprise the complete execution of a single instruction:

1. Load instruction which is located at the memory address stored in the program counter into the instruction register. This sets up the control circuitry to perform the operation specified by the instruction.

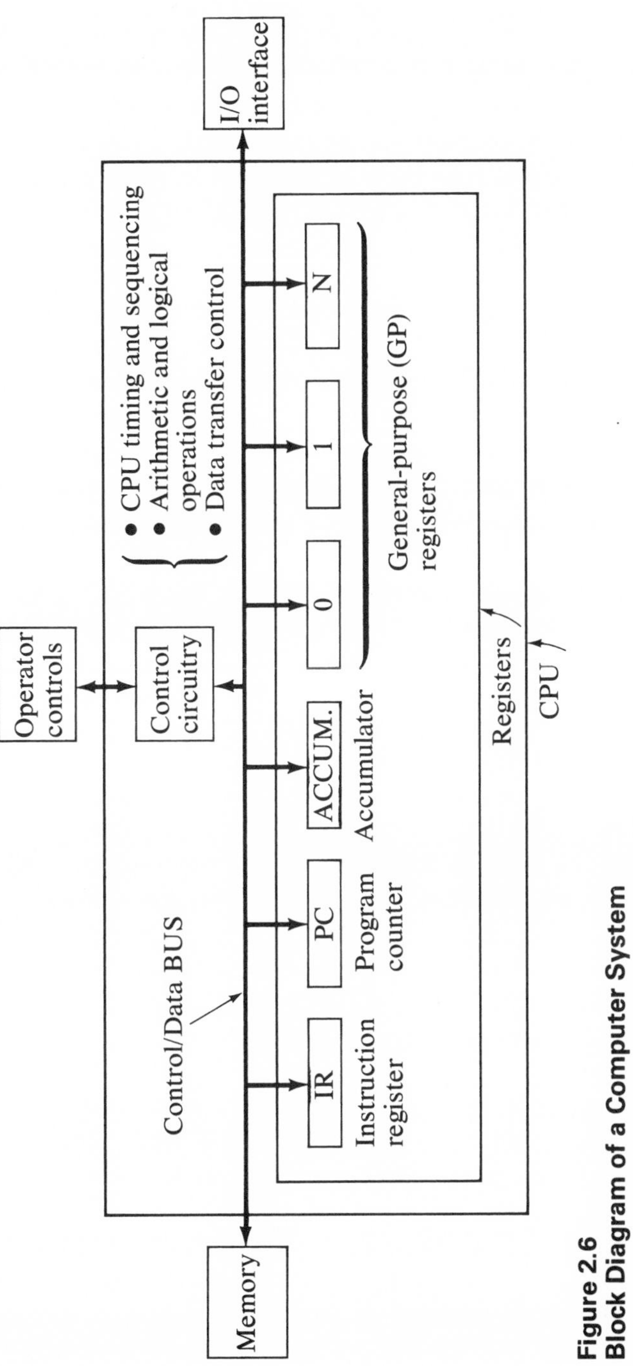

Figure 2.6
Block Diagram of a Computer System

2. Increment (i.e., add 1 to) the contents of the program counter.

3. Execute the instruction. This typically requires several sequential operations, each of which takes place within a single CPU clock cycle. Execution of multiword instructions includes fetching subsequent words of the instruction when required. The program counter is incremented again for each subsequent instruction word fetched.

Branching instructions replace the address in the program counter when executed. In the absence of branching instructions, execution is sequential, as the PC is incremented by 1 for each instruction word fetched.

Program execution is started by loading the PC with the address of the first program instruction and putting the computer into "run" mode. This can be done manually through operator controls or by automatic "boot" hardware which loads and starts an *executive* (monitor) program when the computer power is turned on. Once started, execution continues until stopped by a halt instruction, an error, or operator controls. Computers under the control of an executive program normally run continuously, and individual task programs are loaded and started by the executive. Further detail on systems operating under executive control is developed later in this chapter.

In many modern computers data are transferred between CPU registers, main memory, and the I/O interface on a single set of connectors called a *bus*. A typical 16-bit computer bus has 16 data lines, 16 address lines, and several control lines. The bus operates in a time-shared mode under control of hardware in the control circuitry of the CPU. The address and control lines are used to "connect" the proper source and destination devices to the bus in effecting a single data transfer.

Sequence of Operations and Interrupts

The normal operating mode of a computer requires that an entire program execute, from start to finish, in the order prescribed by the instructions. Other programs cannot be started until the currently executing program has finished. This can lead to very inefficient utilization of the CPU. For example, an output operation that sends a sequence of characters to a typewriter can transfer characters only as quickly as the typewriter can receive them, which can be as slow as 10 characters per second. Once a character has been sent, the CPU must wait, effectively doing nothing, until the output device is ready to receive another character.

To allow more-efficient CPU utilization and provide more flexibility in operation, computers are provided with *interrupt* capability. When a CPU interrupt line is enabled by a peripheral or other external device, normal execution stops (on completion of the currently executing instruction), and the CPU branches to a prespecified memory location for its next instruction, storing the current instruction location in the link. An *interrupt service program* is located at the specified transfer address, which executes and returns to the interrupted program on completion. An interrupt is, in effect, a hardware-initiated subroutine call (i.e., branch and link operation).

In order to successfully interrupt and later return to an executing program, the contents of all CPU registers (called the *state* of the CPU) must be stored in memory (saved) prior to beginning execution of the interrupt service program and reloaded into the same registers (restored) when control is returned to the interrupted program. This is done as part of the interrupt service program.

Every peripheral or external device capable of generating a CPU interrupt has an individual interrupt service routine. In many computers all such routines are grouped together into a single program, which first identifies the interrupting device and then branches to the proper service routine. More advanced designs can branch directly to a number of assignable memory addresses, often called *vectors*, one of which is assigned to each interrupting device.

The ability to interrupt and later resume execution of the currently running program is what enables a computer to monitor and control processes effectively. A program can run immediately in response to an external event (for example, a critical temperature detector) that occurs at a random time unrelated to internal computer operation. In addition, the time in which the CPU would be idle, waiting for an I/O device to be ready, for example, can be used to run other programs, returning to the waiting program on a "ready" interrupt from the peripheral device.

Most computers have an interrupt structure that allows the assignment of different *priority levels* to different devices. An interrupt service routine for a device of a given priority can itself be interrupted by a device of higher priority. An interrupt from a lower or equal priority device would, however, not be serviced until the current interrupt service is completed. Such a structure provides great flexibility in sequence of operations and permits the most efficient utilization of the CPU with satisfactory response to external events.

A *real-time clock* is an important hardware component in all computer systems that need to keep track of elapsed time or time intervals. It is a simple device that interrupts the CPU on a periodic basis.

Many clocks run off AC line supply and generate interrupts 60 times per second, or every 16.67 milliseconds. The interrupt service routine for a real-time clock simply counts interrupts, that is, increments a specified memory location each time the routine runs. By reading this memory location, other programs can easily determine time of day or can track elapsed time.

Hardware Configurations

Our introduction to computer hardware, structure, and principles of operation is now complete. The diagram of Figure 2.7 expands Figure 2.6 to include the many devices that are connected to the computer through the I/O interface. All real-time computer process-control systems would have some, but not necessarily all, of the peripheral devices shown in this diagram. The exact hardware configuration of any given computer system is highly dependent on the application.

2.4 Computer System Software Concepts

Software Classification

No computer system can operate without software, the programs necessary to execute the tasks performed by the system. Software can be classified into three categories:

1. *Application software* consists of programs for tasks directly related to the primary functions of the system. In a real-time process control system, the application software might include:

—A program to read analog inputs into memory.

—A program to convert these "raw" analog inputs to engineering units.

—A program to compute control outputs based on input values.

—A program to set analog outputs.

—A program to print an operating log.

Also included in this category are utility programs for functions such as calculating the square root of a number. Most application programs are written by the user, although vendor-supplied software generally includes common utility programs such as a square root program.

2. *System-support software* consists of programs that aid the user in the development of application programs. Programs in this category are almost always vendor-supplied. Included are:

—Language processors and linkers which convert programs written in high-level languages such as FORTRAN into machine-language programs.

—Editors, which facilitate the creation or modification of user-written programs.

—Programs which aid in debugging (finding program errors).

3. *Executive software*, often called the *operating system* of the

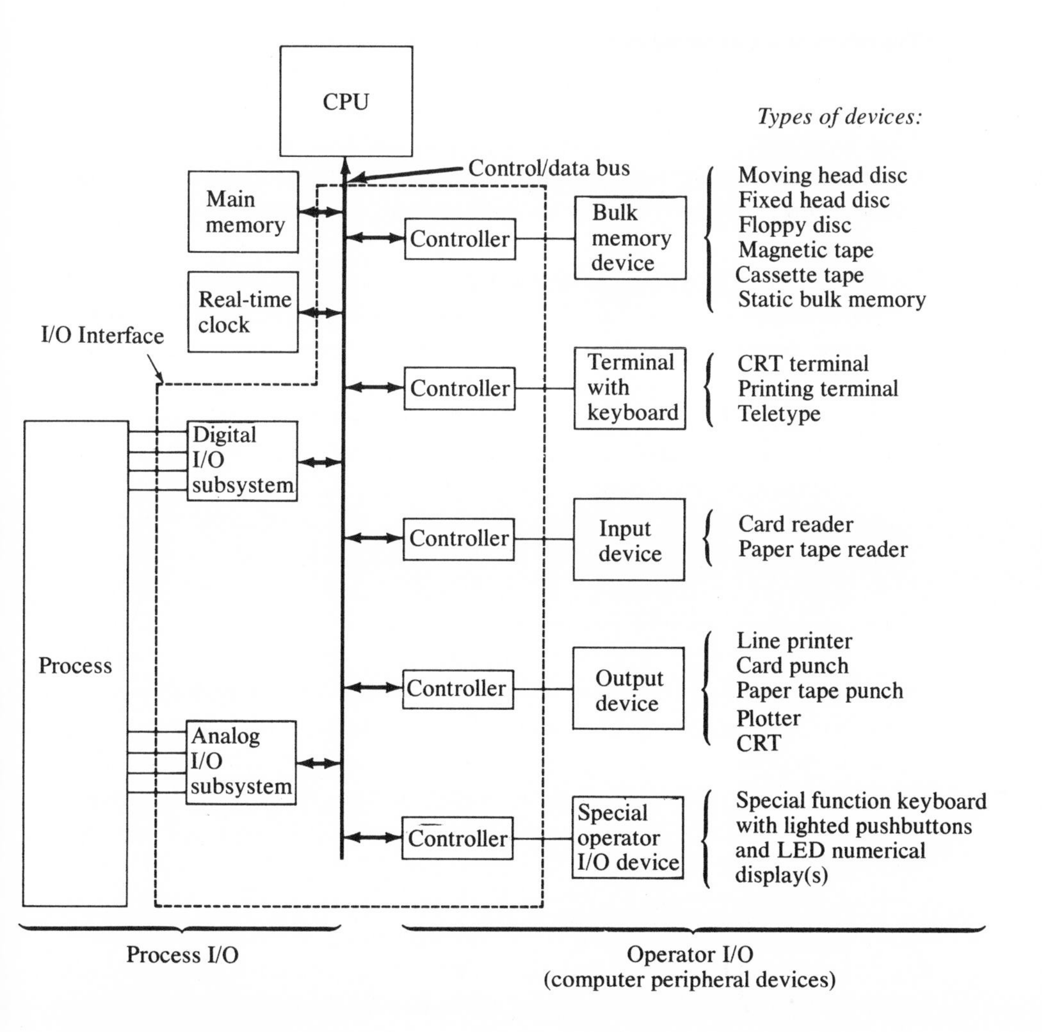

Figure 2.7
Hardware Configurations

computer, consists of programs that oversee or supervise the actual operation of the system while it is running, performing such functions as:

—Scheduling and starting the execution of system-application programs.

—Allocating main memory and loading programs into main memory from bulk memory (e.g., disc).

—Overseeing I/O operations.

—Servicing interrupts.

Programming Languages

Although it is possible to write computer programs directly in machine language (i.e., binary instructions), the extensive bookkeeping required to assign and keep track of memory locations for all data and instructions required makes this a very tedious and error-prone procedure.

To simplify the programming task, all computer manufacturers assign a mnemonic symbol of two or three letters to each instruction type and provide a system-support program called an *assembler* which translates these symbols into machine language. Symbols (names) assigned by the programmer, sometimes called *variables*, are used to designate memory locations to be used for data. The assembler assigns memory locations to such symbols. An *assembly language statement* combines an instruction mnemonic symbol with variable names to produce the symbolic equivalent of a single machine instruction. Examples of such statements are included in the task example developed in the next section.

Writing programs in assembly language enables a programmer to take maximum advantage of the capabilities of an individual computer. It requires, however, a thorough knowledge and understanding of the instruction set and assembly language of the computer being programmed. In addition, programs written for one computer cannot be used on another machine, because all instruction sets and assembly languages are different.

High-level languages such as FORTRAN and BASIC have been developed to further simplify the programming task and provide a degree of standardization that allows programs written in a given language to be run on any computer which has a compiler for that language. High-level language programs are sequences of *source statements* that specify, for example, basic arithmetic or logical operations to be performed on data, I/O operations, branching (conditional or unconditional), and subroutine calls. As in assembly language, symbols defined by the programmer are used to indicate specific variables or memory locations. High-level languages require no specific knowledge of computer architecture or instruction sets. They are easy to learn and have been routinely and widely taught for many years in university and high school technical programs. It is expected that most readers are familiar with one or more high-level languages. The process of translating a program written in a high level language into an executable machine language program is developed later in this chapter.

Example

In this subsection a very simple task is used to compare machine language, assembler language, and high-level language programming. The task is as follows:

Task: Add numbers in memory location 1121 and 1122. Store the sum in memory location 2322.

Steps:

1. Load contents of memory location 1121 into register 1.
2. Add contents of memory location 1122 to contents of register 1. (Register 1 now contains the sum).
3. Store register 1 contents in memory location 2322.

A. MACHINE LANGUAGE

Step	*Memory Locations for Instruction*	*Instruction* Word 1			*Instruction* Word 2
1 (Load)	121,122	1110 0101	0001	0000	0000 0100 0100 0001
2 (Add)	123,124	1110 1101	0001	0000	0000 0100 0110 0010
3 (Store)	125,126	1110 0110	0001	0000	0000 1001 0001 0010
		OP-Code	First operand (Reg. 1)	Second operand (memory)	Memory address of second operand

Three two-word instructions are required for this task. In each instruction, the left 8 bits of word 1 contain the OP code, and the remaining bits specify operand addresses. In these instructions the first operand is register 1, and the second operand is the memory location at the address of the second instruction word.

B. ASSEMBLY LANGUAGE

Symbols (Variables):
I represents contents of memory location 1121
J represents contents of memory location 1122
K represents contents of memory location 2322

Assembler Program:

Step	*Assembly*	*Language*	*Source Statement*
1 (Load)	LDM	1	I
2 (Add)	ADM	1	J
3 (Store)	STM	1	K
	OP Code mnemonic symbol	First operand (register 1)	Second operand (symbolic memory location)

Each assembly-language statement is translated into a corresponding machine-language instruction by the assembler program. The assembler normally assigns and keeps track of memory addresses for symbols. Specific memory locations, if desired, can be assigned by the programmer with additional assembly-language statements.

C. HIGH-LEVEL LANGUAGES

Symbols: Same as assembly-language example.

	Source statement
FORTRAN:	K = I + J
BASIC:	LET K = I + J

In high-level languages, only a single statement is required for this task. The statement is translated into machine-language instructions by the language *compiler* program.

Language Compilers and Interpreters

A system-support program that converts, or translates, high-level language programs into machine-language instructions is called a *compiler*. Writing a high-level language program and converting it to an executable machine-language program actually involves several steps:

1. The *source program* is the set of high-level language statements

created by the programmer. It must be in a form which can be read as input by the computer, for example, punched cards, paper tape, floppy disc, or cassette tape. A system support program called a *source editor* is available in many systems to aid in creating or modifying source programs. Using a source editor, source program statements can be typed into a terminal, and the resulting source program is stored in a file on a bulk memory device. Such source files can also be easily modified using the source editor program.

2. The compiler program is run and reads the source program as input. The compiler produces machine instructions for the program in what is called *object* form. As noted earlier, many machine instructions contain memory addresses which locate data or other instructions as branching targets. Therefore, a machine-language program cannot be completely written until all its instructions and data are assigned specific memory addresses. A compiler generates instructions assuming that the programs will be located in memory beginning at memory address 0. The *object module* produced by the compiler contains, in addition to these instructions, data tables that can be used by another program to *relocate* the program; that is, modify all its memory references so the program can run in another assigned memory location, different from 0.

Many compilers are two-step programs which first translate the source program into assembly language. The system assembler is then run to produce the final object module. All compilers also check for and report programming errors. In addition to the object module, they can also produce printed listings of the source statements, memory "maps" for all program variables, and generated code listings, usually in assembly language.

3. The final step in conversion is to assign a specific memory location to the program and modify the object module addresses for that location. This is done by a system support program called a *linker*. The linker output is the "memory image" of executable machine code*. Actually, the linker is a powerful program that can accept many separately compiled object modules and "link" them into a single executable program, assigning memory addresses to and relocating each module. This capability allows a programmer to develop and compile individual program pieces separately, for example, a main program and the subroutines it calls. It also permits grouping

*The term *code* is a frequently used synonym for program instructions in various formats. Thus we speak of source code, assembler code, object code, and machine code to refer to programs, or subsets thereof, in these formats.

commonly used programs (such as square root) into a *library* in object format. The linker automatically extracts all such required programs from the library and includes them in the final run module (i.e., executable program). Linkers also produce printed memory maps of all program locations.

The oldest and most widely used high-level "scientific" language is FORTRAN IV. Virtually all computers, except the smallest microcomputer systems, have a FORTRAN compiler available. Although FORTRAN was originally created and standardized for batch processing (i.e., sequential program execution) applications, many vendors provide real-time extensions, a set of subroutines that work in conjunction with the computer's operating system to perform functions required by real-time systems, such as:

- Analog and digital process input and output.
- External interrupt handling.
- Scheduling task execution and other timekeeping functions.

Programs written in FORTRAN or other high-level languages generally require more memory and time to execute than equivalent programs written in assembly language by a skilled programmer. Usually, however, this is an acceptable price to pay for the great increase in flexibility and ease of programming provided by the high-level language.

A different approach to the use of high-level languages is a program called an *interpreter*. Unlike a compiler, an interpreter does not convert high-level source statements into machine language; rather, it reads the source statements during actual program execution, "interprets" the statements to determine actual functions required, and calls subroutines (which are part of the interpreter) to carry out the functions. Memory is assigned to variables "dynamically" (i.e., during actual execution) the first time a symbol (i.e., variable name) is encountered. The popular high-level language BASIC is usually implemented with an interpreter. BASIC is similar in structure to FORTRAN but simpler, particularly for I/O operations.

Interpreter-based programs run much more slowly than compiled programs because of the time required to interpret source statements. They can be easily modified, however, simply by changing the source code. Changes to a *compiled* program require that the program be recompiled and relinked after the source code is changed. Interpreters are advantageous, therefore, in applications such as pilot plants where programs are changed frequently, whereas compiler-based languages are more advantageous for applications that, once developed and debugged, require little or no change, or where the higher speed is required by the application.

Real-Time Operating Systems

Executive programs, usually called *operating systems*, are used in many computers to control program execution and permit more efficient utilization of the CPU. The operating system (OS) is the master program in the system. Once it is loaded into memory and started, it runs continuously. All other programs run, in effect, as subroutines that are called by the OS and return to the OS on completion. Many computers have automatic "boot" hardware that loads the OS into memory from bulk storage and starts execution when the power is turned on.

The primary function of any operating system is to schedule and execute programs, which are often called *tasks* when run under OS control. Tasks can be scheduled in two ways:

1. On *command* from an operator or system user. Commands are statements read by the OS from its primary input device, usually a console terminal into which commands are typed by the user. Although each operating system has its own command structure (called *syntax*), many vendors use a command syntax adapted from IBM's JCL (job control language) developed in the early 1960s for System/360 operating systems. For example, the following command (from DEC's RT-11 operating system) is used to compile a FORTRAN program and produce a source listing on the system printer as well as an object module. The source program is stored on bulk memory device DX1 (a floppy disc) in a file named PGM1.FOR, and the object module is to be stored on the same device in a file called PGM1.OBJ. The command is:

```
FORTRAN/LIST/OBJECT:DX1:PGM1.OBJ DX1:PGM1.FOR
```

In response to this command the operating system loads the FORTRAN compiler into memory from bulk storage, opens the specified files on device DX1, and starts execution of the compiler program. When the compiler finishes, it returns control to the OS, which then closes the specified files and reads the next command.

2. On request from a running program. Real-time operating systems include a series of subroutines that can be called to schedule other programs, either immediately, at some later time, or whenever a specified event such as an external interrupt occurs. A call to a scheduling routine makes an entry in a table of pending tasks maintained by the operating system. The task-manager portion of the OS, which runs whenever the CPU is interrupted, executes these tasks as scheduled if the CPU is free. When there is a conflict for the CPU between two or more tasks, the task manager uses task priorities, dis-

cussed in more detail later, to arbitrate the conflict. As an example, consider a program to read analog inputs that must run periodically, once per second. If the first statement of this program is a subroutine call that schedules *itself* to run after 1 second has elapsed, a single execution of the program (scheduled, say, by an initialization program) achieves the desired periodic execution.

Another important operating system function, closely related to task scheduling, is called *multitasking* or *multiprogramming*. In a multitasking system it is possible to have more than one program in memory and in some stage of execution. Although to the user the programs seem to be running concurrently, only one program is executing at any given time. To change from one executing program to another, the OS saves the state of the running program (CPU register contents) in memory and transfers to another program. Execution of the suspended program can be resumed at a later time by restoring the CPU state and restarting execution at the point where the program was suspended. This operation, called *context switching*, is identical in concept to interrupt handling described earlier.

Time-sharing systems are simple examples of multitasking operations, where each user of the system is allocated a portion of CPU time in a "round-robin" fashion. Context switches also occur when the running program is held up (bound), waiting for some event to occur, such as completion of an I/O operation. In this way the CPU is used with maximum efficiency, and is, in fact, never idle whenever there are pending or uncompleted tasks to be run.

The simplest multitasking systems can handle only two concurrent programs, generally called the *foreground* and *background* tasks. The foreground task has priority, so the background runs only when the foreground is suspended for I/O or some other reason. More elaborate systems can handle several (as many as 256) concurrent tasks. Each program is assigned a *priority level* of execution, and the program running at any given time will be the highest-priority program that is not bound. A program can, for example, schedule another program with higher priority than itself to run immediately. This would cause an immediate context switch from the calling to the called program, with the calling program resuming execution after completion of the called task. This is, in effect, identical to a subroutine call within a single task, and allows tasks to interact in a manner specified by the tasks themselves.

Two other key operating system functions are *interrupt handling* and managing *I/O operations*. In most operating systems interrupts transfer control to the OS which then calls the required interrupt service routine. Routines to service peripheral device interrupts are usually vendor-supplied and often part of the device drivers. User-written interrupt service routines, when required, are "connected" to the operating system with OS commands or program calls to vendor-supplied subroutines that perform this function.

I/O operations involving peripheral or process devices are managed by the operating system, which includes a series of I/O subroutines. Generally, a single call to the appropriate subroutine is all that is required in a user-written program to execute I/O operation. The operating system loads and calls the device driver and services device interrupts as required to complete the specified operation. I/O is a complex procedure in many computers, and a single I/O call from a user program will frequently initiate a sequence of several thousand instructions, all under control of the operating system.

Real-time operating systems (RTOS) are complex and often very large programs that require years of development by computer vendors. A complete RTOS may contain hundreds of individual programs or modules. Not all these modules will be required by most systems. The operating system for any given system is created by selecting from the total set the programs needed for desired OS functions and support of the system devices. These programs are linked into a single operating system tailored to the specific computer system in a process known as *system generation.*

Today's sophisticated real-time operating systems have many capabilities beyond the basic functions described here. Such powerful executive programs substantially ease the development of on-line computer applications.

2.5 Configurable Digital Systems and Networks

In spite of the powerful support software and operating systems available for digital computers today, it is an expensive and time-consuming task to develop application software for real-time computer control and information systems. In custom, one-of-a-kind applications it is not unusual for the software development costs to exceed the cost of the computer system hardware, sometimes by a factor of two or three. Writing such software also requires personnel with a very high level of specialized computer skills.

In an attempt to reduce software costs and permit engineers without a high level of computer skills to produce application software, several manufacturers have created software packages consisting of generalized subroutines to perform functions commonly found in computer control systems, for example, process I/O control algorithms, operator communication, and generation of reports and logs. Such systems are "programmed" by entering data and parameters into *tables* specifying which functions are to be used as well as which process or internal variables are to be used by each function. The systems run under the control of an internal, interpreter-like executive program that reads the data tables and calls the required subroutines to perform the specified functions.

The process of entering data into the tables begins in many such systems with filling in a standard form supplied by the manufacturer to specify each function; consequently, they are often called *fill-in-the-forms* packages. More generally, they are referred to as *control-oriented languages* or *table-driven software*. Generating application software using such packages does not require writing programs but rather *configuring* prewritten programs together for a specific application. Thus the combination of a digital computer system with a table-driven application software package is referred to as a *configurable digital system*. Applications are limited to those which require only functions available in the package, although some packages permit user-written programs to be added for specialized functions.

The cost of computer hardware has dropped so dramatically in recent years that several manufactuers of process-control equipment have developed configurable digital systems built around microprocessors that are cost-competitive with conventional analog control hardware. Such systems are *distributed*, which means that several individual computers are connected in a *network* and communicate with one another (i.e., transfer data and commands) on a high-speed, time-shared communications line called a *data highway*. Such systems include not only microprocessor-based controllers with built-in process I/O but also *operator stations* built around color CRTs and keyboards which replace the conventional control panel as an operator interface. They are easily configured by engineers with no specialized computer knowledge. Their total capability includes all conventional control system functions plus many capabilities not easily implemented with analog hardware, including such advanced control strategies (developed in later chapters) as feedforward, dead-time compensation, and multivariable control. Redundancy is widely used in such systems to provide reliability which often exceeds that of the conventional systems they replace.

The advent of such cost-effective and easy-to-apply systems has caused a dramatic upturn in the use of digital systems for process control as well as the application of advanced control strategies. It is likely that by 1985 or so, the great majority of new process control systems will be implemented with networks of digital computers.

References

1. Mellichamp, D.A., Ed., *CACHE Monograph Series in Real-Time Computing*, Monographs I–VI, CACHE Publication Committee, c/o Brice Carnahan, Dept. of Chemical Engineering, University of Michigan, Ann Arbor, MI 48019, 1977
2. Harrison, T. J., Ed., *Minicomputers in Process Control—An Introduction*, Instrument Society of America, Pittsburgh, 1978.

CHAPTER 3

Single-Loop Computer Control

Having read the description of the hardware and software required for computer-control applications, let us now see how we might go about implementing computer control on a typical process loop. For simplicity, let us consider a single-loop system. The single-loop system is a flow control loop currently on conventional control. The pneumatic controller of this loop is the proportional + integral (PI) type. It is desired to place this loop under computer control, utilizing the discrete equivalent of the PI controller.

3.1. The Present System

A schematic of the analog flow control loop is shown in Figure 3.1.

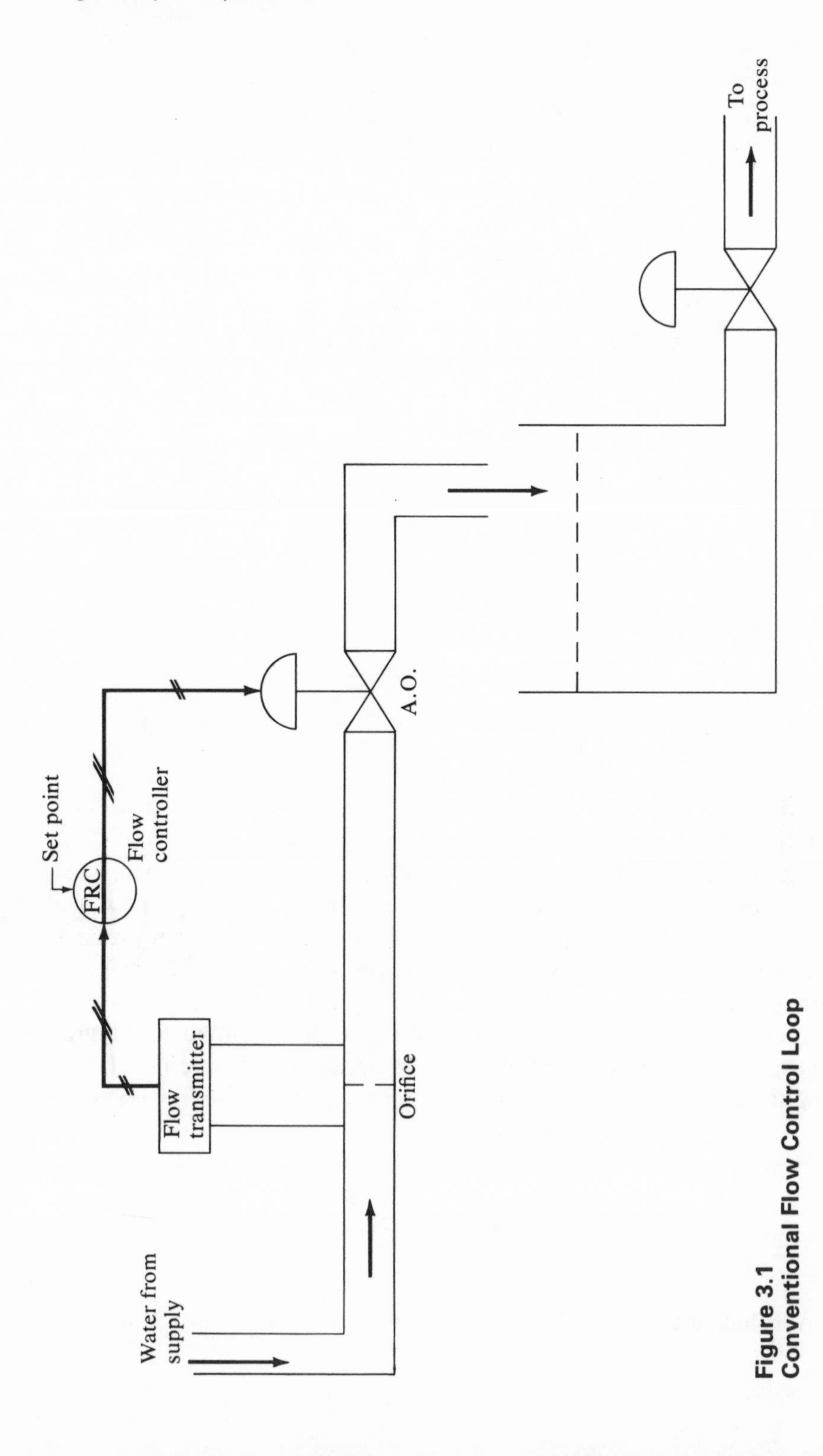

Figure 3.1
Conventional Flow Control Loop

The flow transmitter measures the pressure differential across an orifice and converts it into a 3 to 15 psig pneumatic signal. The flow controller compares this signal with the desired value, the set point. If an error exists, the controller outputs a signal (also between 3 and 15 psig) which manipulates the valve position so as to eliminate the error. Note that all instrument signals in this pneumatic loop are continuous signals. If some of the instruments were of the electronic type, their inputs and outputs would be current or voltage signals, but all would still be continuous signals.

The operation of the ideal PI controller may be described in terms of the following equation:

$$v = v_0 + \frac{100}{PB}\left\{e + \frac{1}{\tau_I}\int_0^t e\,dt\right\} \tag{3.1}$$

where

v = output signal from the controller at time t, percentage of full-scale
v_0 = output signal from the controller at time $t = 0$ (at the time when the controller is switched from manual to automatic), percentage of full-scale
e = error signal (i.e., $R - c_m$), percentage of full-scale (c_m is the measured variable, percentage of full-scale, and R is the set point, percentage of full-scale)
PB = proportional band, percent (defined as the proportional scale change in input necessary to cause a full-scale change in controller output)
τ_I = integral time, minutes

Some literature describes Equation (3.1) in terms of the gain k_c and reset T_I. The relationship between k_c, PB, and τ_I, T_I is

$$k_c = \frac{100}{PB} \quad \text{dimensionless} \tag{3.2}$$

and

$$T_I = \frac{1}{\tau_I} \quad \text{repeats per minute} \tag{3.3}$$

Thus if PB is 50% and τ_I is 2 minutes, then

$$k_c = \frac{100}{50} = 2 \tag{3.4}$$

and

$$T_I = \frac{1}{2} = 0.5 \text{ repeats/minute}$$

Let us assume that the desired flow rate is 3.78 gallons per minute (gpm). Then the value of R and v_0 can be calculated from the instrument calibrations as follows:

1. From orifice calibration shown in Table 3.1, at 3.78 gpm, the flow transmitter output is 7.26 psig (i.e., 35.5% of full-scale range of 3 to 15 psig). Therefore, $R = 35.5$.

2. From valve characteristics shown in Table 3.1, at 3.78 gpm, the signal to the valve must be 8.9 psig (i.e., 49.17% of full-scale range of 3 to 15 psig).

3. $v_0 = 49.17$ (from Table 3.1, this corresponds to 6 on the flow recorder/controller).

Let us also assume that tests have been conducted at the desired operating level of 3.78 gpm and the following tuning constants have been found to be satisfactory:

$$PB = 50\% \tag{3.5}$$

$$\tau_I = 1 \text{ minute}$$

To start up and operate this loop, the following procedure may be used:

1. Turn on instrument air and water supply.

Table 3.1
Flow Transmitter Calibrations and Valve Characteristics

Flow (gpm)	Indicated Flow on FRC (arbitrary units)	Output Signal from Flow Transmitter, (percentage of full-scale)	Signal to Valve (percentage of full-scale)
0	0	0 (= 3 psig)	0 (= 3 psig)
0.79	1	1	14.58
1.29	2	3	21.67
1.92	3	8	29.17
2.54	4	15	34.2
3.20	5	24	40.5
3.78	6	35.5 (= 7.26 psig)	49.2 (= 8.9 psig)
4.44	7	49	59.2
4.92	8	64	65.8
5.61	9	82.5	79.6
6.31	10	100 (= 15 psig)	100 (= 15 psig)

2. Move the auto/manual knob on flow controller to manual.
3. Set the proportional band at 50 and reset dial at 1.
4. Move set-point needle to 6 on FRC (i.e., 35.5% of full-scale input).
5. Adjust the output of the controller to get 49.17% (8.9 psig) output signal. This will be achieved when the set point and measurement needles on FRC match.
6. Switch FRC to automatic.

3.2. Switchover to Computer Control

Figure 3.2 shows the schematic of the computer-controlled flow loop. The output of the flow transmitter is fed to an air-to-current (P/I) transducer. The resulting 4 to 20 mA electrical signal is connected to the terminals on the control computer which represent one of the analog-to-digital (A/D) converter channels. The discrete output of the A/D converter is available to the computer on demand. The discrete output from the computer is converted to a continuous signal, also on demand, by one of the digital-to-analog (D/A) converter channels. The D/A output, 4 to 20 mA, is available at the analog output terminals of the computer. It is connected to an I/P transducer, which produces a proportionate 3 to 15 psig pneumatic signal. This signal then operates the valve.

The control computer is instructed to sample the A/D channel every T seconds (where T is the sampling period). The computer program operates on this measurement (which represents the value of the measured variable at the sampling instant) using the discrete equivalent of the PI controller equation and computes the desired control-algorithm output. The computer is then asked to forward this output to the D/A converter and on to the valve. This procedure is repeated every T seconds so as to achieve closed-loop computer control. Now, let us develop a computer program that will accomplish this task. First, we must derive an equation that represents the discrete equivalent to the PI controller.

The output of the control equation at the nth sampling instant, using numerical integration for the integral of Equation (3.1), is

$$v_n = v_0 + \frac{100}{PB}\left\{e_n + \frac{1}{\tau_I}\sum_{i=0}^{n} e_i T\right\} \tag{3.6}$$

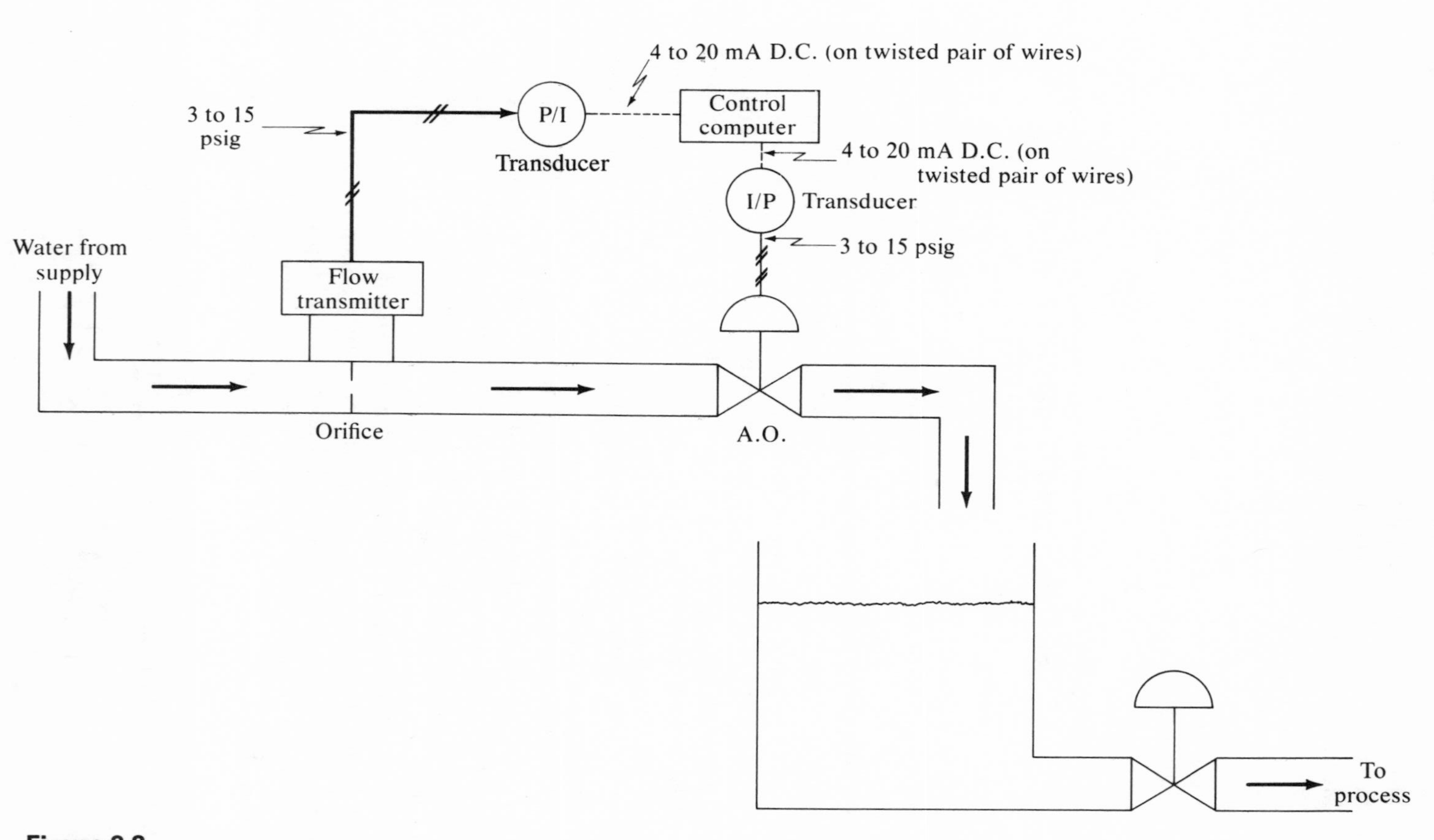

Figure 3.2
Computer-Controlled Flow Loop

Similarly, the output at the $(n - 1)$th sampling instant is

$$v_{n-1} = v_0 + \frac{100}{PB}\left\{e_{n-1} + \frac{1}{\tau_I}\sum_{i=0}^{n-1} e_i T\right\} \tag{3.7}$$

Subtracting Equation (3.7) from Equation (3.6) gives

$$v_n = v_{n-1} + \frac{100}{PB}\left\{(e_n - e_{n-1}) + \frac{T}{\tau_I} e_n\right\} \tag{3.8}$$

where

v_n = control-law output at the nth sampling instant
v_{n-1} = control-law output at the $(n - 1)$th sampling instant
e_n = error at the nth sampling instant
e_{n-1} = error at the $(n - 1)$th sampling instant
T = sampling period, seconds

The output v and error e are in percentages, as in the case of the conventional control system. Equation (3.8) is the digital equivalent to the conventional PI controller and is referred to as the proportional + integral (PI) control algorithm.

The output of the algorithm one sampling period after the loop is placed under computer control is computed from the equation

$$v_1 = v_0 + \frac{100}{PB}\left(e_1 + \frac{T}{\tau_I} e_1\right) \tag{3.9}$$

where v_0 is initialized to a value which is estimated to cause the measured variable to equal its set point at steady state. The algorithm output at subsequent sampling instants is computed from Equation (3.8).

The flowchart of a computer program to implement computer control using the PI control algorithm is shown in Figure 3.3. From this flowchart the control program can be developed using a suitable programming language.

The start-up and operational procedure for the computer control loop consists of the following steps.

1. Connect the equipment as shown in Figure 3.2.
2. Turn on instrument air and water supply.
3. Turn on computer and the I/O device (e.g., teletypewriter).
4. Initialize the computer (the procedure will vary from computer to computer).

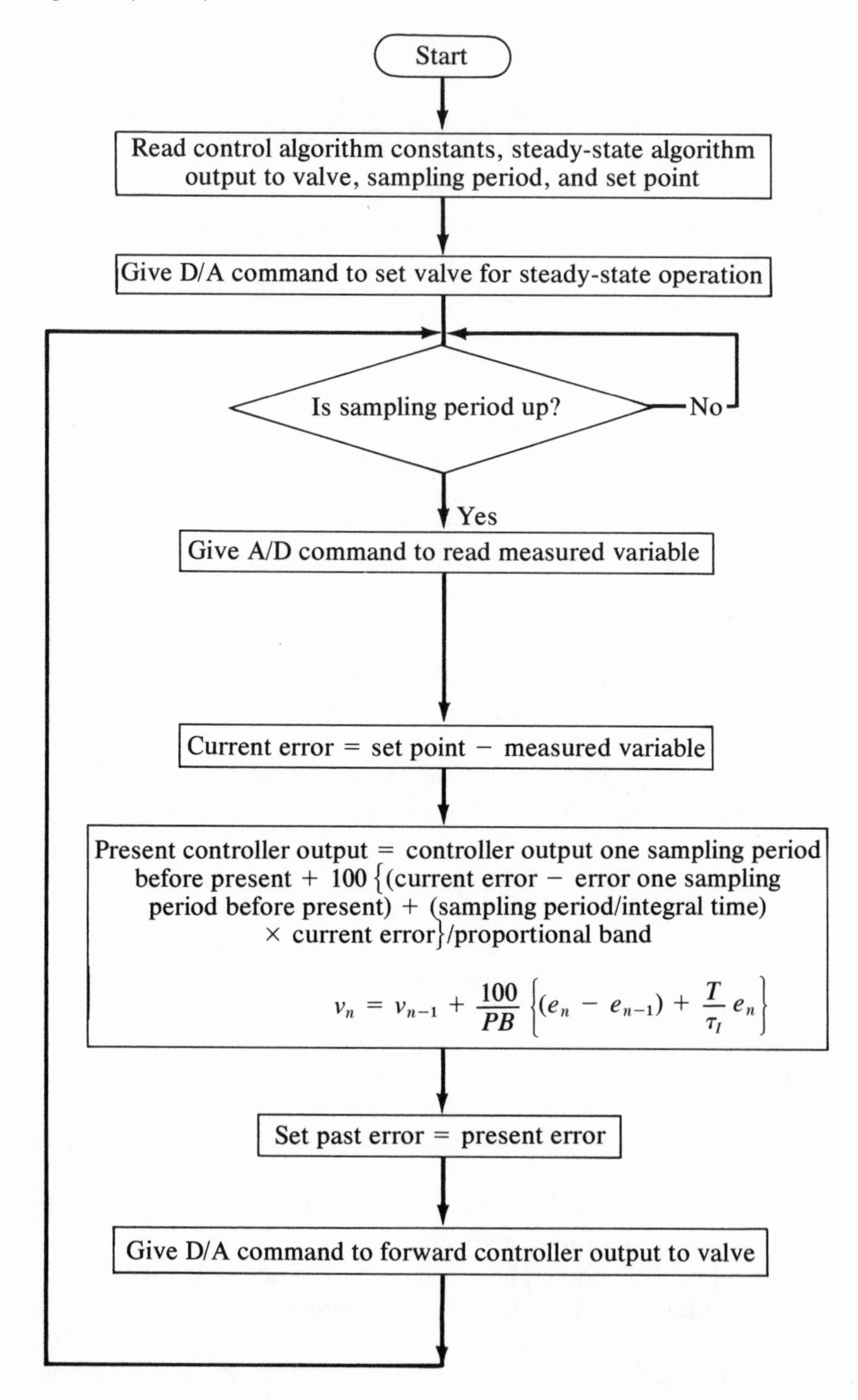

Figure 3.3
Flowchart of Computer Program to Implement Control

5. Follow the manufacturer's procedure to enter the control program into the computer.
6. Execute the program.

The procedure described in this chapter applies to single-loop computer control using the PI control algorithm. This approach can be readily extended to computer control of multiple loops using the PI (or P, or PID) control algorithm by suitable modifications using repetitive program structures. The real power of a computer in control applications lies in its ability to facilitate the implementation of improved or advanced control strategies that would be impractical with analog hardware. The development of such strategies requires that we be able mathematically to represent and analyze the computer-control loops. This type of analysis is also needed to determine the stability characteristics of computer-control loops. The next several chapters are, therefore, devoted to the mathematical analysis of the computer-control loops.

CHAPTER 4

Mathematical Representation of the Sampling Process

In a computer-control loop, the process variable is sampled every T seconds, where T is the sampling period. This function is performed by an A/D converter. For the purpose of analyzing and designing computer-control systems, it is necessary to develop a mathematical representation of the analog-to-digital converter or the *sampler*, and this is the subject of this chapter.

The basic purpose of the sampler is to obtain the values of the process variable, which is a continuous function, at regular intervals of time. Figure 4.1 is a pictorial representation of this concept.

The sampler permits the input signal to pass through only during the short interval τ but blocks it during the remaining portion of the sampling period. The signal transmission interval τ is very short in comparison with the sampling period. The values of the output function during the interval τ are equal to the corresponding values of the input function during the interval. In other words, the output from the sampler can be thought of as a train of very narrow pulses, the enve-

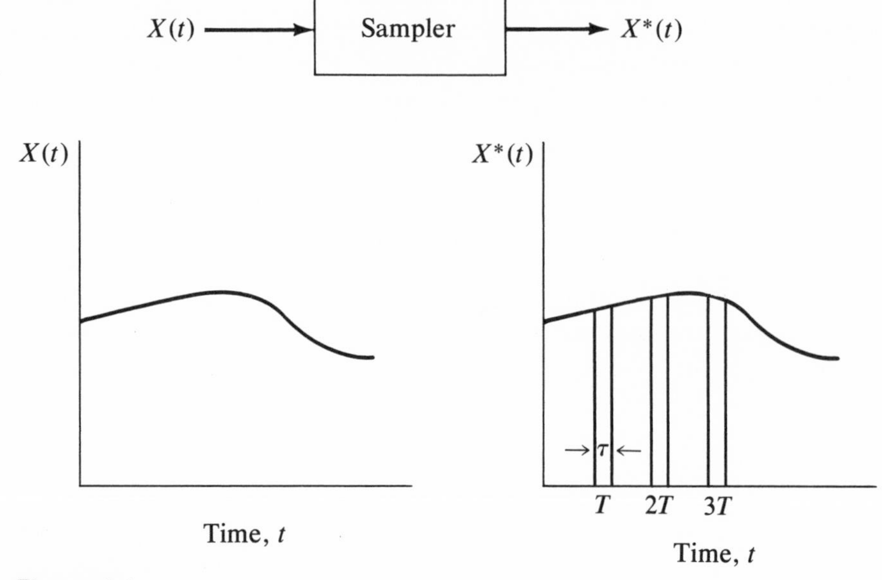

Figure 4.1
Input-Output Signals from Sampler

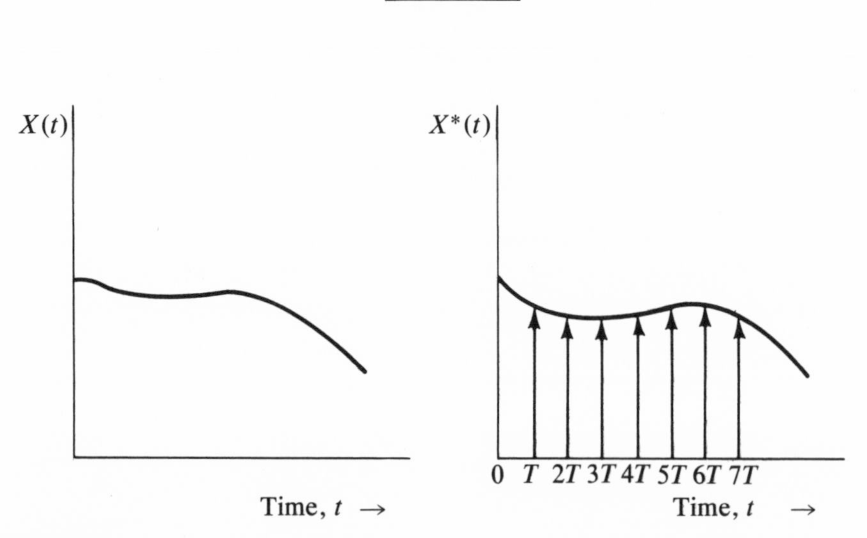

Figure 4.2
Idealized Sampling Operation

lope of which is identical to the input signal. The quantity $f_s = 1/T$ is called the *sampling frequency*.

Because the sampling duration τ (i.e., the width of the sampled pulse) is small compared with the most significant time constant of the control system as well as the sampling period T, the sampler output can be considered as a train of impulses whose strengths are equal to the values of the continuous function at the respective sampling instants, as shown in Figure 4.2. This assumption greatly simplifies the representation of the sampler. To fix the concept of representing the sampled function by impulses, let us review the mathematical representation of a rectangular pulse and the impulse.

A rectangular pulse function $f(t)$ is represented (see also Figure 4.3) as

$$f(t) = \begin{cases} 0 & t < 0 \\ \dfrac{1}{h} & 0 \leqslant t \leqslant h \\ 0 & t > h \end{cases} \tag{4.1}$$

Now, let $h \to 0$. Then we obtain a new function that is zero everywhere except at the origin, where it is infinite. However, the area under this function always remains unity. This new function is called the *delta function* $\delta(t)$. The fact that its area remains unity means that

$$\int_{-\infty}^{\infty} \delta(t)\, dt = 1 \tag{4.2}$$

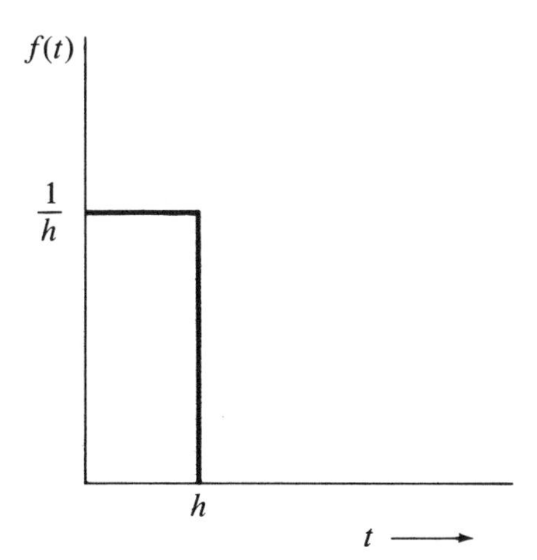

Figure 4.3
Pulse Function

The delta function is also called the *unit impulse function* (see Figure 4.4).

Mathematically, a *train of unit impulses* (whose areas or strengths are unity) can be represented (see also Figure 4.5) as

$$\delta_T(t) = \sum_{n=0}^{\infty} \delta(t - nT) \tag{4.3}$$

In this equation $\delta(t)$ is the unit impulse function at $t = 0$, and $\delta(t - nT)$ is the unit impulse function occurring at $t = nT$.

What we are interested in as the output of the sampler is a train of impulses whose areas numerically equal the values of the continuous function at the respective sampling instants. Such a relationship can be readily written with the aid of $\delta_T(t)$ as

$$\begin{aligned} X^*(t) &= X(t)\ \delta(t - nT) \\ &= X(t) \sum_{n=0}^{\infty} \delta(t - nT) \end{aligned} \tag{4.4}$$

Note that

$$X(t) = 0 \text{ for } t < 0$$

Because the function $\delta(t - nT)$ is zero everywhere except at $t = nT$, the only values of $X(t)$ needed are those at $t = nT$. Thus Equation (4.4) can be rewritten as

$$X^*(t) = \sum_{n=0}^{\infty} X(nT)\ \delta(t - nT) \tag{4.5}$$

Note that although Equations (4.4) and (4.5) represent the relation-

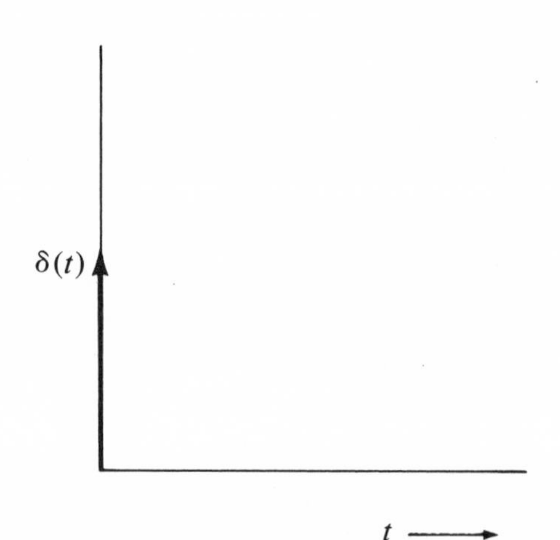

Figure 4.4
Impulse Function

ship between the continuous function and the sampled function, we cannot substitute numerical values into the right-hand side of the equation to obtain the values of the sampled function at a particular sampling instant. This is because the delta function has meaning only when it appears as an integral. The value of the Kth sample is given by

$$X(KT) = \int_0^{\infty} X(t)\ \delta(t - KT)\ dt \tag{4.6}$$

The validity of Equation (4.6) should be intuitively clear. At $t = KT$ the integral $\int_0^{\infty} \delta(t - KT)\ dt$ has a value of unity at which instant $X(t)$ has a value of $X(KT)$.

On the basis of the discussion in this chapter, then, we are able to represent the sampled function as

$$\begin{aligned} X^*(t) &= X(t) \sum_{n=0}^{\infty} \delta(t - nT) \\ &= \sum_{n=0}^{\infty} X(nT)\ \delta(t - nT) \end{aligned} \tag{4.7}$$

and the value of the Kth sample is given by

$$X(KT) = \int_0^{\infty} X(t)\ \delta(t - KT)\ dt \tag{4.8}$$

A pictorial representation of the various functions is summarized in Figure 4.6.

It should be noted that there are some well-defined limits on how

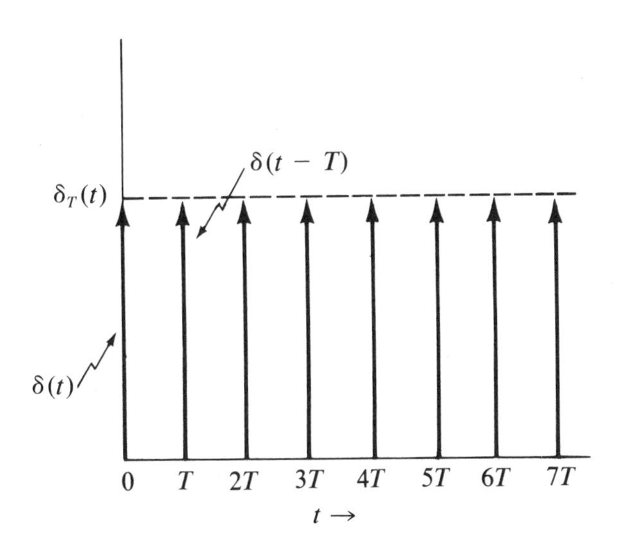

Figure 4.5
Train of Unit Impulses

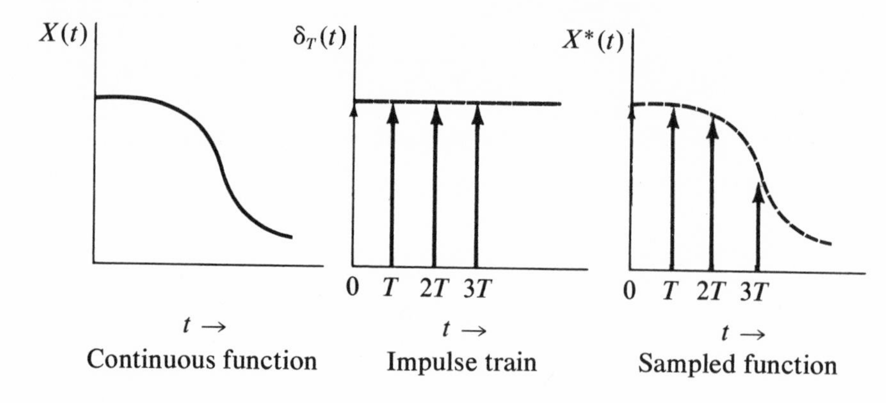

Figure 4.6
Pictorial Summary of Sampling Operation

frequently the process variable should be sampled. If we do not sample at all or sample very infrequently, the quality of control will be very poor. In fact, there is a theorem that specifies a minimum sampling rate for proper signal recovery. On the other hand, if we sample infinitely fast, the performance of the computer-control system will approach that of the conventional-control system, but this is a hypothetical concept. Somewhere, between these extremes, is the desirable sampling frequency for computer-control applications. In Chapter 7 we consider these aspects of sampling-frequency selection.

CHAPTER 5

The Z-Transformation

We found Laplace transforms to be helpful in the analysis of conventional-control systems. In subsequent chapters we will see that the analysis of sampled-data control problems is conveniently handled in terms of Z-transforms. In this chapter we shall develop some of the important concepts of Z-transforms. To begin, recall that the Laplace transform of a function $f(t)$ is defined as

$$l\{f(t)\} = f(s) = \int_0^{\infty} f(t)\ e^{-st}\,dt \tag{5.1}$$

Also recall that the sampled function $f^*(t)$ and the continuous function $f(t)$ are related by

$$f^*(t) = f(t) \sum_{0}^{\infty} \delta(t - nT) = \sum_{n=0}^{\infty} f(nT)\ \delta(t - nT) = f(0)\,\delta(t) + f(T)\,\delta(t - T) + f(2T)\,\delta(t - 2T) + f(3T)\ \delta(t - 3T) + \cdots \tag{5.2}$$

where $f(T)\ \delta(t - T)$ represents an impulse at time $1T$ whose area is $f(T)$. Now take the Laplace transform of Equation (5.2)

$$
\begin{aligned}
l\{f^*(t)\} = l\{f(0)\ \delta(t)\} + l\{f(T)\ \delta(t - T)\} + l\{f(2T)\ \delta(t - 2T)\} \\
+ \cdots = f(0)l[\delta(t)] + f(T)l[\delta(t - T)] \\
+ f(2T)l[\delta(t - 2T)] + \cdots = f(0) + f(T)\ e^{-sT}\ l\{\delta(t)\} \\
+ f(2T)\ e^{-s2T}\ l\{\delta(t)\} + \cdots \qquad (5.3)
\end{aligned}
$$

because $l\{\delta(t)\} = 1$

$$l\{f^*(t)\} = f^*(s) = \sum_{n=0}^{\infty} f(nT)\ e^{-nsT}$$

If we introduce a new variable $Z = e^{sT}$ into Equation (5.3), we get

$$f^*(s)\Big|_{Z = e^{Ts}} = \sum_{n=0}^{\infty} f(nT)\ Z^{-n} \qquad (5.4)$$

This result is defined to be the Z-transform[1] of $f(t)$ and is denoted as $F(Z)$. Thus

$$F(Z) = Z\{f(t)\} = f^*(s)\Big|_{Z = e^{Ts}} = \sum_{n=0}^{\infty} f(nT)\ Z^{-n} \qquad (5.5)$$

5.1. *Z*-Transform of Various Functions

Let us apply the definition and evaluate the Z-transform of some common functions.

Unit Step Function

$$f(t) = \begin{cases} 0 & t < 0 \\ u(t) & t \geqslant 0 \end{cases} \qquad (5.6)$$

By definition,

$$Z\{u(t)\} = F(Z) = \sum_{n=0}^{\infty} u(nT)\ Z^{-n}$$

$$= \sum_{n=0}^{\infty} Z^{-n} = 1 + Z^{-1} + Z^{-2} = \frac{1}{1-r} \tag{5.7}$$

where

$r < 1$ is the ratio of successive terms
$= Z^{-1}$

Therefore,

$$Z\{u(t)\} = \frac{1}{1 - Z^{-1}} \qquad |Z^{-1}| < 1 \tag{5.8}$$

Exponential Function

$$f(t) = \begin{cases} 0 & t < 0 \\ e^{-at} & t \geqslant 0 \end{cases}$$

$$\begin{aligned} Z\{e^{-at}\} &= \sum_{n=0}^{\infty} e^{-anT} Z^{-n} = \sum_{n=0}^{\infty} (e^{-aT} Z^{-1})^n \\ &= \frac{1}{1 - Z^{-1} e^{-aT}} \qquad |Z^{-1}| < e^{aT} \end{aligned} \tag{5.9}$$

Ramp Function

$$f(t) = \begin{cases} 0 & t < 0 \\ kt & t \geqslant 0 \end{cases}$$

$$\begin{aligned} F(Z) = Z\{kt\} &= \sum_{n=0}^{\infty} knT Z^{-n} \\ &= kT\{Z^{-1} + 2Z^{-2} + 3Z^{-3} + \cdots\} \\ &= kTZ^{-1}\{1 + 2Z^{-1} + 3Z^{-2} + 4Z^{-3} + \cdots\} \\ &= \frac{kT Z^{-1}}{(1 - Z^{-1})^2} \\ &= \frac{kT Z}{(Z-1)^2} \qquad |Z^{-1}| < 1 \end{aligned}$$

$$F(Z) = \frac{kTZ}{(Z-1)^2}$$

Sine Function

$$f(t) = \begin{cases} 0 & t < 0 \\ \sin at & t \geqslant 0 \end{cases}$$

because

$$f(t) = \sin at, \quad f(nT) = \sin anT$$

and

$$\sin anT = \frac{e^{jnaT} - e^{-jnaT}}{2j}$$

We get upon substituting in the definition

$$Z\{\sin at\} = \sum_{n=0}^{\infty} \sin anT\, Z^{-n} = \sum_{n=0}^{\infty} \frac{1}{2j} (e^{jnaT} - e^{-jnaT})\, Z^{-n}$$

$$= \frac{1}{2j}\left[\sum_{n=0}^{\infty} \left(e^{jaT} Z^{-1}\right)^n - \sum_{n=0}^{\infty} \left(e^{-jaT} Z^{-1}\right)^n\right]$$

Therefore,

$$Z\{\sin (at)\} = \frac{1}{2j}\left(\frac{1}{1 - e^{jaT} Z^{-1}} - \frac{1}{1 - e^{-jaT} Z^{-1}}\right)$$

Putting the right side over a common denominator, we have

$$Z\{\sin at\} = \frac{1}{2j}\left(\frac{1 - e^{-jaT} Z^{-1} - 1 + e^{jaT} Z^{-1}}{1 + Z^{-2} - Z^{-1}(e^{jaT} + e^{-jaT})}\right)$$

$$= \frac{Z^{-1}(e^{jaT} - e^{-jaT})/2j}{1 + Z^{-2} - 2Z^{-1}\,\dfrac{(e^{jaT} + e^{-jaT})}{2}}$$

Therefore,

$$Z\{\sin at\} = \frac{Z^{-1} \sin aT}{1 - 2Z^{-1} \cos aT + Z^{-2}}$$

$$= \frac{Z \sin aT}{Z^2 - 2Z \cos aT + 1}$$

By similar procedure, Z-transforms of other functions can be obtained. For convenience, Z-transforms are tabulated with Laplace transforms and the corresponding $f(t)$ in Appendix A.

5.2. Properties of Z-Transforms

1. Z-transforms Are Linear

$$\begin{aligned} Z\{f(t) + g(t)\} &= \sum_{n=0}^{\infty} [f(nT) + g(nT)]\, Z^{-n} \\ &= \sum f(nT)\, Z^{-n} + \sum_{n=0}^{\infty} g(nT)\, Z^{-n} \\ &= Z\{f(t)\} + Z\{g(t)\} \end{aligned} \tag{5.10}$$

that is, the Z-transform of the sum of two functions is the sum of their individual transforms.

$$\begin{aligned} Z\{cf(t)\} &= \sum_{n=0}^{\infty} cf(nT)\, Z^{-n} \\ &= c \sum_{n=0}^{\infty} f(nT)\, Z^{-n} \\ &= cZ\{f(t)\} \end{aligned} \tag{5.11}$$

Thus the Z-transform of the product of a constant and a function equals the product of the constant and the Z-transform of the function.

2. Initial Value-Theorem. If the Z-transform of $f(t)$ is $F(Z)$, then

$$\lim_{t \to 0} f(t) = \lim_{Z \to \infty} F(Z)$$

Proof. By definition,

$$F(Z) = \sum_{n=0}^{\infty} f(nT)\, Z^{-n} = f(0) + f(T)\, Z^{-1} + f(2T)\, Z^{-2} + \cdots \tag{5.12}$$

Let us take the limit as $Z \to \infty$

$$\lim_{Z \to \infty} F(Z) = f(0) = \lim_{t \to 0} f(t) \tag{5.13}$$

which proves the theorem.

3. Final-value Theorem. If the Z-transform of $f(t)$ is $F(Z)$, then

$$\lim_{Z \to 1} \left[F(Z)\,(1 - Z^{-1}) \right] = \lim_{t \to \infty} f(t) \tag{5.14}$$

Proof

$$Z\left[f(t + T) - f(t)\right] \equiv \sum_{n=0}^{\infty} \left\{ f\left[(n + 1)T\right] - f(nT) \right\} Z^{-n} \tag{5.15}$$

Also, by definition

$$F(Z) = \sum_{n=0}^{\infty} f(nT)\, Z^{-n} = f(0) + f(T)\, Z^{-1} + f(2T)\, Z^{-2} + \cdots$$

If we multiply both sides of this equation by Z, we get

$$ZF(Z) = Zf(0) + f(T) + f(2T)\, Z^{-1} + f(3T)\, Z^{-2} + \cdots \tag{5.16}$$

Now,

$$\begin{aligned} Z\left\{ f(t + T) \right\} &= \sum_{n=0}^{\infty} f(nT + T)\, Z^{-n} \\ &= f(T) + f(2T)\, Z^{-1} + f(3T)\, Z^{-2} + \cdots \\ &= ZF(Z) - Zf(0) \end{aligned} \tag{5.17}$$

Substituting in Equation (5.15), we have

$$ZF(Z) - Zf(0) - F(Z) = \sum_{n=0}^{\infty} \left[f(nT + T) - f(nT) \right] Z^{-n} \tag{5.18}$$

Now, divide this equation by Z and take the limit as $Z \to 1$

$$\begin{aligned} &\lim_{Z \to 1} \left[(1 - Z^{-1})\, F(Z) \right] - f(0) = \lim_{k \to \infty} \sum_{n=0}^{k} \left\{ f(nT + T) - f(nT) \right\} \\ &= \lim_{k \to \infty} \left\{ \left[f(T) - f(0) \right] + \left[f(2T) - f(T) \right] + \left[f(3T) - f(2T) \right] \right. \\ &\qquad \left. + \cdots \left[f(kT + T) - f(kT) \right] \right\} \\ &= \lim_{k \to \infty} f\left[(k + 1)T \right] - f(0) \end{aligned} \tag{5.19}$$

Therefore,

$$\lim_{Z \to 1} \left[(1 - Z^{-1})\, F(Z) \right] = \lim_{t \to \infty} f(t) \tag{5.20}$$

which establishes the result.

4. Translation of the Function. If

$$f^*(t) = \begin{cases} f^*(t - kT) & t \geqslant kT \\ 0 & t < kT \end{cases} \tag{5.21}$$

$$= u(t - kT) f^*(t - kT)$$

The function $f^*(t)$ is delayed k sampling periods. Then

$$Z\{f(t - kT)\, u(t - kT)\} = Z^{-k} F(Z) \qquad (5.22)$$

Proof

$$Z\{f(t - kT)\, u(t - kT)\} = \sum_{n=0}^{\infty} f(nT - kT)\, u(nT - kT)\, Z^{-n} \qquad (5.23)$$

Let us make a variable substitution $m = n - k$. Then, Equation (5.23) becomes

$$\begin{aligned} Z\{f(t - kT)\, u(t - kT)\} &= \sum_{m=-k}^{\infty} f(mT)\, u[mT]\, Z^{-(m+k)} \\ &= Z^{-k} \sum_{m=0}^{\infty} f(mT)\, Z^{-m} \qquad (5.24) \\ &= Z^{-k} F(Z) \end{aligned}$$

which proves the result.

5. Complex Translation Theorem. The complex translation theorem states that if the Z-transform of $f(t)$ is $F(Z)$, then,

$$Z\{e^{\mp at} f(t)\} = Z\{f(s \pm a)\} = F(Ze^{\pm aT}) \qquad (5.25)$$

Proof By definition

$$\begin{aligned} Z\{e^{\mp at} f(t)\} &= \sum_{n=0}^{\infty} f(nT)\, e^{\mp anT}\, Z^{-n} \\ &= \sum_{n=0}^{\infty} f(nT)\, (e^{\pm aT} Z)^{-n} \\ &= F(Z\, e^{\pm aT}) \qquad (5.26) \end{aligned}$$

Suppose we wish to use this theorem to obtain the Z-transform of $u(t)e^{-at}$. First, we note that the Z-transform of $u(t)$ is given by

$$Z\{u(t)\} = \frac{Z}{Z - 1} \qquad (5.27)$$

To get $Z\{u(t)\, e^{-aT}\}$ we substitute in Equation (5.27) the product $Z\, e^{aT}$ for Z. Thus

$$\frac{Z\, e^{aT}}{Z\, e^{aT} - 1} = \frac{1}{1 - Z^{-1} e^{-aT}} = Z\{u(t)\, e^{-at}\}$$

5.3. The Inverse Z-Transformation[2]

Three methods are outlined here for inverting the Z-transform:

1. Partial fraction expansion.
2. Long division.
3. Tables of Z-transform.

The operation of inverting the Z-transform is denoted as

$$f^*(t) = Z^{-1}\{F(Z)\}$$

Note that the inverse yields the sampled function $f^*(t)$ and not the continuous function $f(t)$. The $f^*(t)$ so obtained is unique, but it is conceivable that the sampled function $f^*(t)$ could be derived from two different continuous functions $f_1(t)$ and $f_2(t)$. It follows, therefore, that the inverse of $F(Z)$ does not necessarily yield a unique continuous function $f(t)$. Thus we should not expect any additional information about $f(t)$ from the inverse other than the values at the sampling instants. Tables of inverse transforms are available. Some simple functions $F(Z)$ for which the inverse is simply the continuous function $f(t)$ are shown in Appendix A.

Most often the inverses we encounter cannot be obtained directly from the tables, and some mathematical manipulations are necessary. One way to accomplish inversion is to expand the function in *partial fractions*. The method is best illustrated by an example.

Example 1.

Find the inverse transform of

$$F(Z) = \frac{Z^2(Z^2 + Z + 1)}{(Z - 0.5)(Z - 1)(Z^2 - Z + 0.8)}$$

The method requires the expansion of $F(Z)/Z$ into partial fractions so that the inverse Z-transform of each of the component term multiplied by Z is recognizable from the tables of Z-transforms.

$$F(Z) = \frac{Z^2(Z^2 + Z + 1)}{(Z - 0.5)(Z - 1)(Z^2 - Z + 0.8)} = ZF_1(Z)$$

Therefore

$$F_1(Z) = \frac{Z(Z^2 + Z + 1)}{(Z - 0.5)(Z - 1)(Z^2 - Z + 0.8)} = \frac{A}{Z - 0.5} + \frac{B}{Z - 1}$$

$$+ \frac{C}{Z - (0.5 + j.74)} + \frac{D}{Z - (0.5 - j.74)}$$

Multiply the entire equation by $(Z - 0.5)$ and set $Z = 0.5$. Then

$$A = \frac{(0.5)\,(0.25 + 0.5 + 1)}{-(0.5)\,(0.25 - 0.5 + 0.8)} = -3.18$$

Multiply by $Z - 1$ and set $Z = 1$. Then,

$$B = \frac{(1)\,(3)}{(0.5)\,(0.8)} = 7.5$$

Next, multiply the same equation by $Z - (0.5 + j.74)$ and set $Z = 0.5 + j.74$. Then, we will get

$$C = -1.67 + 0.52j$$

Similarly,

$$D = -1.67 - 0.52j$$

Thus $F_1(Z)$ becomes

$$F_1(Z) = \frac{7.5}{Z - 1} - \frac{3.18}{Z - 0.5} + \frac{-1.67 + 0.52j}{Z - (0.5 + j.74)}$$

$$+ \frac{-1.67 - 0.52j}{Z - (0.5 - j.74)}$$

Now, recognize that the presence of a real number other than one in the denominator of the Z-transform expression will give rise to an exponential function in its inverse. Therefore, we must express the number 0.5 in exponential form. Similarly, complex numbers give rise to exponential and sine/cosine terms. Thus, by converting from rectangular to polar coordinates,

$$e^{-0.693} = 0.5$$

and

$$e^{-0.11 \pm 0.98j} = 0.5 \pm 0.74j$$

Substitution and multiplication by Z yields

$$F(Z) = Z\,F_1(Z)$$

$$= 7.5\,\frac{Z}{Z - 1} - 3.18\,\frac{Z}{Z - e^{-0.693}}$$

$$+ (-1.67 - 0.52j)\,\frac{Z}{Z - e^{-.11+0.98j}}$$

$$+ (-1.67 - 0.52j)\,\frac{Z}{Z - e^{-.11-0.98j}}$$

Now we may look up the inverse of each term in tables. Thus

$$f^*(t) = 7.5 - 3.18e^{-0.693t/T} + (-1.67 + 0.52j)e^{-(0.11-0.98j)t/T}$$
$$- (1.67 + 0.52j)\, e^{-(0.11+0.98j)t/T}$$

which reduces to

$$f^*(t) = 7.5 - 3.18e^{-0.693t/T} - 2e^{-0.11t/T}\{1.67\cos(0.98t/T)$$
$$+ 0.52\sin(0.98t/T)\}$$

Several values of the function $f(t)$ at the sampling instants are:

t	0	T	$2T$	$3T$	$4T$	$5T$	$6T$
$f(t)$	1	3.5	7.0	9.3	9.3	7.6	6.1

Another method of inverting $F(Z)$ is by *long division*. By this method $F(Z)$ is expanded into a power series of Z^{-1} by long division. The coefficient of the Z^{-n} term corresponds to the value of the function $f(t)$ at the nth sampling instant. This can be seen easily if we write the equation that defines the Z-transform of a function as

$$F(Z) = Z[f(t)] = \sum_{n=0}^{\infty} f(nT)\, Z^{-n}$$
$$F(Z) = f(0) + f(T)\, Z^{-1} + f(2T)\, Z^{-2} + f(3T)\, Z^{-3} + \cdots$$

If $F(Z)$ is expanded into a power series, we have $F(Z) = a_0 + a_1Z^{-1} + a_2Z^{-2} + a_3Z^{-3} + \cdots$ By comparison $a_0 = f(0)$; $a_1 = f(T)$ and $a_n = f(nT)$, which is the value of $f(t)$ at the nth sampling instant.

Example 2. Evaluate the inverse of

$$F(Z) = \frac{Z^2(Z^2 + Z + 1)}{(Z - 0.5)(Z - 1)(Z^2 - Z + 0.8)}$$

$$F(Z) = \frac{Z^4 + Z^3 + Z^2}{Z^4 - 2.5Z^3 + 2.8Z^2 - 1.7Z + 0.4}$$

$$\begin{array}{l}
\qquad 1 + 3.5Z^{-1} + 6.95Z^{-2} + 9.28Z^{-3} + 9.29Z^{-4} + 7.68Z^{-5} + 6.19Z^{-6} \\
Z^4 - 2.5Z^3 + 2.8Z^2 - 1.7Z + 0.4 \;\big)\; Z^4 + Z^3 + Z^2 \\
\qquad Z^4 - 2.5Z^3 + 2.80Z^2 - 01.70Z + 00.40 \\
\hline
\qquad 3.5Z^3 - 1.80Z^2 + 01.70Z - 00.40 \\
\qquad 3.5Z^3 - 8.75Z^2 + 09.80Z - 05.95 + 01.40Z^{-1} \\
\hline
\qquad 6.95Z^2 - 08.10Z + 05.55 - 01.40Z^{-1} \\
\qquad 6.95Z^2 - 17.38Z + 19.46 - 11.82Z^{-1} + 02.78Z^{-2} \\
\hline
\qquad 9.28Z - 13.91 + 10.42Z^{-1} - 02.78Z^{-2} \\
\qquad 9.28Z - 23.20 + 25.98Z^{-1} - 15.78Z^{-2} + 03.71Z^{-3} \\
\hline
\qquad 9.29 - 15.56Z^{-1} + 13.00Z^{-2} - 03.71Z^{-3} \\
\qquad 9.29 - 23.24Z^{-1} + 26.01Z^{-2} - 15.79Z^{-3} + 03.72Z^{-4} \\
\hline
\qquad 7.68Z^{-1} - 13.01Z^{-2} + 12.08Z^{-3} - 03.72Z^{-4} \\
\qquad 7.68Z^{-1} - 19.20Z^{-2} + 25.50Z^{-3} - 13.06Z^{-4} + 3.07Z^{-5} \\
\hline
\qquad 6.19Z^{-2} - 13.42Z^{-3} + 09.34Z^{-4} - 3.07Z^{-5}
\end{array}$$

Thus

$$F(Z) = 1 + 3.5Z^{-1} + 7Z^{-2} + 9.3Z^{-3} + 9.3Z^{-4} + 7.7Z^{-5} + 6.2Z^{-6} + \cdots$$

$f(t) =$	1	3.5	7	9.3	9.3	7.7	6.2
t	0	T	$2T$	$3T$	$4T$	$5T$	$6T$

In summary, the stepwise procedure to determine the Z-transform of a function $f(t)$ is:

1. Substitute nT for t in $f(t)$ to get $f(nT)$.
2. Evaluate the sum $F(Z) = \sum_{n=0}^{\infty} f(nT)\, Z^{-n}$.
3. Express the result in closed form.

To invert the Z-transform $F(Z)$ to get $f^*(t)$, the procedure is

1. Divide $F(Z)$ by Z to get $F_1(Z)$.
2. Expand $F_1(Z)$ in partial fractions.
3. Multiply $F_1(Z)$ by Z to obtain $F(Z)$.
4. Look up the inverse transform of the terms in $F(Z)$ in the tables of Z-transforms.

Or

1. Obtain a power series of Z^{-n} from $F(Z)$ by long division.
2. Note that the coefficient in front of Z^{-n} is the value of $f(t)$ at the nth sampling instant.

Remember that the inversion procedure gives correct information about $f(t)$ only at the sampling instants. We should also keep in mind that because Z-transforms are derived from Laplace transforms, their use is restricted to the solution of linear-control problems. This in turn means that we can seek only information about the control system in the vicinity of a nominal steady-state operating level. From our work in conventional control this concept of linearizing the process equations around some steady state should be familiar to us. The Laplace-transform methods could then be applied to the linearized equations so as to design the control system. This procedure works well so long as the range of linearity is not exceeded. In studying the subsequent material we should remember that the restrictions of linearity apply to the solution of sampled-data control systems as well.

The problem of Z-transform inversion is conveniently handled by a digital computer. A Fortran-based computer program to perform the inversion is shown in Appendix C1. To use this program, arrange the function to be inverted in the form

$$C(Z) = \frac{P_0 + P_1Z^{-1} + P_2Z^{-2} + P_3Z^{-3} + \cdots}{Q_0 + Q_1Z^{-1} + Q_2Z^{-2} + Q_3Z^{-3} + \cdots} \tag{5.28}$$

where

$P_0, P_1, \ldots, P_n$ and $Q_0, Q_1, \ldots, Q_n$ are constants

When provided with these constants as inputs, the program outputs $C(nT)$ according to the equation

$$C(Z) = C(0) + C(T)Z^{-1} + C(ZT)Z^{-2} + C(3T)Z^{-3} + \cdots \tag{5.29}$$

where $C(nT)$ represents the value of $C(t)$ at the nth sampling instant.

References

1. Ragazzini, J. R., Zadeh, L. A., The Analysis of Sampled-Data Systems, *Trans. AIEE*, **71**, Pt II, (1952) 225–234.
2. Barker, R. H., The Pulse Transfer Function and Its Application to Sampling Servo Systems, *Proc. IEE, London*, **99**, Pt IV, (December 1952) 302–317.

Short Bibliography on Mathematics In Sampled Data Control Systems

1. Kuo, B. C., *Analysis and Synthesis of Sampled-data Control Systems*, Prentice-Hall, Englewood Cliffs, N.J., 1963.
2. Tou, J. T., *Digital and Sampled-data Control Systems*, McGraw-Hill, New York, 1959.
3. Smith, C. L., *Digital Computer Process Control*, Intex Publishers, Scranton, Pa., 1972.

CHAPTER 6

Pulse Transfer Functions

In conventional control systems the Laplace transform of an input function $X(s)$ is related to the Laplace transform of the output function $Y(s)$ by the transfer function of the system $G(s)$ according to the Equation

$$\frac{Y(s)}{X(s)} = G(s) \tag{6.1}$$

Schematically, the transfer function is represented as

input $X(s)$ → $\boxed{G(s)}$ → output $Y(s)$

In sampled-data systems we must relate the pulsed input to the pulsed output. We have seen that pulsed signals can be conveniently handled by Z transforms. By analogy with the transfer function representation of the conventional control systems, it is tempting to postulate that the Z transform of the pulsed input can be related to the Z transform of the pulsed output according to the expression

$$\frac{Y(Z)}{X(Z)} = G(Z) \tag{6.2}$$

It will be shown that Equation (6.2) is indeed valid. The term $G(Z)$ is called the pulse-transfer function[1] or the Z-transfer function of the system. To show that Equation (6.2) is correct we have to derive an alternate expression for the sampling process.

6.1. Complex Series Representation of the Sampler

Recall that the sampling operation gives us the values of the output at each sampling instant and can be represented as

continuous input $X(t) \rightarrow$ [Sampler] $\rightarrow X^*(t)$ sampled output

We have said that the sampled output is related to the continuous input by the expression

$$X^*(t) = \delta_T(t)\, X(t) \tag{6.3}$$

where

$$\delta_T(t) = \sum_{n=-\infty}^{\infty} \delta(t - nT)$$

Since $\delta_T(t)$ is a periodic function, it can be expanded into a complex Fourier series:

$$\delta_T(t) = \sum_{n=-\infty}^{\infty} C_n\, e^{jn\omega_s t} \tag{6.4}$$

where

$$\omega_s = \text{sampling frequency, in radians per unit time}$$

$$= \frac{2\pi}{T}$$

and the Fourier coefficients are given by

$$C_n = \frac{1}{T} \int_{-T/2}^{+T/2} \delta_T(t)\, e^{-jn\omega_s t}\, dt \tag{6.5}$$

Substituting for $\delta_T(t)$ from Equation (6.3) into Equation (6.5) gives

$$C_n = \frac{1}{T}\int_{-T/2}^{T/2} e^{-jn\omega_s t}\left(\sum_{n=-\infty}^{\infty}\delta(t-nT)\right)dt$$

$$= \frac{1}{T}\int_{-T/2}^{T/2} e^{-jn\omega_s t}\sum_{n=-\infty}^{-1}\delta(t-nT)\,dt \rightarrow 0 \; +$$

$$\frac{1}{T}\int_{-T/2}^{T/2} e^{-jn\omega_s t}\,\delta(t)\,dt$$

$$+\frac{1}{T}\int_{-T/2}^{+T/2} e^{-jn\omega_s t}\left(\sum_{n=1}^{\infty}\delta(t-nT)\right)dt \rightarrow 0 \tag{6.6}$$

Therefore,

$$C_n = \frac{1}{T}\int_{-T/2}^{+T/2} e^{-jn\omega_s t}\,\delta(t)\,dt \tag{6.7}$$

Now recall that the theorem of the means for integrals of continuous functions states that

$$\int_a^b g(t)\,dt = (b-a)\,g(t_1) \qquad a \leqslant t_1 \leqslant b \tag{6.8}$$

Therefore,

$$C_n = \frac{1}{T}e^{-jn\omega_s t}\Big|_{t=0} = \frac{1}{T} \tag{6.9}$$

Substituting this result in Equation (6.4) gives

$$\delta_T(t) = \frac{1}{T}\sum_{n=-\infty}^{\infty} e^{jn\omega_s t}$$

and

$$X^*(t) = \frac{1}{T}\sum_{n=-\infty}^{\infty} X(t)\,e^{jn\omega_s t} \tag{6.10}$$

Taking the Laplace transform of this equation we have

$$X^*(s) = l\{X^*(t)\} = \frac{1}{T}\sum_{n=-\infty}^{\infty} X(s+jn\omega_s) \tag{6.11}$$

6.2. Development of the Pulse Transfer Function

We wish to relate the sampled input $X^*(t)$ to the sampled output $Y^*(t)$ as shown in the following figure:

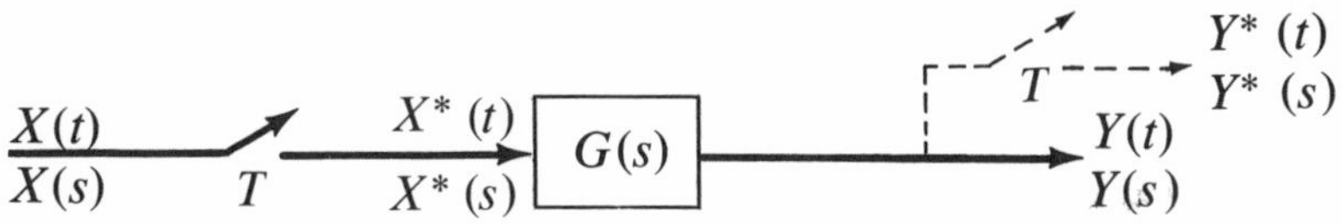

$X(t)$, $X(s)$ → T → $X^*(t)$, $X^*(s)$ → $G(s)$ → $Y(t)$, $Y(s)$; T → $Y^*(t)$, $Y^*(s)$

From this figure,

$$Y(s) = X^*(s)\,G(s) \tag{6.12}$$

By analogy to Equation (6.11)

$$Y^*(s) = \frac{1}{T}\sum_{n=-\infty}^{\infty} Y(s + jn\omega_s) \tag{6.13}$$

In view of Equation (6.12),

$$Y(s + jn\omega_s) = G(s + jn\omega_s)\,X^*(s + jn\omega_s) \tag{6.14}$$

Substitution in Equation (6.14) yields

$$\begin{aligned} Y^*(s) &= \frac{1}{T}\sum_{n=-\infty}^{\infty} Y(s + jn\omega_s) \\ &= \frac{1}{T}\sum_{n=-\infty}^{\infty} G(s + jn\omega_s)\,X^*(s + jn\omega_s) \end{aligned} \tag{6.15}$$

But because

$$\begin{aligned} X^*(s + jn\omega_s) &= l\{X^*(t)\,e^{jn\,\omega_s t}\} \\ &= l\{X^*(t)\,e^{jn\,2\pi t/T}\} \end{aligned} \tag{6.16}$$

substituting $t = nT$ gives

$$X^*(s + jn\omega_s) = l\{X^*(t)\,e^{j2\pi n^2}\}$$

Since

$$e^{-j2\pi n^2} = \cos 2\pi n^2 - j \sin 2\pi n^2 = 1 \text{ for all } n,$$

$$X^*(s + jn\omega_s) = l\{X^*(t)\} = X^*(s)$$

Equation (6.15) becomes

$$Y^*(s) = \frac{1}{T}\sum_{n=-\infty}^{\infty} G(s + jn\omega_s)\, X^*(s)$$

$$= \frac{1}{T} X^*(s) \sum_{n=-\infty}^{\infty} G(s + jn\omega_s) \tag{6.17}$$

By the same reasoning which led to Equation (6.11), this can be written as

$$Y^*(s) = X^*(s)\, G^*(s) \tag{6.18}$$

Taking the Z transform of this equation gives

$$Y(Z) = X(Z)\, G(Z)$$

or (6.19)

$$G(Z) = \frac{Y(Z)}{X(Z)}$$

Note the similarity of Equation (6.19) to the familiar transfer function $G(s)$ given in Equation (6.1), which would be the transfer function of the system if the samplers were removed. This important similarity enables us to treat the Z transforms and the pulse transfer function in much the same manner as the Laplace transforms and the conventional transfer functions. A step-by-step procedure for developing pulse transfer functions is given below.[2]

1. Derive the transfer function $G(s)$ of the system by conventional techniques.
2. For $X(t) = \delta(t)$ (i.e., the impulse function)

$$G(s) = \frac{Y(s)}{X(s)} = \mathrm{Y(s)} \tag{6.20}$$

since $l\{\delta(t)\} = 1$.

3. Invert $G(s)$ to get $g(t)$ which is numerically equal to the impulse response $y(t)$ according to Equation (6.20)
4. Note that for $X(t) = \delta(t)$, $X(Z) = Z\{\delta(t)\} = 1$. Therefore,

$$G(Z) = \frac{Y(Z)}{X(Z)} = Y(Z)$$

5. Thus $G(Z)$ can be obtained from $y(t)$ [or $g(t)$] by substituting nT for t and evaluating the sum of the series as

$$G(Z) = \sum_{n=0}^{\infty} y(nT)\, Z^{-n} \tag{6.21}$$

For physically realizable systems, convergence of this series is assured. Frequently, Z-transfer functions, with $G(S)$ available as the starting point, are desired, in which case steps 3 and 4 suffice. When $G(s)$ is a complicated function, it may be split into partial fractions; then steps 2 through 5 can be applied to the individual partial fractions.

Example

Find the pulse transfer function of the system shown below:

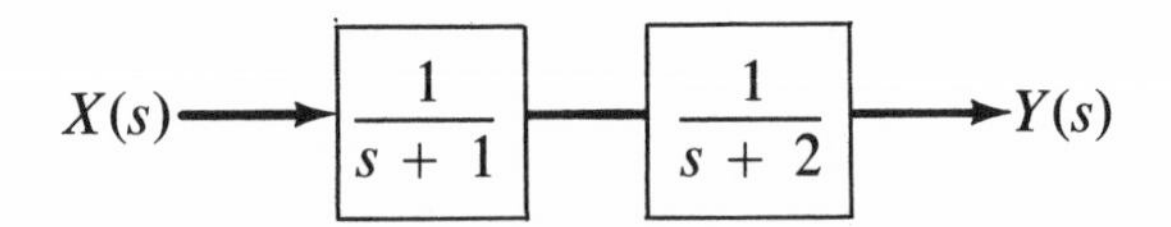

Solution:

$$G(s) = \frac{1}{(s+1)(s+2)}$$

$$= \frac{1}{s+1} - \frac{1}{s+2}$$

Therefore,

$$g(t) = e^{-t} - e^{-2t} = y(t)$$

$$g(nT) = e^{-nT} - e^{-2nT} = y(nT)$$

Now,

$$G(Z) = \sum_{n=0}^{\infty} y(nT)\, Z^{-n} = \sum_{n=0}^{\infty} (e^{-nT} - e^{-2nT})\, Z^{-n}$$

$$= \frac{1}{1 - e^{-T} Z^{-1}} - \frac{1}{1 - e^{-2T} Z^{-1}}$$

After simplification we get

$$G(Z) = \frac{Z(e^{-T} - e^{-2T})}{(Z - e^{-T})(Z - e^{-2T})}$$

References

1. Barker, R. H., The Pulse Transfer Function and Its Application to Sampling Servo Systems, *Proc. IEE, London*, **99**, (4), 302–317, December 1952.
2. Tou, J. T., *Digital and Sampled-data Control Systems*, McGraw-Hill, New York, 1959.

CHAPTER 7

Data Holds

In the sampled-data control loop the continuous signal representing the measured value of the controlled variable is sampled by means of a sampler (i.e., an A/D converter) and compared to the discrete form of the set point to produce an error. An appropriate computer program is executed to produce a control action which consists of discrete data (i.e., a train of impulses of varying strengths). The function of the *holding device* is to reconstruct the continuous signal from the discrete data. This device is called a *digital-to-analog converter.* In this chapter we develop mathematical expressions that describe the operation of some of the common holding devices.

The schematic of the holding device is shown in Figure 7.1. The holding device converts the pulsed-input signal into a continuous signal by interpolation or extrapolation of the input pulses so that the input signal can be approximately reproduced. The smoothing of the pulsed data by a holding device is essentially an extrapolation problem. The extrapolated time function between two consecutive sampling instants nT *and* $(n + 1)T$ depends on its values at the preceding sampling instants nT, $(n - 1)T$, $(n - 2)T$, ... , and can generally be described by a power series expansion of the output between the interval $t = NT$ and $t = (n + 1)T$ as shown in Figure 7.1.

Let $Y(t)$ be the output function and $Y_n(t)$ be its value between sampling instants nT and $(n + 1)T$, that is,

$$Y_n(t) = Y(t) \qquad nT \leq t \leq (n + 1)T. \tag{7.1}$$

Then,

$$Y_n(t) = Y(nT) + Y^{(1)}(nT)(t - nT) + \frac{Y^{(2)}(nT)}{2!}(t - nT)^2 + \cdots + \frac{Y^k(nT)}{k!}(t - nT)^k \tag{7.2}$$

where

$Y(nT)$ = value of $Y(t)$ at $t = nT$

$Y^k(nT)$ = value of the kth-order derivative of $Y(t)$ evaluated at $t = nT$

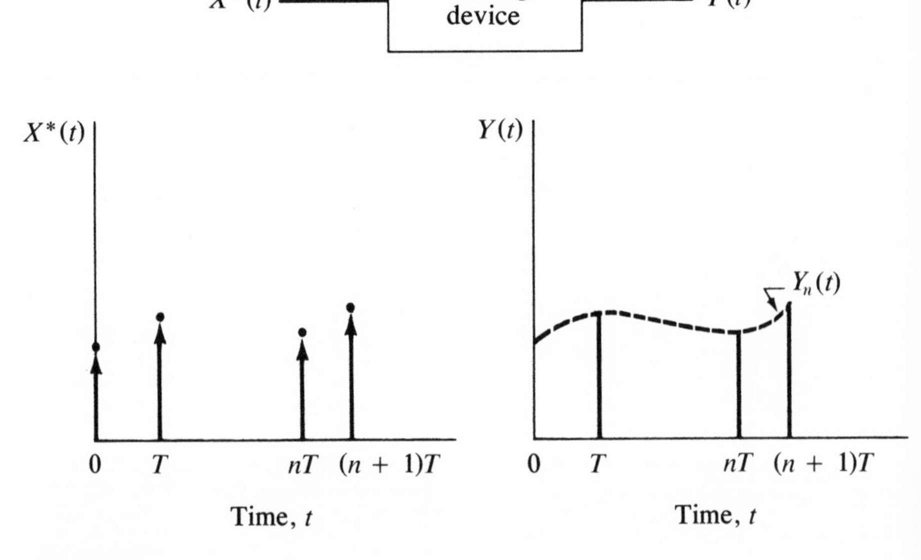

Figure 7.1
Representation of Holding Devices

When a holding device approximates the time function between two consecutive sampling instants by a zero-order polynomial, it is referred to as a *zero-order hold*. The equation for the zero-order hold is

$$Y_n(t) = Y(nT) \qquad nT < t < (n+1)T \tag{7.3}$$

When the device approximates the time function between two successive sampling instants by a first-order polynomial

$$Y_n(t) = Y(nT) + Y^{(1)}(nT)(t - nT) \tag{7.4}$$

it forms a first-order holding device. Similarly, when the time function is described by a kth-order polynomial, the device is referred to as a kth-order holding device. In general, a higher-order holding device produces a better approximation of the desired time function from the input-data pulses.

Since the holding device receives information only at the sampling instants, the values of these derivatives [see Equation (7.2)] can only be estimated from the sampled-input data. These estimates are given by the following expressions:
First-order derivative:

$$Y^{(1)}(nT) = \left.\frac{dY}{dt}\right|_{nT} \cong \frac{Y(nT) - Y[(n-1)T]}{T} \tag{7.5}$$

Second-order derivative:

$$Y^{(2)}(nT) = \left.\frac{d^2Y}{dt^2}\right|_{T=nT} = \frac{Y^{(1)}(nT) - Y^{(1)}[(n-1)T]}{T} \tag{7.6}$$

and for the kth-order derivative, we have

$$Y^k(nT) = \frac{Y^{k-1}(nT) - Y^{k-1}[(n-1)T]}{T} \tag{7.7}$$

These derivatives can be expressed in terms of $Y(nT)$ as

$$Y^k(nT) \cong \frac{1}{T^k}\left\{Y(nT) - kY[(n-1)T] + \frac{k(k-1)}{2!}Y[(n-2)T] - \cdots + (-1)^k Y[(n-k)T]\right\} \tag{7.8}$$

From these equations it can be seen that, to obtain an estimated value of a derivative of $Y(t)$, the minimum number of data pulses required is equal to the order of the derivative plus one. Thus first-order derivative calculations require two pulses. To estimate a second-order derivative at a sampling instant nT requires three consecutive pulses at nT, $(n - 1)T$, and $(n - 2)T$. Thus to estimate a higher-order derivative requires a greater delay before a reliable value of that derivative can be obtained. The delay introduced by a higher-order holding device may be detrimental to system stability. On the other hand, to obtain a better reproduction of the desired time function from the input data pulses and to reduce the ripple content it is advisable to employ a higher-order holding device. Consequently, to design a holding device for a sampled-data control system we must compromise between the tolerable ripple content and the specified system stability and dynamic performance. Because of the relatively high cost and constructional complexity of higher-order holding devices and the amount of phase lag introduced by them, the most common holding devices used in computer control systems are the zero-order holding devices, although the first-order holding devices are occasionally used.

7.1 Transfer Function of the Zero-Order Hold

The schematic of a zero-order hold is shown in Figure 7.2. The input to the holding device is a train of impulses of varying strengths. For example, at the first time instant $t = 0$ the input of strength k_0 to the holding device can be expressed as

$$X^*_0(t) = k_0\,\delta(t) \tag{7.9}$$

During the first time interval the output from the zero-order holding device can be represented as (see Figure 7.3)

$$Y(t) = k_0\left[u(t) - u(t - T)\right] \tag{7.10}$$

Similar expressions can be written for input and output over N time intervals. Thus the input train can be represented by

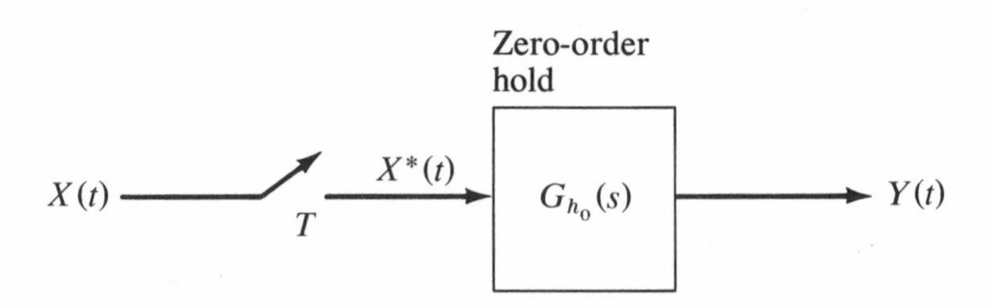

Figure 7.2
Zero-Order Holding Device

$$X^*(t) = \sum_{n=0}^{N} k_n \, \delta(t - nT) \tag{7.11}$$

and the output by

$$\begin{aligned} Y(t) &= k_0 \left[u(t) - u(t - T)\right] + k_1 \left[u(t - T) - u(t - 2T\right] \\ &\quad + \cdots + k_N \left[u(t - (NT) - u(t - (N + 1)T\right] \\ &= \sum_{n=0}^{N} k_n \left[u(t - nT) - u(t - (n + 1)T)\right] \end{aligned} \tag{7.12}$$

Taking the Laplace transform of the input train gives

$$X^*(s) = \sum_{n=0}^{N} k_n \, e^{-snT} \tag{7.13}$$

The Laplace transform of the output is

$$\begin{aligned} Y(s) &= \sum_{n=0}^{N} k_n \left\{ \frac{e^{-snT}}{s} - \frac{e^{-(n+1)sT}}{s} \right\} = \frac{1}{s} \sum_{n=0}^{N} k_n e^{-nsT}(1 - e^{-sT}) \\ &= \frac{1 - e^{-sT}}{s} \sum_{n=0}^{N} k_n e^{-nsT} \end{aligned} \tag{7.14}$$

Therefore, the transfer function of the zero-order hold (which is the ratio of the Laplace transform of the output to that of the input) is

$$\frac{Y(s)}{X(s)} = G_{h_0}(s) = \frac{\left(\dfrac{1 - e^{-sT}}{s}\right) \sum_{n=0}^{N} k_n e^{-nsT}}{\sum_{n=0}^{N} k_n e^{-nsT}}$$

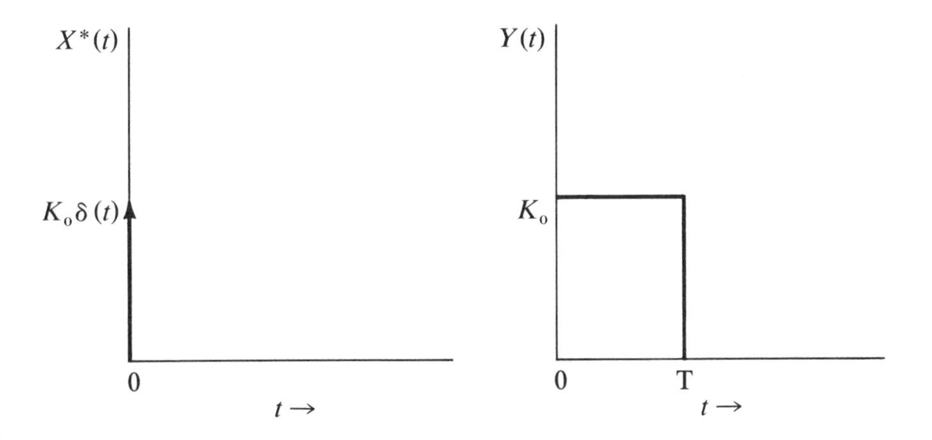

Figure 7.3
Input/Output From Zero-Order Hold During the First Interval

$$G_{h_0}(s) = \frac{1 - e^{-sT}}{s} \tag{7.15}$$

The input-output relationship of the zero-order hold for a typical input function is shown in Figure 7.4.

7.2. Transfer Function of the First-Order Hold

The operation of the first-order hold is shown in Figure 7.5. The extrapolated time function $Y_n(t)$ between two successive sampling instants nT and $(n + 1)T$ is assumed to be a linear function given by

$$Y_n(t) = Y(nT) + Y^{(1)}(nT)(t - nT) \tag{7.16}$$

where

$$Y^{(1)}(nT) = \frac{Y(nT) - Y[(n - 1)T]}{T}$$

or

$$Y_n(t) = Y(nT) + \frac{1}{T}\left[Y(nT) - Y(n - 1)T\right](t - nT) \tag{7.17}$$

Equation (7.17) gives the value of the output between $nT \leqslant t \leqslant (n + 1)T$. Thus if at $t = (n - 1)T$, $X^*_{n-1} = k_{n-1}\,\delta(t - (n - 1)T)$ and at

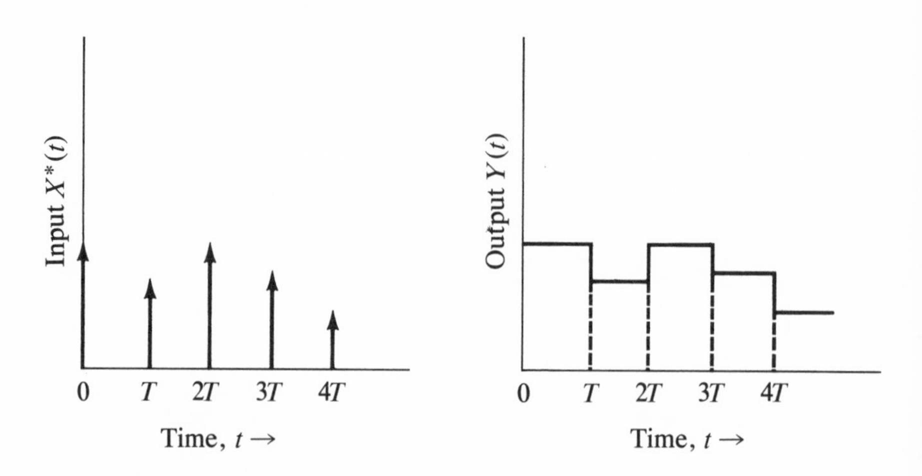

Figure 7.4
Input/Output of Zero-Order Hold

$t = nT$, $X_n^* = k_n\,\delta(t - nT)$, then, $Y(nT)$ will be equal to k_n, and during the time interval $nT \leq t \leq (n + 1)T$ the output will be given by

$$Y_n(t) = k_n + \frac{k_n - k_{n-1}}{T}(t - nT) \tag{7.18}$$

Now, let us derive the transfer function of the first-order hold. Recall that the conventional transfer function can be developed from the impulse response of the system as follows:

$$\text{Transfer function } G(s) = \frac{Y(s)}{X(s)} \tag{7.19}$$

If the input to the system is an impulse function $k_0\,\delta(t)$ then

$$X(s) = l\left[k_0 \delta(t)\right] = k_0 \tag{7.20}$$

Therefore, from Equation (7.19),

$$G(s) = \frac{1}{k_0} Y(s) \tag{7.21}$$

This suggests that if the response of the system to an impulse function can be obtained in the time domain, the Laplace transform of the response gives the transfer function of the system according to Equation (7.21). For the first-order hold, if the input is $k_0\,\delta(t)$ (i.e., an im-

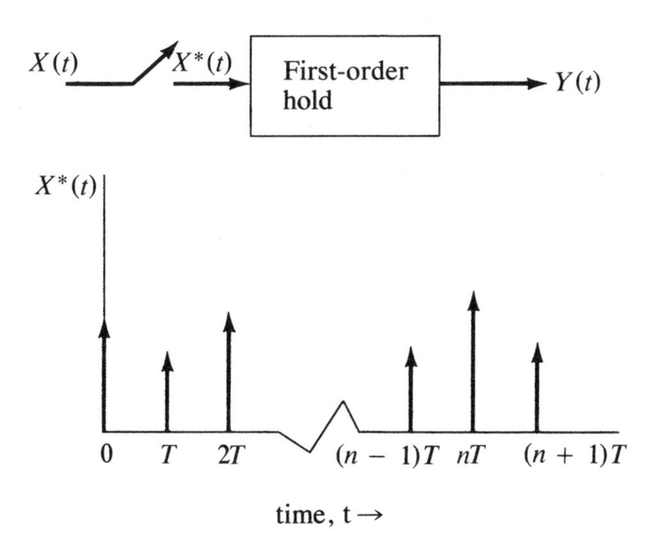

Figure 7.5
First-Order Hold Operation

pulse of strength k_0 at the origin), the output of the holding device will be, from Equation (7.17),

$$Y_0(t) = Y(0) + \frac{Y(0) - Y(-T)}{T}t \qquad 0 < t < T \tag{7.22}$$

$$= k_0 + \frac{k_0 - 0}{T}t = k_0\left(1 + \frac{t}{T}\right)$$

$$Y_1(t) = Y(T) + \frac{Y(T) - Y(0)}{T}(t - T) \qquad T < t < 2T \tag{7.23}$$

$$0 + \frac{0 - k_0}{T}(t - T) = k_0\left(1 - \frac{t}{T}\right)$$

$$Y_2(t) = Y(2T) + \frac{Y(2T) - Y(T)}{T}(t - 2T) \qquad 2T < t < 3T \tag{7.24}$$

$$0 + \frac{0 - 0}{T}(t - 2T) = 0$$

The schematic of the input-output representation for the first-order hold is shown in Figure 7.6. Also, from Equation (7.17), $Y_n(t) = 0$ for $n > 2$. Thus the impulse response of the system can be expressed as

$$Y(t) = k_0\left(1 + \frac{t}{T}\right)u(t) - k_0\left(1 + \frac{t}{T}\right)u(t - T)$$
$$+ k_0\left(1 - \frac{t}{T}\right)u(t - T) - k_0\left(1 - \frac{t}{T}\right)u(t - 2T) \tag{7.25}$$

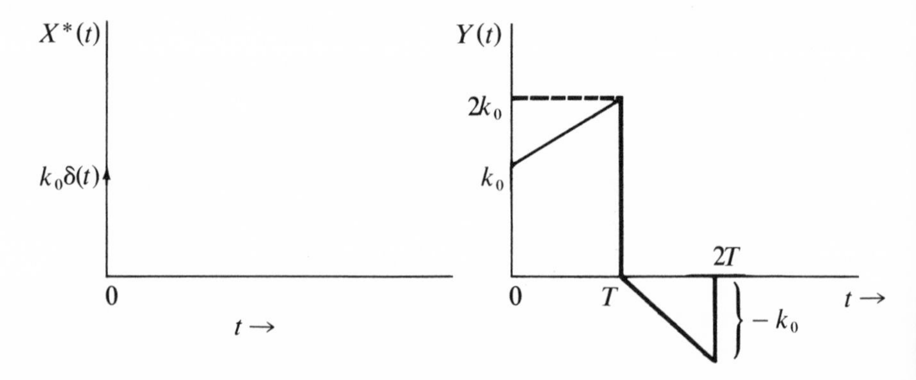

Figure 7.6
Input-Output from First-Order Hold

The reader should verify that this equation gives the results of Equations (7.22) through (7.24). Recall that

$$u(t - nT) = \begin{cases} 1 & t \geqslant nT \\ 0 & \text{otherwise} \end{cases} \tag{7.26}$$

Equation (7.25) can be simplified to give

$$\begin{aligned} Y(t) = k_0\left(1 + \frac{t}{T}\right)u(t) - 2k_0\frac{t}{T}u(t - T) \\ - k_0\left(1 - \frac{t}{T}\right)u(t - 2T) \end{aligned} \tag{7.27}$$

This equation may be rearranged so that we may readily look up its Laplace transform. Thus

$$\begin{aligned} Y(t) = k_0\left(1 + \frac{t}{T}\right)u(t) - 2k_0\left(1 + \frac{t - T}{T}\right)u(t - T) \\ + k_0\left(1 + \frac{t - 2T}{T}\right)u(t - 2T) \end{aligned} \tag{7.28}$$

Taking the Laplace transform of Equation (7.28) and dividing by k_0 gives the transfer function of the first-order hold.

$$\begin{aligned} G_{h_1}(s) = \frac{1}{k_0}l\{Y(t)\} = \frac{1}{s} + \frac{1}{s^2T} - \frac{2}{s}e^{-Ts} - \frac{2}{s^2T}e^{-Ts} \\ + \frac{1}{s}e^{-2Ts} + \frac{e^{-2Ts}}{s^2T} \end{aligned} \tag{7.29}$$

which can be simplified to give

$$G_{h_1}(s) = \left(\frac{1 + Ts}{T}\right)\left(\frac{1 - e^{-Ts}}{s}\right)^2 \tag{7.30}$$

7.3. Sampling Frequency Considerations[1]

This is perhaps an appropriate place to develop the necessary background that can answer the questions about the minimum sampling frequency in sampled-data control systems. It should be clear that if the sampling frequency approaches infinity, the performance of the sampled data system approaches that of the continuous or analog control system. On the other hand, if we do not sample at all or sample very infrequently, the control-loop performance will be unacceptably poor. Somewhere between these extremes is a minimum sampling rate required for proper signal recovery and an optimum sampling frequency based on economics.

To illustrate the need for the minimum rate, let us begin with the schematic of the sampler followed by a hold circuit as shown in Figure 7.7. The continuous signal $x(t)$ is sampled every T seconds or minutes. The hold circuit reconstructs the continuous signal $y(t)$ from the sampled train $X^*(t)$. The question is: Is there a minimum sampling rate that is required to ensure that adequate information about x will be present in y? The answer to this question may be developed by considering the signals x, X^*, and y in the frequency domain, as described in the following paragraphs.

Consider a signal $x(t)$ that has an amplitude spectrum* as shown in Figure 7.8a. Such a signal is said to be band limited, meaning that $|X(j\omega)|$ is zero for all $\omega > \omega_c$ and $\omega < -\omega_c$. The knowledge of $X(j\omega)$ at each value of ω is equivalent to knowing $x(t)$ for each value of t. Now recall that the sampled train $X^*(t)$ can be expressed in the frequency domain as

$$X^*(j\omega) = \frac{1}{T}\sum_{n=-\infty}^{\infty} X[j(\omega + n\omega_s)] \tag{7.31}$$

where

$$\omega_s = \text{sampling frequency, } 2\pi/T$$

The frequency spectrum of the sampled train is shown in Figure 7.8b. Note from Figure 7.8 that $\omega_c < \omega_s/2$ [i.e., the sampling fre-

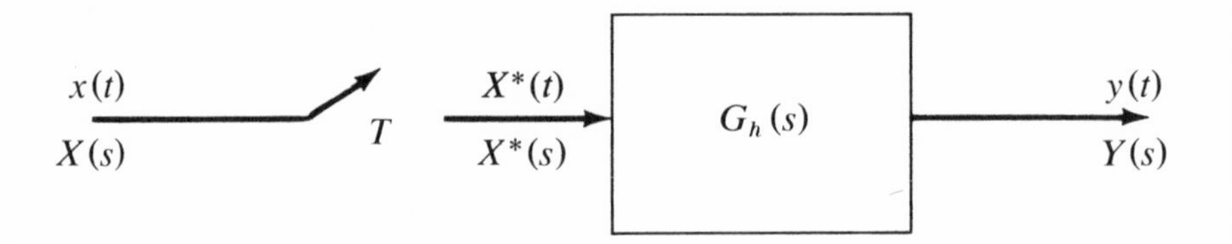

Figure 7.7
Sampler and Hold Operation

*The frequency spectrum of a signal $f(t)$ is a plot of its Fourier transform $F(j\omega)$ in the form of amplitude $|F(j\omega)|$ versus frequency and phase angle of $F(j\omega)$ versus frequency.

quency is greater than twice the highest significant frequency in the input signal $x(t)$].

The amplitude spectrum of the sampled train $|X^*(j\omega)|$ consists of an infinite number of spectra, each identical to that of the continuous signal $X(j\omega)$ but reduced in amplitude by a factor $1/T$. The central band in Figure 7.8*b* is called the primary band, and each of the displaced amplitude spectra is called a side band. From these figures it should be clear that if we are to successfully reconstruct $y(t)$ from $X^*(t)$ such that $y(t)$ has adequate information originally contained in $x(t)$, we must be able to eliminate the side bands (the multiplication by T can be easily accomplished by an operational amplifier).

Now let us see what happens to X^* as it passes through the hold circuit. From Figure 7.7

$$Y(j\omega) = G_h(j\omega)\, X^*(j\omega) \tag{7.32}$$

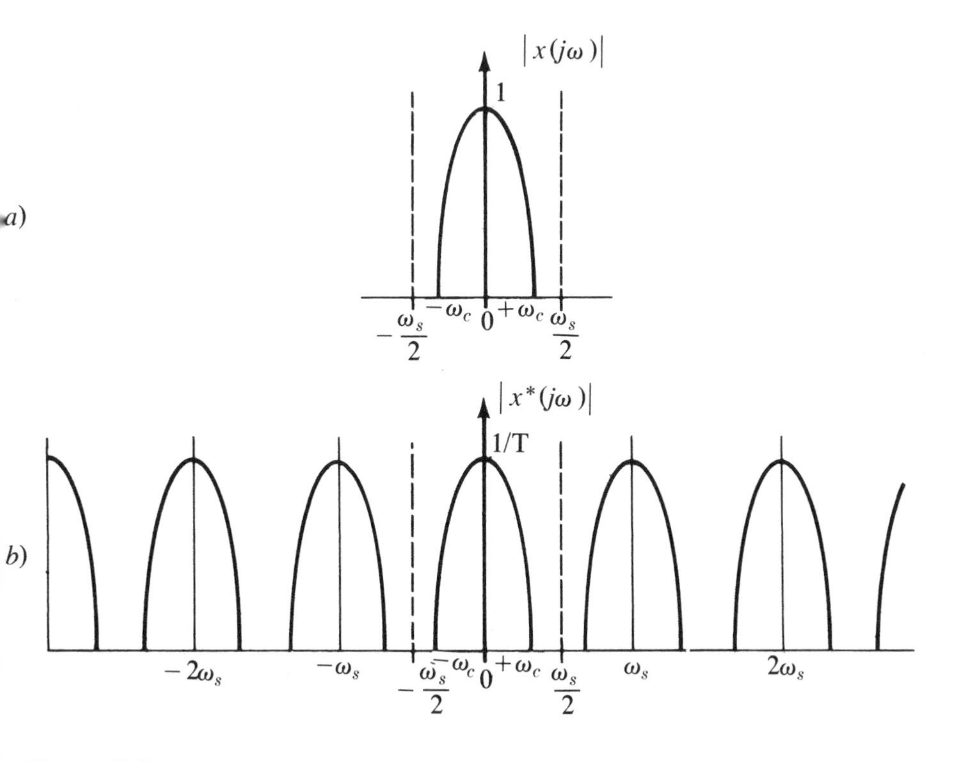

Figure 7.8
Amplitude Spectra of Input and Output Signals from the Sampler (*a*) Pertains to the Input Function and (*b*) to the Sampled Output: Note that the Sampling Frequency is Higher than the Maximum Frequency of the Input Signal (i.e., $\omega_s > 2\omega_c$) (Reprinted by Permission from Ref. 1)

If we select the hold circuit having amplitude spectra as shown in Figure 7.9*a*, it is easy to see that this circuit will act like a filter. It will allow the frequency content of the spectra to pass through, without attenuation, for $|\omega| < |\omega_c|$ but will filter out completely the frequency content for $|\omega| > |\omega_c|$ in accordance with Equation 7.32. This description, shown in Figure 7.9*b*, depicts idealized filter operation, in that it is not possible to construct a physical hardware which has the amplitude spectra shown in Figure 7.9*a*. Practical filters can only approach the performance of the ideal filter.

Now let us see what happens if the sampling frequency is less than twice the highest significant frequency of the continuous signal (i.e., $\omega_s < 2\omega_c$), as shown in Figure 7.10*a*. Again, when we sample this signal we will get an infinite number of spectra, as shown in Figure 7.10*b*. However, in this case, since $\omega_c > \omega_s/2$, a portion of the spectrum in the interval $\omega_s/2 < \omega < \omega_c$ overlaps onto the spectrum of the adjacent side band. It should be clear that even if we use the ideal filter, we will not be able to reconstruct the signal of Figure 7.10*a* from the spectra shown in Figure 7.10*b*. In this case, it will be impossible to recover all the information contained in the original signal. This discussion leads to Shannon's sampling theorem,[2] which states that if

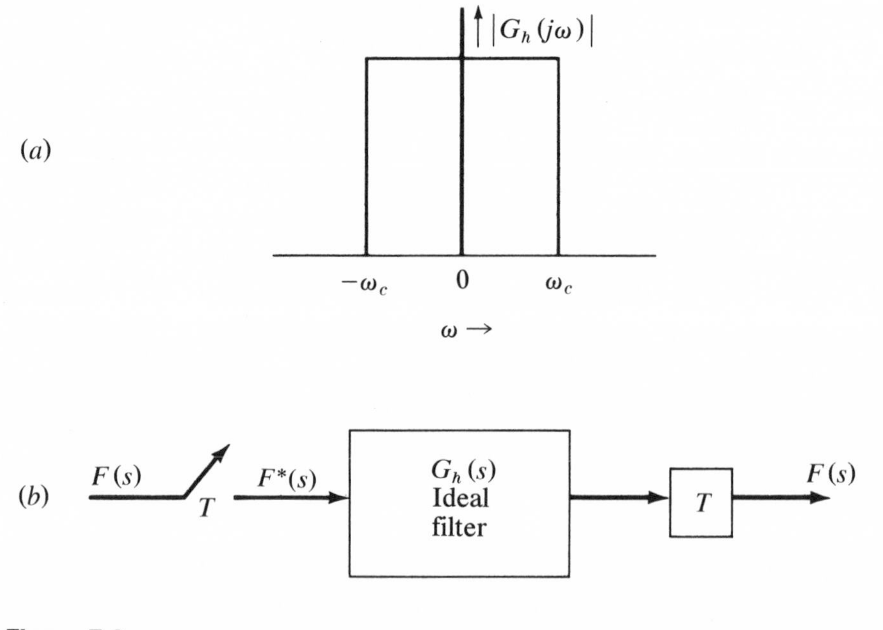

Figure 7.9
Sampler with an Ideal Filter: (*a*) Amplitude Spectra of Ideal Filter and (*b*) Schematic for Complete Signal Recovery

a signal contains no frequency higher than ω_c radians per second, it is completely characterized by the values of the signal measured at instants of time separated by $T = \frac{1}{2}(2\pi/\omega_c)$ second.

Recall that we began this discussion by considering the frequency spectra of a band-limited signal. All physical signals contain components over a wide range of frequencies, although the magnitude of the high-frequency components is generally small. Therefore, sampling will result in a certain amount of overlapping, even if an ideal filter were employed. Further, the amplitude spectra of the ideal filter cannot be reproduced exactly by a practical filter (for example, see Figure 7.11 for amplitude spectra of a zero-order hold). Thus the exact reproduction of the continuous signal from the sampled signal is impossible. However, the sampling theorem gives us a useful guide

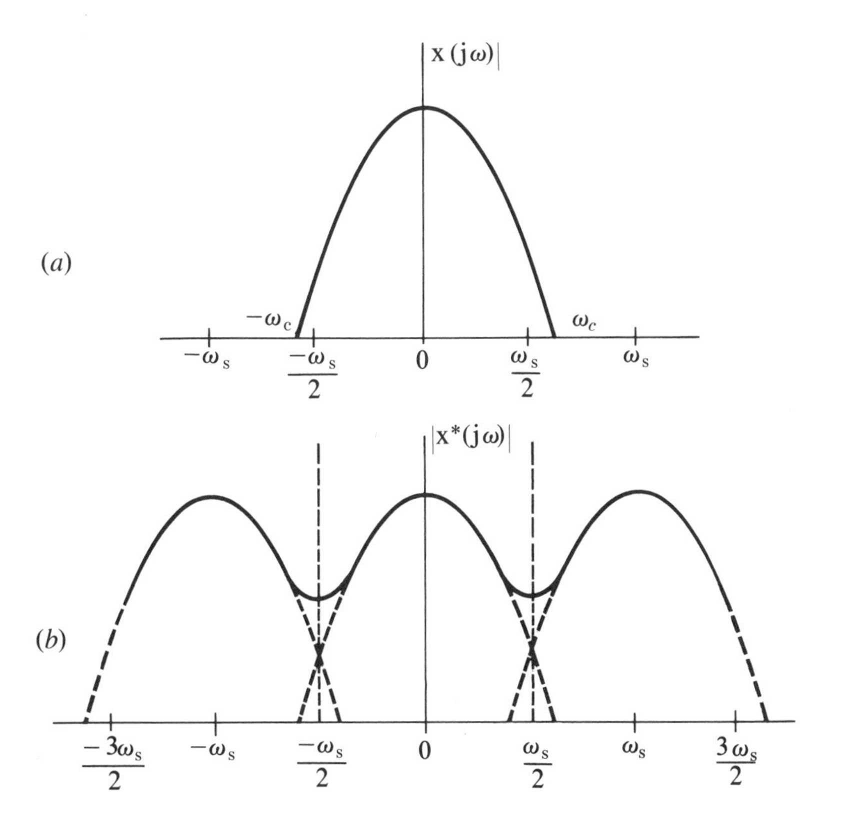

Figure 7.10
Amplitude Spectra of Input and Output Signals when $\omega_s < 2\omega_c$:
(*a*) Input Spectra (*b*) Output Spectra
Reprinted by Permission from C. L. Smith, *Digital Computer Process Control*, Intext Educational Publishers, 1972.

on the minimum sampling frequency in sampled-data control systems.

7.4. Selection of Optimum Sampling Period

The sampling theorem establishes the lower limit on sampling frequency. Somewhere above this minimum is a value for the sam-

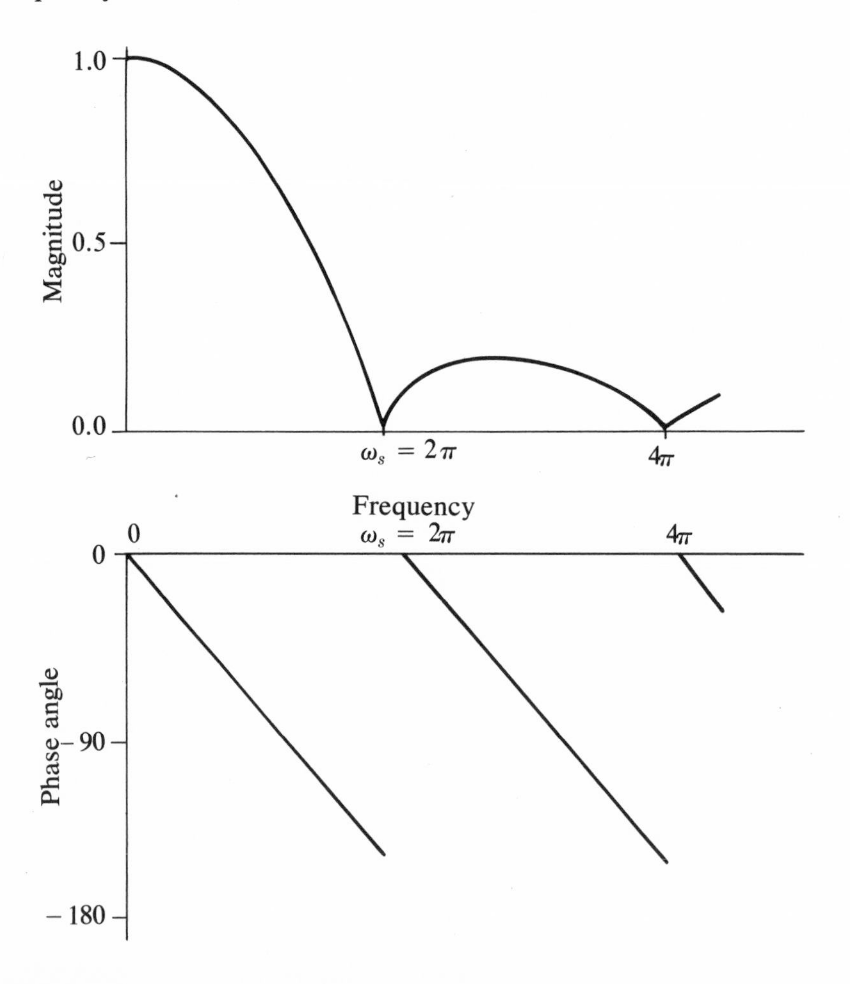

Figure 7.11
Frequency Characteristics of Zero-Order Hold (T = 1)
Reprinted by Permission from C. L. Smith, *Digital Computer Process Control*, Intext Educational Publishers, 1972

pling period that is economically optimum. In this section we present a discussion of how this optimum might be selected.

The simplest method for selecting the sampling period is the one recommended in the DDC guidelines established by the 1963 Users Conference.[3] These guidelines recommended the sampling period of 1 sec for flow loops, 5 sec for level and pressure loops, and 20 sec for temperature and composition loops. Although this procedure is simple to use, the designer may wish to undertake a more detailed analysis to ensure that the sampling period selected is as large as possible, consistent with good control.

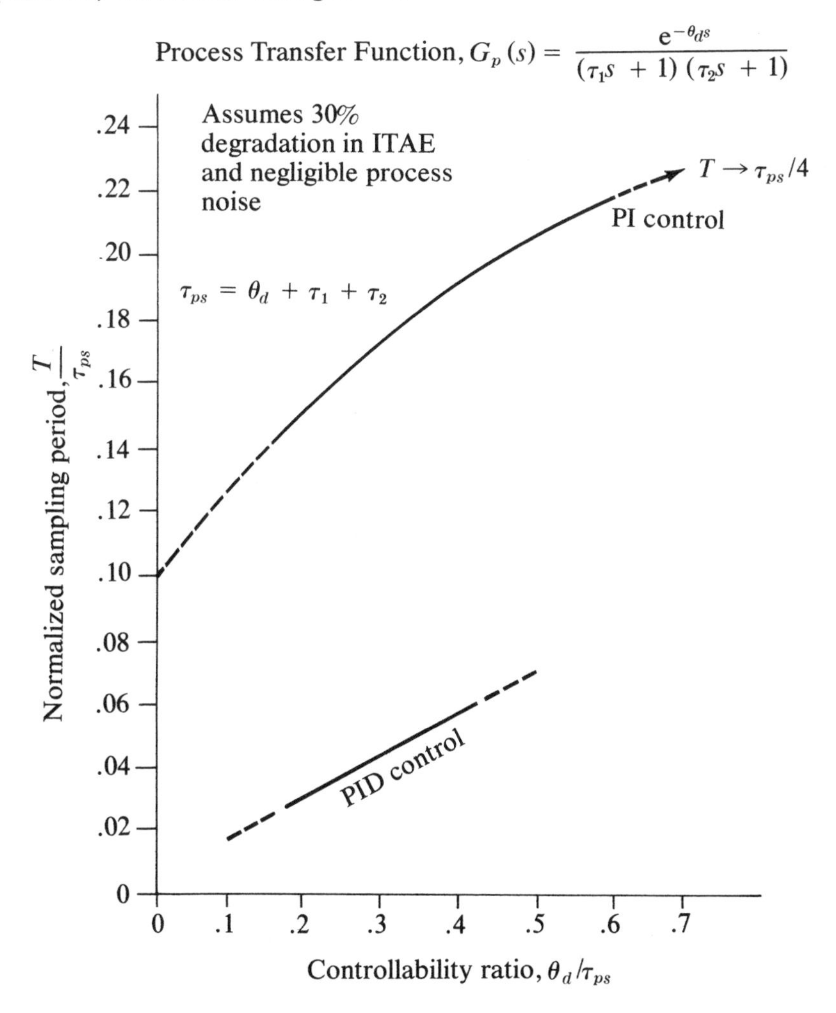

Figure 7.12
Selection of Sampling Period For PI Controllers
(Reprinted by Permission of Ref. 4)

A somewhat more refined estimate of the required sampling period may be obtained if the dynamic parameters of the process model are known. The basis of the approach here is the fact that the transient closed-loop response of a sampled-data control system is inferior as compared to that of the equivalent continuous control system because of the dead-time effects of sampling. By fixing the upper limit on the acceptable degradation, the maximum allowable sampling period can be back calculated through simulation. Figure 7.12 shows a plot of normalized sampling period versus controllability for PI and PID control of overdamped second-order processes with dead time.[4]

When a digital computer is used to execute the two- and three-mode (PI and PID) control equations, there exists a lower limit on the sampling period, because if T is too small, a reset deadband may result when a fixed point calculation is used to implement the control algorithm. With correct binary point selection and a 16-bit word length, the lower limit on the sampling period for both PI and PID control is[4]

$$T > \frac{\tau_I}{100}$$

where τ_I is integral time.

References

1. Kuo, B. C., *Analysis and Synthesis of Sampled-Data Control Systems*, Prentice-Hall, Englewood Cliffs, N.J. 1963.
2. Oliver, R. M., Pierce, J. R., Shannon, C. E., The Philosophy of Pulse Code Modulation, *Proc. IRE*, **36**, 11, November 1948, 1324–31.
3. Guidelines and General Information on User Requirements Concerning Direct Digital Control, First Users Workshop on Direct Digital Control, Princeton, N.J., April 3–4, 1963.
4. Fertik, H. A., Tuning Controllers for Noisy Processes, *ISA Trans.* **14**, 4, 1975.

CHAPTER 8

Open-Loop Response of Sampled-Data Systems

The open-loop system consists of the final control element, the process, and the measuring element in series and can be represented as shown in Figure 8.1. We have studied Z-transforms and their use in representing the sampled data; we have derived the transfer function of the zero-order hold, and in fact we now have the necessary background for evaluating the open-loop response of sampled-data control systems. Referring to Figure 8.1, let us first combine the four blocks into a single block, $G(s)$, which is merely the product of the transfer functions $G_{h_0}(s)$, $G_1(s)$, $G_p(s)$, and $H(s)$. Thus the open-loop system simplifies to

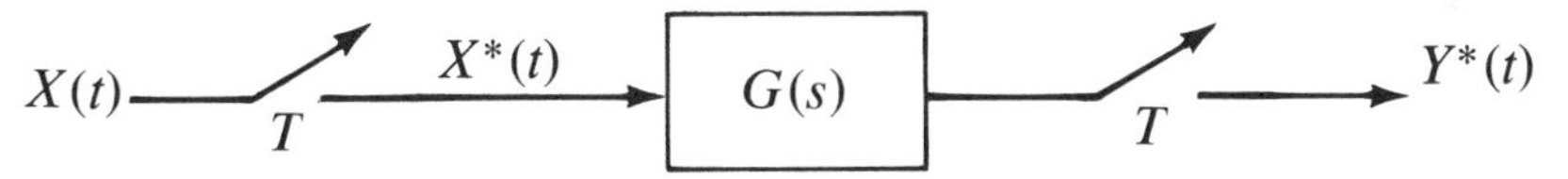

The Z-transform representation of this block diagram is

$$X(Z) \rightarrow \boxed{G(Z)} \rightarrow Y(Z)$$

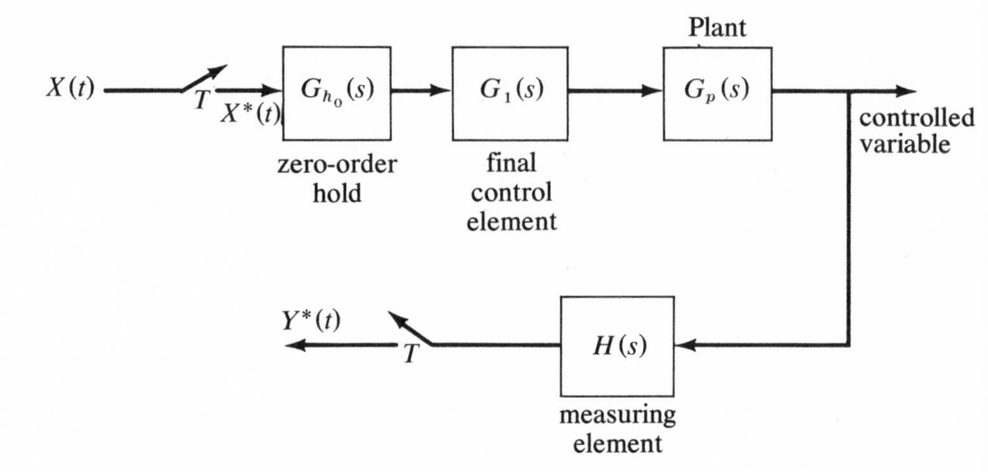

Figure 8.1
Open-Loop Sampled-Data System

that is,

$$\frac{Y(Z)}{X(Z)} = G(Z) \tag{8.1}$$

where $G(Z)$ is the Z transform of the function whose Laplace transform is $G(S)$.

To obtain the open-loop response of the system to a specified input $X(t)$ we take the Z transform of $X(t)$ to get $X(Z)$ and then combine it with $G(Z)$ to get $Y(Z)$. The inverse Z transform of $Y(Z)$ then gives us the open-loop response at the various sampling instants. The procedure is best illustrated by an example.

8.1. Example of Open-Loop Response

Determine the open-loop response of the sampled-data system shown below to a unit step change in input $X(t)$.

$X(t)$ — T — $X^*(t)$ → $G(s) = \dfrac{9}{s(s^2 + 9)}$ — T → $Y^*(t)$

The sampling period T is 1 sec.

Solution

The block diagram is represented as follows:

$$X(Z) \rightarrow \boxed{G(Z)} \rightarrow Y(Z)$$

For a unit step change in $X(t)$

$$X(Z) = \frac{Z}{Z-1} \tag{8.2}$$

Now let us expand $G(s)$ in partial fractions to get

$$G(s) = \frac{9}{s(s^2+9)} = \frac{1}{s} - \frac{s}{s^2+9} \tag{8.3}$$

the inverse of which is

$$G(t) = u(t) - \cos 3t \tag{8.4}$$

Now,

$$\begin{aligned}
G(Z) &= \sum_{n=0}^{\infty} G(nT)\, Z^{-n} \\
&= \sum_{n=0}^{\infty} \left[u(nT) - \cos(3nT)\right] Z^{-n} \\
&= \frac{Z}{Z-1} - \frac{1 - Z^{-1}\cos 3T}{1 - 2Z^{-1}\cos 3T + Z^{-2}} \qquad (8.5) \\
&= \frac{Z}{Z-1} - \frac{Z^2 - Z\cos 3T}{Z^2 - 2Z\cos 3T + 1}. \\
&= \frac{Z^3 - 2Z^2\cos 3T + Z - Z^3 + Z^2\cos 3T + Z^2 - Z\cos 3T}{(Z-1)(Z^2 - 2Z\cos 3T + 1)} \\
&= \frac{Z^2 + Z - Z^2\cos 3T - Z\cos 3T}{(Z-1)(Z^2 - 2Z\cos 3T + 1)} \\
&= \frac{Z(Z+1)(1-\cos 3T)}{(Z-1)(Z^2 - 2Z\cos 3T + 1)}
\end{aligned}$$

Thus

$$\begin{aligned}
Y(Z) &= X(Z)\, G(Z) \\
&= \frac{Z}{(Z-1)} \cdot \frac{Z(Z+1)(1-\cos 3T)}{(Z-1)(Z^2 - 2Z\cos 3T + 1)} \qquad (8.6) \\
&= \frac{Z^2(Z+1)(1-\cos 3T)}{(Z-1)^2(Z^2 - 2Z\cos 3T + 1)}
\end{aligned}$$

This can be expanded in partial fractions to give

$$Y(Z) = Z\left\{\frac{1}{(Z-1)^2} + \frac{1}{2}\frac{1}{Z-1} - \frac{1}{2}\frac{Z+1}{Z^2 - 2Z\cos 3T + 1}\right\} \tag{8.7}$$

So that we can look up the inverse, let us rearrange the above equation as

$$Y(Z) = \frac{Z}{(Z-1)^2} + \frac{1}{2}\frac{Z}{Z-1} - \frac{1}{2}\left(\frac{Z^2 - Z\cos 3T}{Z^2 - 2Z\cos 3T + 1}\right) - \frac{1}{2}\frac{1+\cos 3T}{\sin 3T}\left(\frac{Z\sin 3T}{Z^2 - 2Z\cos 3T + 1}\right) \tag{8.8}$$

Now, the inverse can be found from tables of Z-transforms. Thus

$$Y^*(t) = \frac{t}{T} + \frac{1}{2} - \frac{1}{2}\cos 3t - \frac{1}{2}\frac{1+\cos 3T}{\sin 3T}\sin 3t \tag{8.9}$$

For $T = 1.0$ sec, this simplifies to

$$Y^*(t) = t + 0.5 - 0.5\cos 3t - 0.0354\sin 3t \tag{8.10}$$

The response of $Y^*(t)$ is shown in Figure 8.2. It must be remembered that the output values of $Y^*(t)$ are correct only at the sampling instants. Several forms of the time function $Y(t)$ may be obtained from a given $Y(Z)$ through inverse transformation, yet all of them will indicate correct values only at the sampling instants. The complete time response $Y(t)$ can be obtained by the modified Z-transform method, which is discussed later.

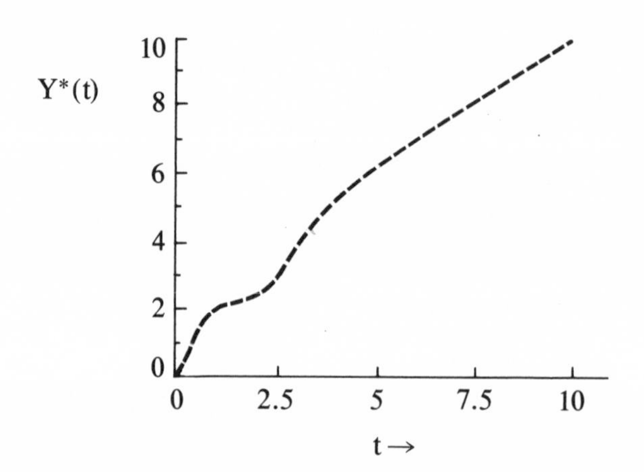

Figure 8.2
Open-Loop Response of Sampled-Data System

In this example we could have used the Z-transform inversion program, Appendix C1, to obtain $Y(t)$. To do this we would substitute $T = 1$ in Equation (8.6) and perform the indicated multiplication. We would then enter the coefficients of the numerator and the denominator polynomials into the computer and execute the program. The program would then output values of $Y(nT)$ at the sampling instants.

CHAPTER 9

Closed-Loop Response of Sampled-Data Control Systems

Let us now learn how to evaluate the transient closed-loop response of computer-control systems to set point and load changes. For the purpose of discussion in this chapter, we assume that the transfer functions of all the elements of the closed loop are known. The first step in evaluating closed-loop response is to obtain the closed-loop pulse transfer function of the system.

9.1. Closed-Loop Pulse Transfer Functions

The block diagram of a single-loop, computer-based, closed-loop control system is shown in Figure 9.1. This is the most general description of the feedback control system in that the dynamics of all the elements are assumed to be significant and therefore are included

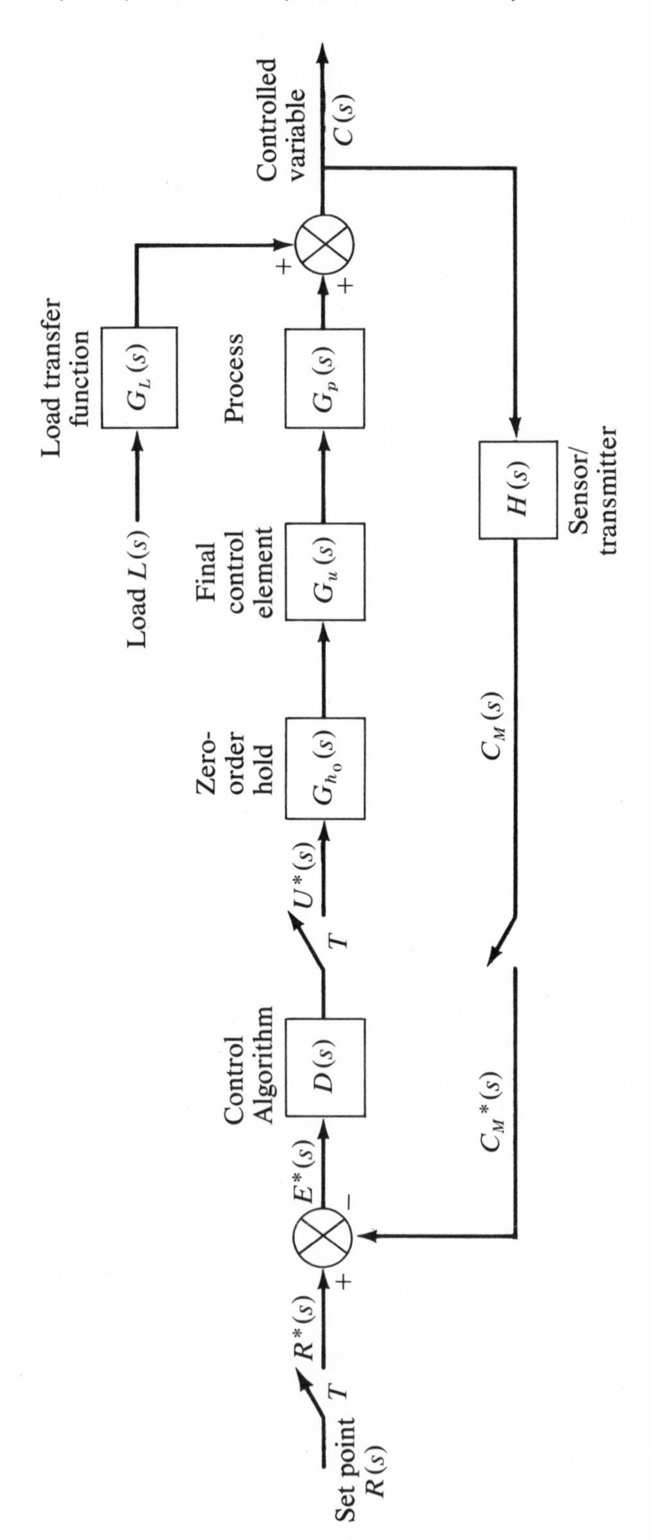

Figure 9.1
Closed-Loop Sampled-Data Control System

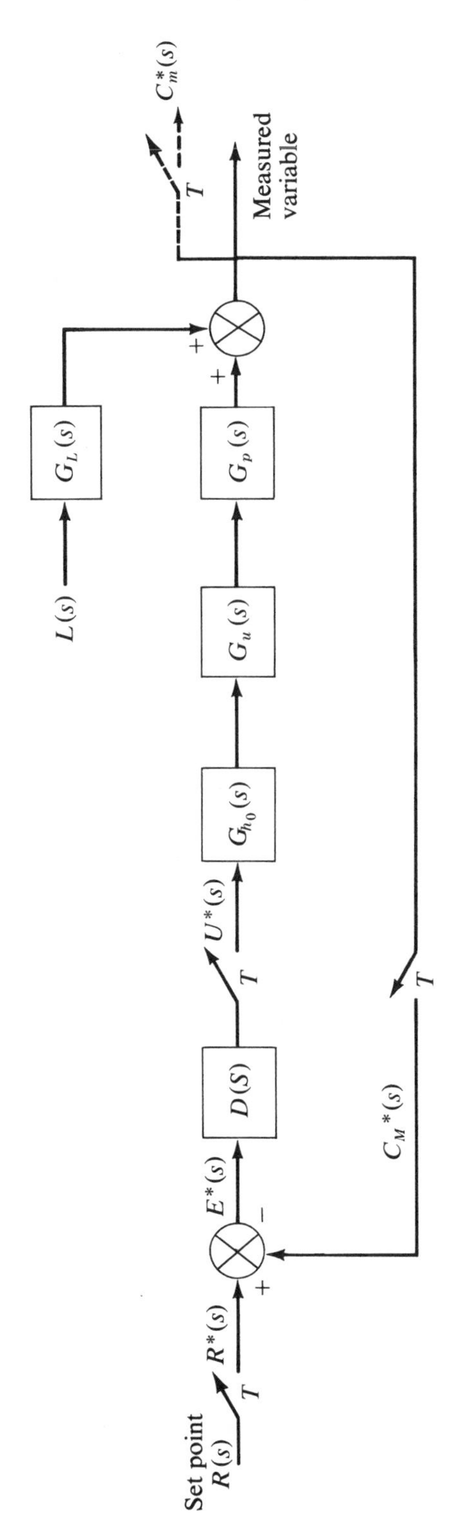

Figure 9.2
Closed-Loop Sampled-Data System with *(a)* Negligible Sensor/Transmitter Lag or *(b)* Sensor/Transmitter Lag Included in $G_p(s)$ and $G_L(s)$ (The Dashed Line Indicates a Fictitious Sampler, Meaning that the *Z*-Transform Analysis Will Give Information about an Output Variable Only at the Sampling Instants)

in the block diagram. Several simplifications of this block diagram are possible: If the sensor/transmitter lag is either negligible or is included in the transfer functions $G_p(s)$ and $G_L(s)$, the block diagram can be simplified as shown in Figure 9.2. On the other hand, if the dynamics are determined from experimental tests, the process transfer function $G_p(s)$ will include the dynamics of the final control element, and the sensor/transmitter and the load transfer function will include the dynamics of the sensor/transmitter. In this instance the block diagram of Figure 9.1 reduces to the one shown in Figure 9.3. Finally, if the load dynamics and the process dynamics are equal, that is, $G_L(s) = G_p(s)$, the block diagram of Figure 9.1 can be simplified as shown in Figure 9.4.

The derivation of the closed-loop pulse transfer function proceeds in a manner analogous to that of the continuous control system. However, note that the sampled-data system has a combination of continuous signals and sampled signals, therefore, we must be very careful in deciding which blocks can be combined into a single block. As an illustration, let us derive the closed-loop pulse transfer function of the block diagram of Figure 9.2.

First, we recall from our work in conventional control that the three blocks $G_{h_0}(s)$, $G_u(s)$, and $G_p(s)$ can be combined into a single block (note that there are no samplers between these blocks). Thus, let

$$G_{h_0}(s)\, G_u(s)\, G_p(s) = G(s)$$

Now, from Figure 9.2 we write

$$C_M(s) = U^*(s)\, G(s) + G_L(s)\, L(s) \tag{9.2}$$

$$R^*(s) - C_M{}^*(s) = E^*(s) \tag{9.3}$$

$$U^*(s) = D(s)\, E^*(s) \tag{9.4}$$

First, we take the starred transform of Equation (9.4) to get

$$U^*(s) = D^*(s)\, E^*(s) \tag{9.5}$$

Now, we substitute this equation for $U^*(s)$ into Equation (9.2) to get

$$C_M(s) = D^*(s)\, E^*(s)\, G(s) + G_L(s)\, L(s) \tag{9.6}$$

Substitution of $E^*(s)$ from Equation (9.3) into this equation gives

$$C_M(s) = D^*(s)\left\{R^*(s) - C^*_M(s)\right\} G(s) + G_L(s)\, L(s) \tag{9.7}$$

Next, we take the starred transform of this equation to get

$$C_M{}^*(s) = D^*(s)\, R^*(s)\, G^*(s) - D^*(s)\, C_M{}^*(s)\, G^*(s) + \overline{G_L(s)\, L(s)}^{\,*} \tag{9.8}$$

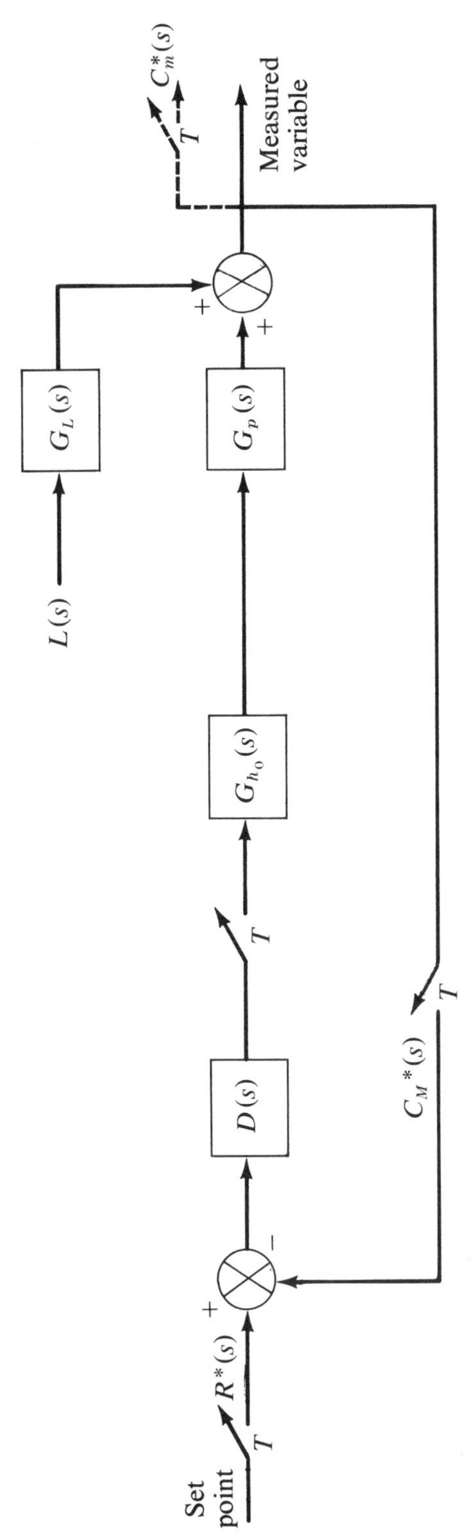

Figure 9.3
Closed-Loop Sampled-Data System where $G_p(s)$ Includes Final Control Element Dynamics and Sensor/Transmitter Dynamics and $G_L(s)$ Includes Sensor/Transmitter Dynamics

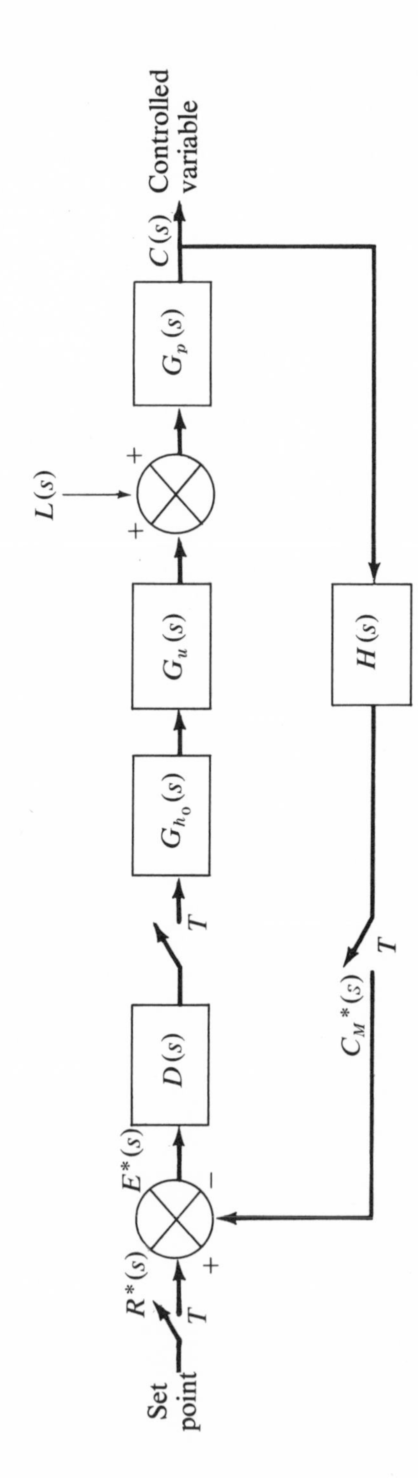

Figure 9.4
Closed-Loop Sampled-Data Control System Having Equal Process and Load Dynamics

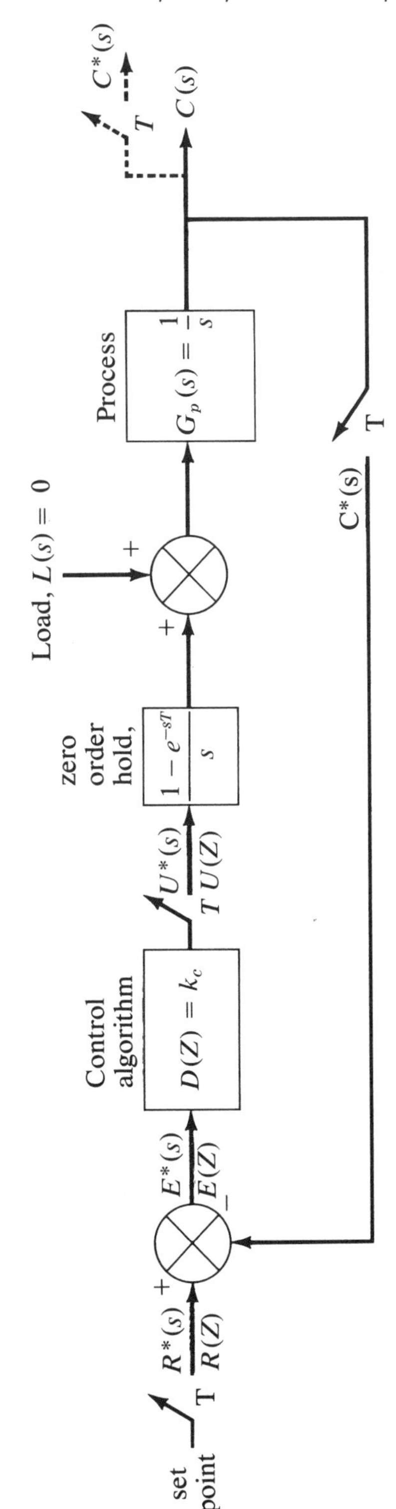

Figure 9.5
Proportional Control of a Sampled-Data System

where

$$G^*(s) = \overline{G_{ho}(s)\, G_u(s)\, G_p(s)}^{\,*}$$

The terms containing $C_M{}^*(s)$ can now be combined to give

$$C_M{}^*(s) = \frac{D^*(s) R^*(s) G^*(s)}{1 + D^*(s) G^*(s)} + \frac{\overline{G_L(s) L(s)}^{\,*}}{1 + D^*(s) G^*(s)} \tag{9.9}$$

The Z transform of this equation may be taken to give

$$C_M(Z) = \frac{D(Z) G(Z)}{1 + D(Z) G(Z)} R(Z) + \frac{G_L L(Z)}{1 + D(Z) G(Z)} \tag{9.10}$$

In taking the starred transform of terms such as those in Equation (9.7) it must be cautioned that (1) a function that is already starred is unaffected and (2) whenever a product of transfer functions appears, we must first multiply the terms and then take the starred transform. Thus the starred transform of $G_L(s)\, L(s)$ is $\overline{G_L(s)\, L(s)}^*$ and not $G_L(s)^*\, L(s)^*$. Consequently, $Z\{G_L(s)\, L(s)\} \neq G_L(Z)\, L(Z)$ in general. The operation of taking the Z-transform of $G_L(s)\, L(s)$ is denoted as $G_L L(Z)$.

Equation (9.10) relates the response of $C_M{}^*(t)$ to set point and load changes. For set point changes only it reduces to

$$\frac{C_M(Z)}{R(Z)} = \frac{D(Z)\, G(Z)}{1 + D(Z)\, G(Z)} \tag{9.11}$$

where

$$G(Z) = Z\{G_{ho}(s)\, G_u(s)\, G_p(s)\}$$

Similarly, we get for load changes alone

$$C_M(Z) = \frac{G_L\, L(Z)}{1 + D(Z)\, G(Z)} \tag{9.12}$$

where

$$G_L\, L(Z) = Z\{G_L(s)\, L(s)\}$$

Equation (9.11) is the closed-loop pulse transfer function of the sampled-data system to set-point changes. Note the similarity of this equation with its counterpart in conventional control systems. Also observe from Equation (9.12) that for load changes the pulse transfer function as classically defined (i.e., $C_M(Z)/L(Z)$) cannot be explicitly obtained. This is because the load enters the process without first being sampled.

For complicated loops (for example, the double cascade systems), this procedure of writing out the equations and solving for $C_M(s)$ to eventually obtain $C_M(Z)$ can get very tedious. In these instances the reader is referred to the method of signal flow graphs[1,2] for arriving at the pulse transfer functions. For single-loop situations the method described in this chapter is adequate.

9.2. Example to Determine Closed-Loop Transient Response

(a). Determine the pulse-transfer function $C(Z)/R(Z)$ for the sampled-data control system of Figure 9.5 with $L(S) = 0$.

(b). Evaluate the transient response to a step change in set point.

The closed-loop pulse transfer function for set point changes is

$$\frac{C(Z)}{R(Z)} = \frac{D(Z)\,G_{h_0}G_p(Z)}{1 + D(Z)\,G_{h_0}G_p(Z)}$$

$$\begin{aligned} G_{h_0}G_p(Z) = Z\{G_{h_0}(s)G_p(s)\} &= Z\left\{\frac{1}{s}\frac{1-e^{-sT}}{s}\right\} \\ &= Z\left\{\frac{1}{s^2}\right\} - Z\left\{\frac{e^{-sT}}{s^2}\right\} \\ &= Z\left\{\frac{1}{s^2}\right\} - Z^{-1}Z\left\{\frac{1}{s^2}\right\} \\ &= (1 - Z^{-1})Z\left(\frac{1}{s^2}\right) \\ &= (1 - Z^{-1})\left(\frac{TZ^{-1}}{(1 - Z^{-1})^2}\right) \\ &= \frac{TZ^{-1}}{(1 - Z^{-1})} \end{aligned}$$

The proportional controller output CO at the nth sampling instant is related to error by the expression

$$(CO)_n = K_c\,e_n + (CO)_{\text{steady state}}$$

similarly, at $(n - 1)$th sampling instant

$$(CO)_{n-1} = K_c\, e_{n-1} + (CO)_{\text{steady state}}$$

Therefore,

$$(CO)_n - (CO)_{n-1} = K_c\,(e_n - e_{n-1})$$

Taking the Z transform of this equation gives

$$CO(Z) - Z^{-1}\,CO(Z) = K_c E(Z) - Z^{-1}\,K_c E(Z)$$

$$CO(Z)\left[1 - Z^{-1}\right] = K_c \mathrm{E}(Z)\left[1 - Z^{-1}\right]$$

Therefore,

$$D(Z) = \frac{CO(Z)}{E(Z)} = K_c$$

Substitution of the expressions for $G_{h_0}\, G_p\,(Z)$ and $D(Z)$ into the pulse-transfer function will give

$$\frac{C(Z)}{R(Z)} = \frac{K_c T\, Z^{-1}/(1 - Z^{-1})}{1 + K_c T\, Z^{-1}/(1 - Z^{-1})}$$

$$= \frac{K_c T}{Z + (K_c T - 1)}$$

b. Evaluation of Transient Closed-Loop Response

For a unit step change in set point

$$R(t) = u(t)$$

Therefore,

$$R(Z) = \frac{1}{1 - Z^{-1}} = \frac{Z}{Z - 1}$$

Substitution of $R(Z)$ in the closed-loop pulse transfer function gives

$$C(Z) = \frac{K_c T Z}{(Z - 1)\left[Z + (K_c T - 1)\right]} = Z C_1\,(Z)$$

Therefore,

$$C_1(Z) = \frac{K_c T}{(Z - 1)\left[Z + (K_c T - 1)\right]}$$

Now we expand $C_1(Z)$ in partial fractions. Thus

$$\frac{K_c T}{(Z-1)\left[Z+(K_c T-1)\right]} = \frac{A}{Z-1} + \frac{B}{Z-(1-K_c T)}$$

Multiply by $Z - 1$ and set $Z = 1$

Then,

$$A = 1$$

Multiply by

$$Z - (1 - K_c T)$$

and set

$$Z = 1 - K_c T$$

Then,

$$B = -1$$

Therefore,

$$\begin{aligned} C_1(Z) &= \frac{K_c T}{(Z-1)\left[Z+(K_c T-1)\right]} \\ &= \frac{1}{Z-1} - \frac{1}{Z-(1-K_c T)} \end{aligned}$$

and

$$\begin{aligned} C(Z) &= Z\, C_1(Z) \\ &= \frac{Z}{Z-1} - \frac{Z}{Z-(1-K_c T)} \\ &= \frac{1}{1-Z^{-1}} - \frac{1}{1-e^{-\ln 1/(1-K_c T)}\, Z^{-1}} \end{aligned}$$

The second term is of the form

$$\frac{1}{1-e^{-aT}\, Z^{-1}}$$

the inverse of which is

$$e^{-at}$$

where

$$-aT = -\ln\left[1/(1 - K_c T)\right]$$

$$a = \frac{1}{T}\ln\left[1/(1 - K_c T)\right]$$

Therefore,

$$\begin{aligned} C(t) &= u(t) - e^{-t/T \ln[1/(1 - K_c T)]} \\ &= u(t) - e^{\ln(1 - K_c T)t/T} \\ &= u(t) - (1 - K_c T)^{t/T} \end{aligned}$$

Remember that this equation is valid only at the sampling instants.

References

1. Ash, R., Kim, W.H., Kranc, G.M., A general Flow Graph Technique for the Solution of Multiloop Sampled Problems, *Trans. A. S. M. E., Journal of Basic Engineering*, June 1960. pp. 360–370.
2. Kuo, B. C., *Analysis and Synthesis of Sampled-data Control Systems*, Prentice Hall, Englewood Cliffs, N.J., 1963, pp. 112–142.

CHAPTER 10

Design of Sampled-Data Control Systems

In this chapter we consider the design of control algorithms via Z transforms. In Chapter 3 we briefly considered the use of conventional (P, PI, or PID) algorithms in computer-control applications. In this chapter we discuss the conventional control algorithms in greater detail. We will shortly see that the algorithms designed by the Z-transform method enable us to specify the desired response characteristics but their development requires the knowledge of the process transfer function. On the other hand, the development of conventional control algorithms does not require the knowledge of Z transforms. To begin the development of control algorithms by Z transforms, consider the block diagram of a typical sampled-data control system shown in Figure 10.1. The process transfer function $G_p(s)$ includes the dynamics of the sensor and the final control element.

Our objective is to design the control algorithm $D(Z)$ such that the desired loop performance is achieved. Once we have developed an expression for $D(Z)$ we may invert it into the time domain to give

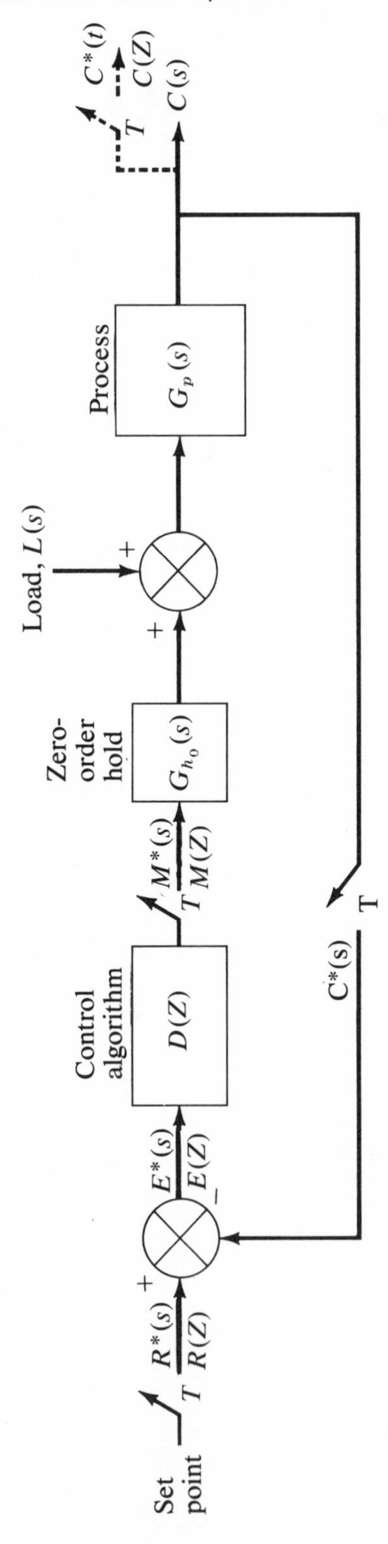

Figure 10.1
Closed-Loop Sampled-Data Control System

an equation suitable for computer programming. It is possible to develop the algorithm for set-point changes or for load changes.

To develop the algorithm for set-point changes, recall from our study of Chapter 9 that the closed-loop pulse transfer function of the system of Figure 10.1 for set-point changes is

$$\frac{C(Z)}{R(Z)} = \frac{D(Z)G_{h_0}G_p(Z)}{1 + D(Z)G_{h_0}G_p(Z)} \tag{10.1}$$

where

$$G_{h_0}G_p(Z) = Z\{G_{h_0}(s)\,G_p(s)\}$$

The solution of Equation (10.1) for $D(Z)$ gives

$$D(Z) = \frac{1}{G_{h_0}G_p(Z)} \cdot \frac{C(Z)/R(Z)}{[1 - C(Z)/R(Z)]} \tag{10.2}$$

The design procedure is to select a suitable set-point input (e.g., step input or ramp) and specify the desired response characteristic (e.g., the controlled variable shall reach the new set point in one sampling period) so that $C(Z)/R(Z)$ can be computed. Equation (10.2) can then be solved for $D(Z)$, provided the process transfer function is known.

The transfer function of the process can be developed from a dynamic mathematical model based on first principles (e.g., heat, mass, or momentum balances), or it can be developed from experimental tests in the plant. We consider the experimental evaluation of transfer functions later in the text. For the purpose of discussion in this chapter, let us assume that the transfer function $G_p(s)$ is known.

The control algorithm may also be designed for load changes. In this case we must use the closed-loop pulse transfer function for load changes, which is given as

$$C(Z) = \frac{G_L\,L(Z)}{1 + D(Z)G_{h_0}G_p(Z)} \tag{10.3}$$

where

$$G_L\,L(Z) = Z\{G_L(s)\,L(s)\}$$

and

$$G_{h_0}G_p(Z) = Z\{G_{h_0}(s)\,G_p(s)\}$$

The solution of Equation (10.3) for $D(Z)$ gives

$$D(Z) = \frac{G_L\,\mathrm{L}(Z) - C(Z)}{G_{h_0}G_p(Z)\,C(Z)} \qquad (10.4)$$

The procedure in this instance is to select a suitable load input L, specify the desired response characteristics, and take the Z transform as indicated in Equation (10.4) to obtain $D(Z)$.

Whether we should design $D(Z)$ for set-point changes or for load changes depends on whether we can measure the load disturbances and whether we can correctly anticipate the form of these disturbances. Based on a thorough understanding of the process dynamics, the designer must decide whether to base the design on set-point changes or on load changes. Fortunately, it has been reported that in many cases the set-point designs work well for load changes as well.[1]

In this chapter we will consider three control algorithms: (1) the deadbeat algorithm, (2) the Dahlin algorithm, and (3) the digital equivalent to the conventional controller algorithm.

10.1 Deadbeat Algorithm[2]

An algorithm that requires the closed-loop response to have finite settling time, minimum rise time, and zero steady-state error is referred to as a deadbeat algorithm. As an example, consider the design of a deadbeat algorithm for a unit step change in set point R. For this case

$$R(t) = u(t)$$

and (10.5)

$$R(Z) = \frac{1}{1 - Z^{-1}}$$

If we specify that the controlled variable C shall reach the new set point in one sampling period and remain at that value thereafter, we will have satisfied the above criteria. Mathematically, these requirements may be stated as

$$\begin{aligned} C(Z) &= 0 + Z^{-1} + Z^{-2} + Z^{-3} + \cdots \\ &= Z^{-1}(1 + Z^{-1} + Z^{-2} + Z^{-3} + \cdots) \qquad (10.6) \\ &= \frac{Z^{-1}}{1 - Z^{-1}} \end{aligned}$$

Now, we compute $C(Z)/R(Z)$ from Equations (10.5) and (10.6). Thus,

$$\frac{C(Z)}{R(Z)} = \frac{Z^{-1}}{1 - Z^{-1}} \cdot \frac{1 - Z^{-1}}{1} = Z^{-1} \tag{10.7}$$

Next, we substitute for $C(Z)/R(Z)$ into Equation (10.2) so as to obtain

$$D(Z) = \frac{1}{G_{h_0}G_p(Z)} \frac{Z^{-1}}{1 - Z^{-1}} \tag{10.8}$$

This is as far as we can proceed with the development of the algorithm without knowing the process transfer function $G_p(s)$. Let us now consider the design of a deadbeat controller for a process whose transfer function is given by

$$G_p(s) = \frac{1}{0.4s + 1} \tag{10.9}$$

Let us assume that the sampling period $T = 1$. For this example

$$\begin{aligned} G_{h_0}G_p(Z) &= Z\{G_{h_0}(s)\, G_p(s)\} \\ &= Z\left\{\frac{1 - e^{-sT}}{s} \cdot \frac{1}{0.4s + 1}\right\} \\ &= Z\left\{\frac{1}{s(0.4s + 1)}\right\} - Z\left\{\frac{e^{-sT}}{s(0.4s + 1)}\right\} \end{aligned} \tag{10.10}$$

Applying the theorem on translation of the function to Equation (10.10) gives

$$\begin{aligned} G_{h_0}G_p(Z) &= Z\left\{\frac{1}{s(0.4s + 1)}\right\} - (Z^{-1})\, Z\left\{\frac{1}{s(0.4s + 1)}\right\} \\ &= (1 - Z^{-1})\, Z\left\{\frac{1}{s(0.4s + 1)}\right\} \end{aligned} \tag{10.11}$$

This equation should be rearranged slightly so that we may look up the Z transform from tables. Thus

$$\begin{aligned} G_{h_0}G_p(Z) &= (1 - Z^{-1})\, Z\left\{\frac{1}{s(0.4s + 1)}\right\} \\ &= (1 - Z^{-1})\, Z\left\{\frac{2.5}{s(s + 2.5)}\right\} \\ &= \frac{(1 - Z^{-1})\, Z(1 - e^{-2.5T})}{(Z - 1)\,(Z - e^{-2.5T})} \\ &= \frac{1 - e^{-2.5T}}{Z - e^{-2.5T}} \end{aligned} \tag{10.12}$$

The next step is to substitute this expression for $G_{h_0}G_p(Z)$ into Equation (10.8) which gives

$$D(Z) = \frac{Z - e^{-2.5T}}{1 - e^{-2.5T}} \cdot \frac{Z^{-1}}{1 - Z^{-1}} \tag{10.13}$$

For $T = 1$ this equation becomes

$$\begin{aligned} D(Z) &= \frac{(Z - 0.082)Z^{-1}}{0.918(1 - Z^{-1})} \\ &= \frac{1 - 0.082\,Z^{-1}}{0.918(1 - Z^{-1})} \end{aligned} \tag{10.14}$$

Equation (10.14) is the deadbeat algorithm for this example in the Z domain. Let us now invert this algorithm into the time domain. To accomplish this we note from Figure 10.1 that

$$D(Z) = \frac{M(Z)}{E(Z)} \tag{10.15}$$

where

$M(Z) = Z$ transform of controller output

$E(Z) = Z$ transform of error $R(Z) - C(Z)$.

Thus

$$D(Z) = \frac{M(Z)}{E(Z)} = \frac{1 - 0.082\,Z^{-1}}{0.918(1 - Z^{-1})} \tag{10.16}$$

The cross multiplication of terms in Equation (10.16) gives

$$0.918\,M(Z) - 0.918Z^{-1}\,M(Z) = E(Z) - 0.082\,Z^{-1}\,E(Z)$$

This equation can be inverted to give the algorithm for computing the controller output. Thus

$$M_n = M_{n-1} + 1.09E_n - 0.089\,E_{n-1} \tag{10.17}$$

where

M_n = controller output at the nth sampling instant

M_{n-1} = controller output at the $(n - 1)$th sampling instant

E_n = error (set point − measurement) at the nth sampling instant

E_{n-1} = error at $(n - 1)$th sampling instant

The Fortran or Basic statements

```
      M = M + 1.09*E - 0.089*E1
      E1 = E                                        (10.18)
C     RETURN FOR NEW SAMPLE AND REPEAT ALGORITHM
```

will accomplish the computations indicated in Equation (10.17).

It must be emphasized that this deadbeat algorithm was designed on the basis of a step change in set point. If the controlled system is subjected to some other kind of input, the response may be unacceptable. Furthermore, the process transfer function is seldom known exactly, and therefore, the controlled response even to a step change in set point will deviate from that specified in Equation (10.6). However, this algorithm is particularly simple to use and may be adequate for specific situations. A more thorough discussion of the limitations of deadbeat control strategies is given by Kuo.[2]

10.2. Dahlin Algorithm[3]

Dahlin's algorithm specifies that the closed-loop sampled-data control system behave as though it were a continuous first-order process with dead time, that is,

$$\frac{C(s)}{R(s)} = \frac{e^{-\theta_d s}}{(\lambda s + 1)} \tag{10.19}$$

where λ is the time constant of the closed-loop response which is selected by trial and error by the designer and θ_d is the process dead time. For a unit step change in $R(t)$

$$R(s) = \frac{1}{s}$$

and therefore, (10.20)

$$C(s) = \frac{e^{-\theta_d s}}{s(\lambda s + 1)}$$

Taking the Z transform of Equation (10.20) gives

$$C(Z) = \frac{(1 - e^{-T/\lambda})\, Z^{-(N+1)}}{(1 - Z^{-1})(1 - e^{-T/\lambda} Z^{-1})} \tag{10.21}$$

where N = largest integral number of sampling periods in θ_d. For a step change in $R(t)$

$$R(Z) = \frac{1}{1 - Z^{-1}} \tag{10.22}$$

and

$$\frac{C(Z)}{R(Z)} = \frac{(1 - e^{-T/\lambda})\, Z^{-(N+1)}}{(1 - e^{-T/\lambda}\, Z^{-1})} \tag{10.23}$$

Therefore, the control algorithm $D(Z)$ should be evaluated according to

$$\begin{aligned} D(Z) &= \frac{C(Z)/R(Z)}{1 - C(Z)/R(Z)} \cdot \frac{1}{G_{h_0}G_p(Z)} \\ &= \frac{(1 - e^{-T/\lambda})\, Z^{-(N+1)}}{1 - e^{-T/\lambda}\, Z^{-1} - (1 - e^{-T/\lambda})Z^{-(N+1)}} \cdot \frac{1}{G_{h_0}G_p(Z)} \end{aligned} \tag{10.24}$$

To select a suitable value for λ, we would assume a trial value and simulate the control system on a computer. By repeatedly varying λ and examining the closed-loop response, the proper selection of λ can be made. Equation (10.24) represents the Dahlin algorithm. As an illustration consider the design of Dahlin algorithm for computer control of a process whose open-loop transfer function is given by[4]

$$G_p(s) = \frac{e^{-0.8s}}{0.4s + 1} \tag{10.25}$$

with $T = 0.4$, and therefore $N = 2$. Now,

$$\begin{aligned} G_{h_0}G_p(Z) &= Z\left\{\frac{1 - e^{-sT}}{s} \cdot \frac{e^{-0.8s}}{0.4s + 1}\right\} \\ &= \frac{0.6321\, Z^{-3}}{1 - 0.3679\, Z^{-1}} \end{aligned} \tag{10.26}$$

Assuming a value of 0.15 for λ and substitution of Equation (10.26) into Equation (10.24) gives

$$D(Z) = \frac{M(Z)}{E(Z)} = \frac{0.9305 - 0.3423\, Z^{-1}}{0.6321 - 0.0439Z^{-1} - 0.5882Z^{-3}} \tag{10.27}$$

Cross multiplying and inverting this equation into the time domain gives the algorithm in a form suitable for programming. Thus

$$M_n = 1.4721 E_n - 0.5421 E_{n-1} + 0.0695 M_{n-1} + 0.9306 M_{n-3} \tag{10.28}$$

The control system of the example was simulated on a digital computer. The response of the system to a unit step change in set point is shown in Figure 10.2.

The deadbeat and Dahlin computer-control algorithms are designed for a specific input, for example, a step change in set point. If a load change occurs in a process for which the control algorithm is based on a change in set point, the response will not be equally good. The usual procedure, therefore, is to design for the worst possible change in either set point or load that is likely to occur.

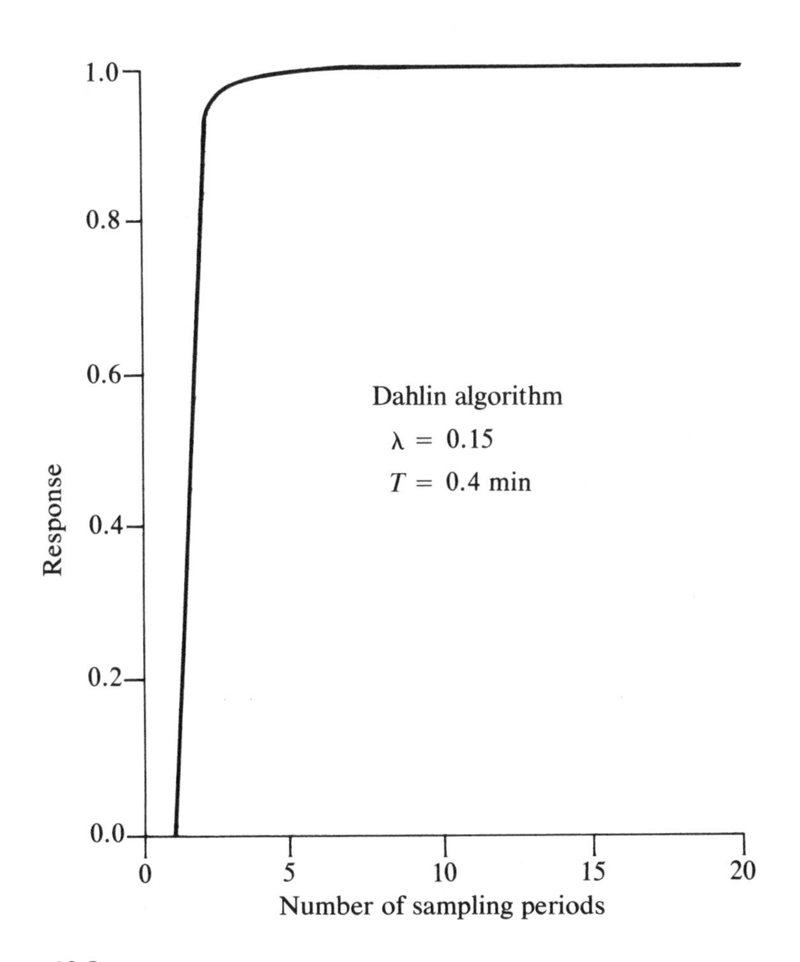

Figure 10.2
Response of the Control System to a Unit Step Change in Set Point

10.3. Digital Equivalent to a Conventional Controller

The operation of an ideal PID controller is described by

$$v = v_0 + K_c\left(e + \frac{1}{\tau_I}\int_0^t e\,dt + \tau_D\frac{de}{dt}\right) \tag{10.29}$$

In conventional control applications, a controller, whose output approximates the right side of Equation (10.29), can be built through the use of pneumatic components or operational amplifiers, integrators, and summers. In computer-control applications a discrete equivalent to Equation (10.29) is employed. In the development of algorithms that are based on Z transforms we specify the nature of the response to be achieved, whereas in the digital equivalent to the PID controller we "adjust" the constants K_c, τ_I, and τ_D so as to achieve desired re-

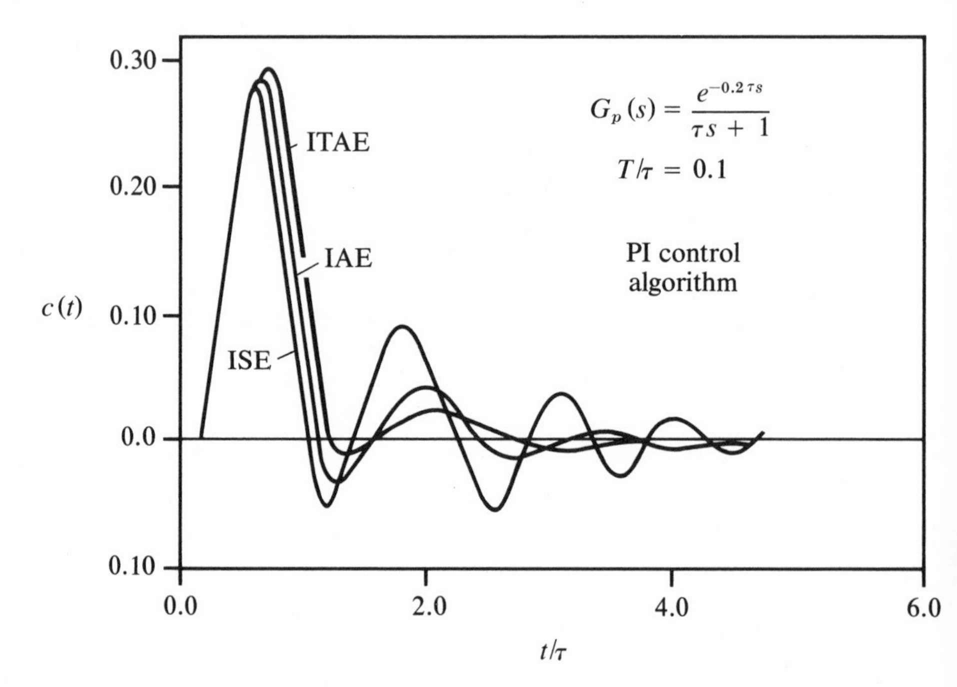

Figure 10.3
Closed-Loop Response of a First-Order Process with Dead Time to a Unit Step Change in Load with Different Performance Criteria (Reproduced with Permission from A. M. Lopez, et al., Tuning PI & PID Digital Controllers, *Instruments and Control Systems,* 42, 2, February 1969. p. 89)

sponse. The computer control system containing the PID control algorithm can be simulated and the constants adjusted so as to minimize the value of an integral of the type

1. Integral of the square error,

$$\mathrm{ISE} = \int_0^\infty [e]^2\, dt$$

2. Integral of the absolute value of the error,

$$\mathrm{IAE} = \int_0^\infty |e|\, dt \tag{10.30}$$

3. Integral of the time multiplied by the absolute value of the error,

$$\mathrm{ITAE} = \int_0^\infty t|e|\, dt$$

Just which criterion to choose depends on the type of response desired.[1]

For example, large errors contribute more heavily to ISE than to IAE, which means that ISE will favor responses with smaller overshoots for load changes, as shown in Figure 10.3. Note that ISE gives a longer settling time. In the ITAE integral, time appears as a factor, and therefore, this criterion, penalizes heavily errors that occur late in time but virtually ignores errors that occur early in time. Figure 10.3 shows that ITAE criterion gives shortest settling time but has the largest overshoot among the three criteria considered.

To obtain the digital equivalent to the PID controller, the derivative and the integral terms of Equation (10.29) are numerically approximated to give an expression for the output of the algorithm at the nth sampling instant. Thus

$$v_n = v_0 + K_c \left[e_n + \frac{T}{\tau_I} \sum_{i=0}^{n} e_i + \frac{\tau_D}{T} (e_n - e_{n-1}) \right] \tag{10.31}$$

where

v_n = controller output at nth sampling instant

e_n = error (set point − measurement) at the nth sampling instant

v_0 = steady state output of the control algorithm that gives zero error

Equation (10.31) is referred to as the "position" form of the control algorithm, since the actual controller output is computed. To derive an alternate form of the algorithm we write the expression for controller output at the $(n - 1)$th sampling instant as

$$v_{n-1} = v_0 + K_c \left[e_{n-1} + \frac{T}{\tau_I} \sum_{i=0}^{n-1} e_i + \frac{T_D}{T} (e_{n-1} - e_{n-2}) \right] \quad (10.32)$$

Then we subtract Equation (10.32) from Equation (10.31) to obtain

$$v_n = v_{n-1} + K_c \left[(e_n - e_{n-1}) + \frac{T}{\tau_I} e_n + \frac{\tau_D}{T} (e_n - 2e_{n-1} + e_{n-2}) \right] \quad (10.33)$$

Equation (10.33) is referred to as the velocity form of the PID algorithm, because it computes the incremental output instead of the actual output of the controller. The velocity form of the algorithm also provides some protection against reset windup, because it does not incorporate sums of error sequences.

Although we have not considered stability aspects of sampled-data control systems as yet, perhaps this is an appropriate place to point out that the performance of the computer-control algorithm depends not only on the tuning constants but also on the sampling period. Indeed, it can be shown[5] that although a second-order conventional control system is stable for all values of the proportional gain constant, computer control of the same system can give unstable response for some specific combinations of the proportional gain and the sampling period. We consider the stability aspects of sampled data control systems later in the text. For now, we should remember from our discussion in Chapter 7 that the technical requirements of the sampling theorem must be satisfied.

Using simulation techniques and the IAE performance criteria, Fertik[6] developed controller parameter charts for conventional PI and PID controllers. The tuning constants are based on a process model of the form

$$G_p(s) = \frac{K_p \, e^{-\theta_d s}}{(\tau_1 s + 1)(\tau_2 s + 1)} \quad (10.34)$$

The charts are shown in Figures 10.4 and 10.5. The abscissa for the charts is the process controllability parameter which is defined as

$$\text{process controllability} = \frac{\theta_d}{\tau_1 + \tau_2 + \theta_d} \quad (10.35)$$

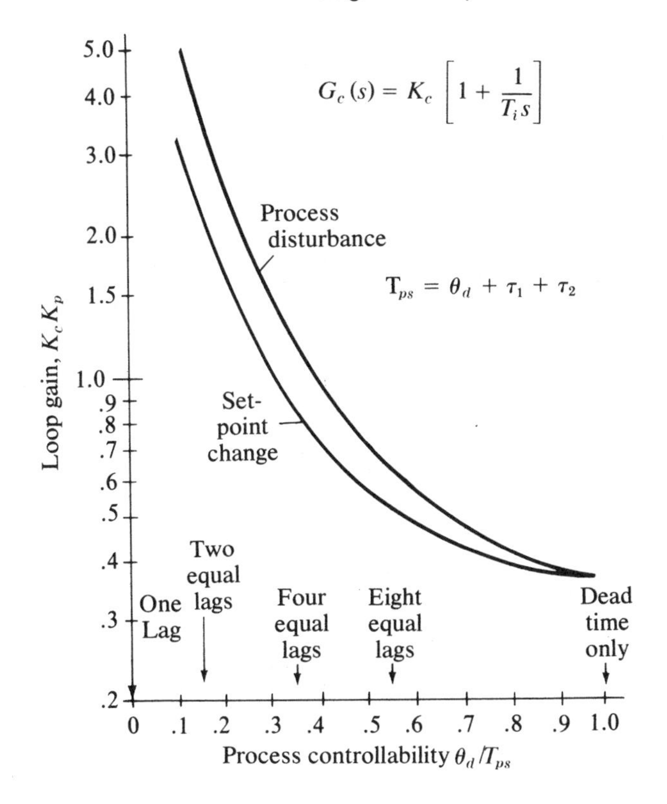

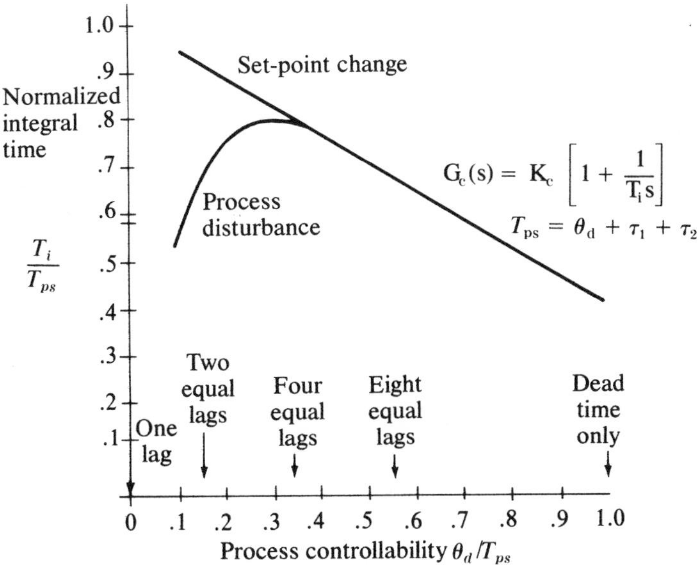

Figure 10.4
Controller Parameter Charts for PI Controllers
(By Permission from Ref. 6)

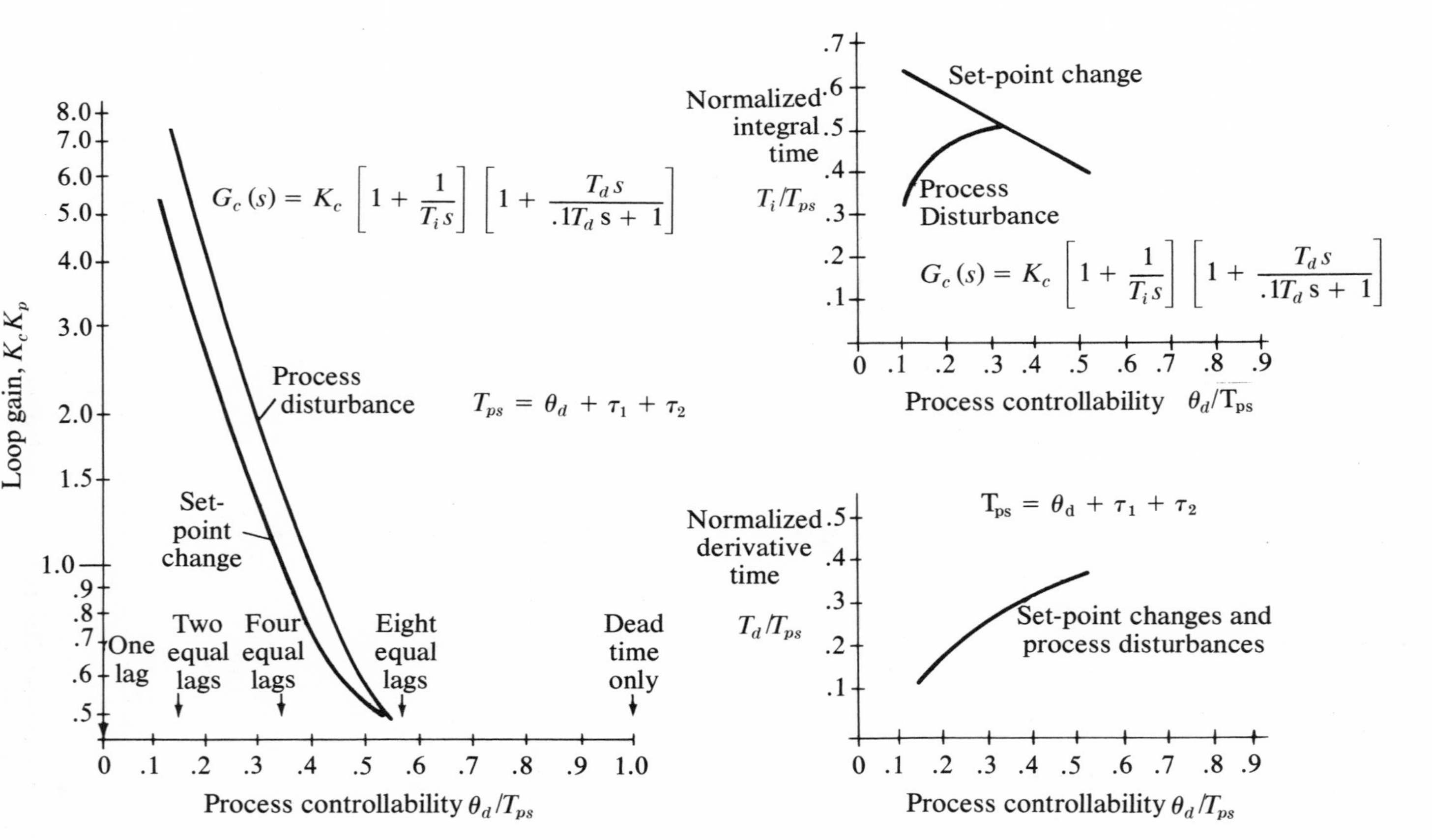

Figure 10.5
Controller Parameter Charts for PID Controllers
(By Permission from Ref. 6)

Thus the values of the parameter range from 0 for processes containing no dead time to 1 for pure dead-time processes.

For computer-control applications the same charts are used, except that the process controllability is calculated according to the equation

$$\text{process controllability} = \frac{\theta_d + T/2}{\tau_1 + \tau_2 + \theta_d + T/2} \tag{10.36}$$

where

$$T = \text{sampling period}$$

The above calculation corrects for the dead-time effects of sampling. Thus the same controller parameter-charts, Figures 10.4 and 10.5, can be used for digital PI or PID controllers.

We have included a computer program in Appendix C2 which determines the transient closed-loop response of sampled-data control systems to set-point or load changes. The program is based on one or two time constants plus a dead-time model. If the parameters of the open-loop process model are available, the reader may enter trial values of the tuning constants, execute the program, plot the load or set point response, and adjust the tuning constants until satisfactory performance (at least by a visual check) is achieved.

10.4. Treatment of Noisy Process Signals

We may define the term noise as those fluctuations in the process signal which do not contain useful control information. In process-control applications noise arises from one or more of the following sources:

Presence of those process disturbances that are too rapid to be reduced significantly by control action.

Turbulence around sensors (e.g., liquid flow or furnace pressure) and instrument noise.

Stray electrical pick up such as from AC power lines.

Process noise resulting from these sources occupies a frequency band that is much wider than that of the controllable process fluctuations. Since control action can only minimize the effects of these low-frequency, controllable fluctuations, process noise always causes a degradation of control performance. Further, the noise can cause aliasing of the signal (sampling-induced low-frequency noise) if the sampling frequency is not carefully chosen (see Figure 10.6). High-frequency noise, which cannot be attenuated by control, also causes excessive actuator wear. It should be clear from this introduction that we must incorporate noise-reduction techniques in those sampled-data control loops that contain noisy process signals. Some process signals are relatively noise-free. For example, when temperature is measured in a large thermal capacity, such as a thermowell, noise reduction is not required.

The frequency bands which the different noise sources occupy are shown in Figure 10.7. Noise reduction is accomplished via pure analog and/or digital filtering as shown in Figure 10.8.

Analog Filters

Stray electrical pickup can be minimized by proper grounding, shielding, and routing of wires. This type of high-frequency noise can be eliminated by an analog (RC) filter, shown schematically in Figure

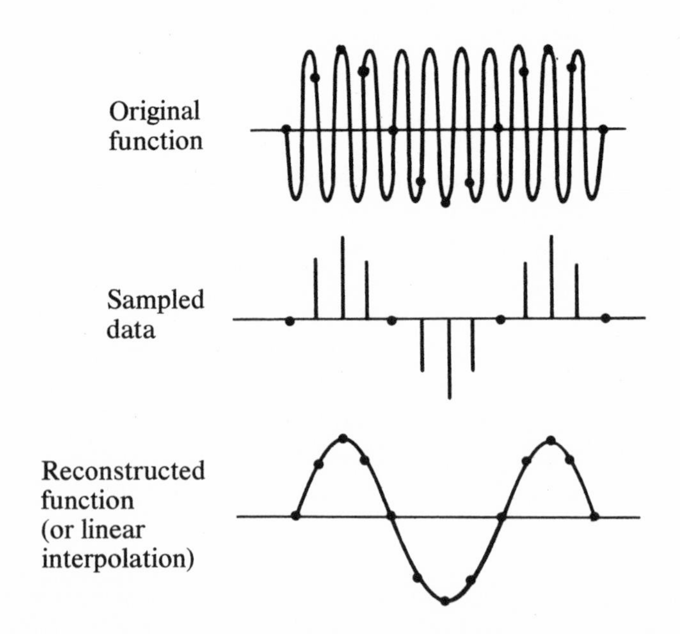

Figure 10.6
Aliasing Error Resulting from Sampling at Rates of 7/8 Samples per Cycle or Original Function (Reprinted by Permission from Ref. 7)

10.9. The time constant of the analog filter τ_b is usually about $T_c/2$. The equation for the RC filter shows that it behaves as a simple first-order lag. Analog filters that are built from passive elements (resistors and capacitors) give time constants up to a few seconds. For filter time constants greater than a few seconds, active elements using operational amplifiers are needed. Since the filter serves only one input signal, this construction becomes relatively expensive. Therefore, combined analog/digital filtering is usually preferred, as shown in Figure 10.8.

Digital Filtering Algorithms

There are two algorithms for digital filtering; one is based

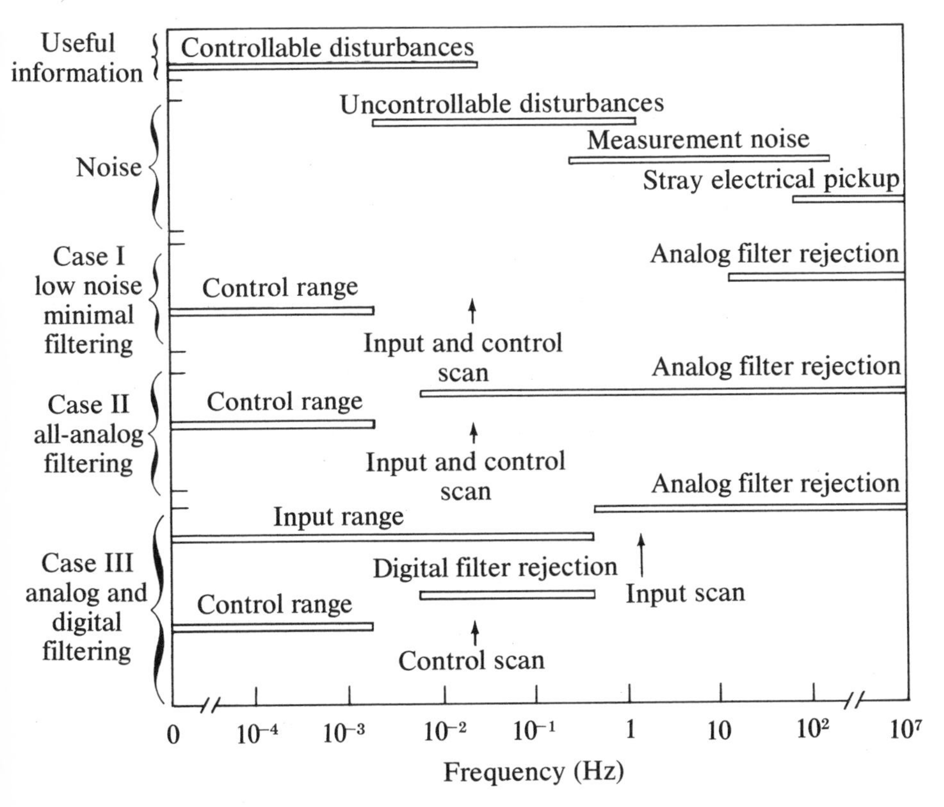

Figure 10.7
Frequency Ranges for Input Signal and Associated Scanning and Filtering Functions for a Typical Process
(Reprinted by Permission from Ref. 7)

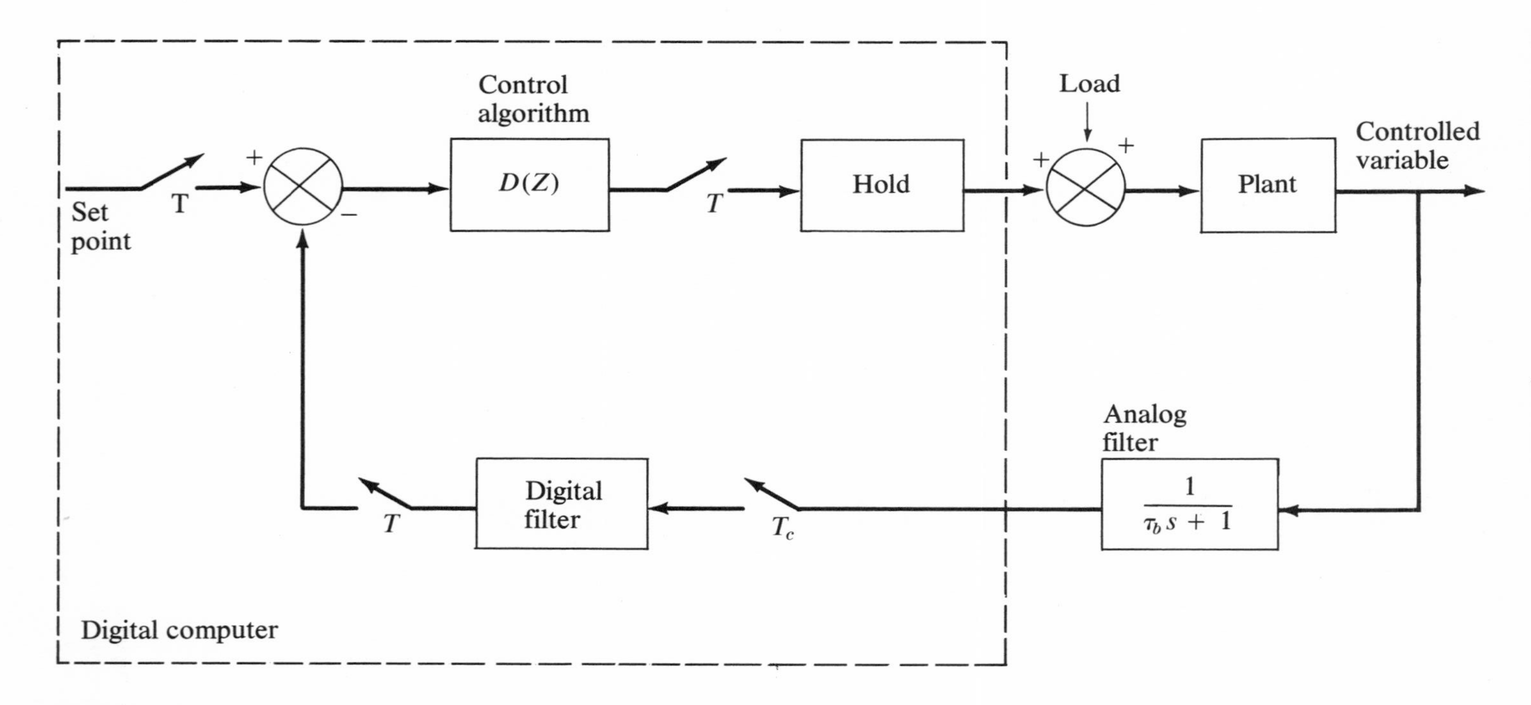

Figure 10.8
Sampled-Data Control System with Noise Reduction Schemes

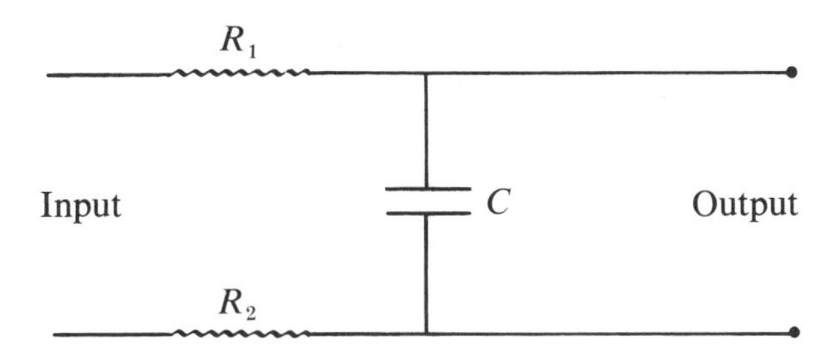

Figure 10.9
Analog RC Filter

simply on the arithmetic average of N samples, whereas the other is the digital equivalent of the first-order lag.

The mean value of the filter output which should be employed in control calculations according to the first method is

$$y_k = \frac{1}{N} \sum_{i=0}^{N-1} x_{k-i} \tag{10.37}$$

where

x_k = kth input to the filter

y_k = filter output

N = number of input samples

Since N samples of x_k are required to produce a filter output y_k, the successive time intervals for control action (i.e., the sampling period T) is $T = NT_c$.

The second algorithm is the digital equivalent to the first-order lag. Recall that the differential equation for the first-order lag is

$$\tau_f \frac{dy}{dt} + y = x(t) \tag{10.38}$$

where x and y are input and output, respectively, and τ_f is the time constant of the first-order system. The numerical approximation of the differential equation is

$$y_n = \frac{1}{Q} x_n + \left(1 - \frac{1}{Q}\right) y_{n-1} \tag{10.39}$$

where Q is a design parameter that is related to the sampling period T and the filter time constant τ_f as

$$Q = \frac{1}{1 - e^{-T/\tau_f}}$$

$$\cong \frac{\tau_f}{T} \text{ for large } \frac{T_f}{T} \text{ (say, greater than 10).} \qquad (10.40)$$

To determine the digital filter time constant, consider the frequency spectrum of a process signal shown in Figure 10.10.[1] If we sample at a frequency of ω_s (equal to $2\pi/T$), the frequency components above $\omega_s/2$ will be folded onto the low-frequency components if no filtering were used. The use of a digital first-order lag filter will greatly attenuate (though not completely eliminate) these components. Since the corner frequency of the first-order lag is simply the reciprocal of its time constant, the filter time constant is given by

$$\tau_f = \frac{1}{\omega_s/2} = \frac{2T}{2\pi} = \frac{T}{\pi}$$

A rule of thumb is to set the digital filter time constant equal to $T/2$.

Characterization and Measurement of Process Noise Parameters.

The process noise characteristics (magnitude and bandwidth) can be estimated from steady-state process data when there are no changes in the manipulated variable. Figure 10.11 shows such data for an illustrative process.[6] It can be assumed that most random noise found in industrial control can be represented by exponentially

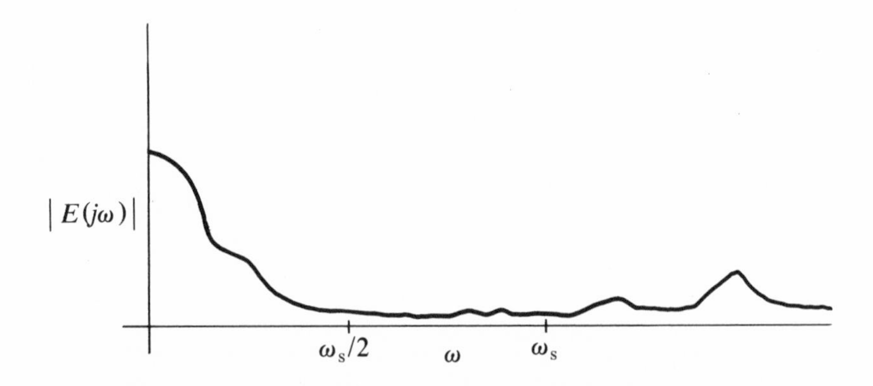

Figure 10.10
Typical Frequency Spectrum of a Process Signal (Reproduced with Permission from Ref. 1)

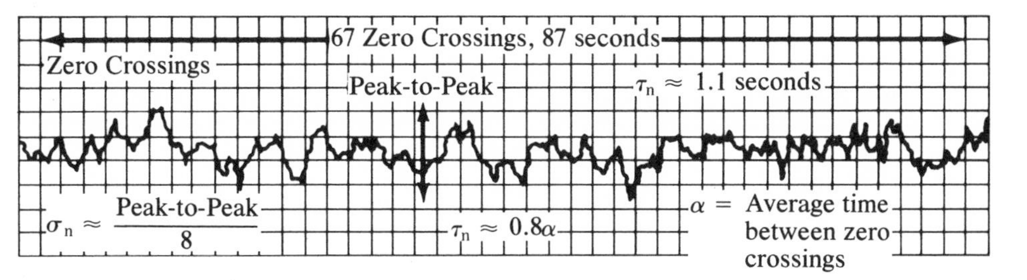

Figure 10.11
Measuring Noise Parameters (By Permission from Ref. 6)

Table 10.1
Selection of Scan and Control Sampling Rates for Processes with Noise

Type of control algorithm	*Scan sampling period, T_c*		*Control Sampling period, T*
	$\tau_n < \tau_b$	$\tau_n > 2\tau_b$	
PI	$\frac{\tau_b}{2} < T_c < \tau_b$	$\frac{1}{30}\ \tau_f < T_c < \frac{\tau_n}{2}$	$\frac{\tau_i}{100} < T < \frac{\tau_f}{2}$
PID (with a filter on the derivative mode)	—Normally not used when significant process noise is present. —May be used when noise bandwidth does not extend into the higher frequencies amplified by the rate term. May also be applied to processes with wide-bandwidth noise provided σ_n is small enough so that the standard deviation of the controller output, σ_m, is acceptable, where $\sigma_m = k_c G \sigma_n$. $G \cong 10$ for analog control. May be reduced for digital control to bring σ_m down to an acceptable level. If σ_m is still too large as indicated by the final control element activity, use the two-mode controller.		$\frac{\tau_d}{10G} < T < \frac{\tau_d}{G}$ $G = \frac{\text{derivative time, } \tau_d}{\text{derivative filter time constant}}$

correlated noise with a dominant time constant τ_n having a magnitude that can be measured by its standard deviation σ_n. From the chart record, Figure 10.11, these parameters are calculated as

$$\tau_n = 0.8\alpha$$

and

$$\sigma_n = \frac{\text{peak-to-peak amplitude}}{8}$$

where α, the average time between zero crossings, equals the total time interval divided by the number of zero crossings in that interval. Zero crossings refer to the intersection of the process variable with the estimated mean as the process variable cuts back and forth across the mean. The estimation of σ_n is based on finding the peak-to-peak amplitude for the measured interval and dividing by 8. For digitally executed PI and PID control algorithms, these noise parameters will assist in the selection of the scan sampling period T_c and the control sampling period T, as shown in Table 10.1.

As an illustration, a chart record of control with and without filtering is shown in Figure 10.12. The beneficial effects of filtering are clearly evident.

References

1. Smith, C. L., *Digital Computer Process Control,* Intext Educational Publishers, Scranton, PA, 1972.
2. Kuo, B. C., *Analysis and Synthesis of Sampled-data Control Systems*, Prentice-Hall, Englewood Cliffs, N.J., 1963.
3. Dahlin, E. B., Designing and Tuning Digital Controllers, *Instruments and Control Systems*, **41**, 6, June 1968.
4. Schork, F. J., private communication, Department of Chemical Engineering, University of Wisconsin, Madison, WI, June 1978.
5. Tou, J. T., *Digital and Sampled-data Control Systems*, McGraw-Hill, New York, 1959, p. 242.
6. Fertik, H. A., Tuning Controllers for Noisy Processes, *ISA Trans.*, **14**, 4, 1975.
7. Goff, K. W., Dynamics in Direct Digital Control, Part I, *ISA J.*, **13**, 11, November 1966, 44–49; Part II, *ISA J.*, **13**, 12, December 1966, 44–54.

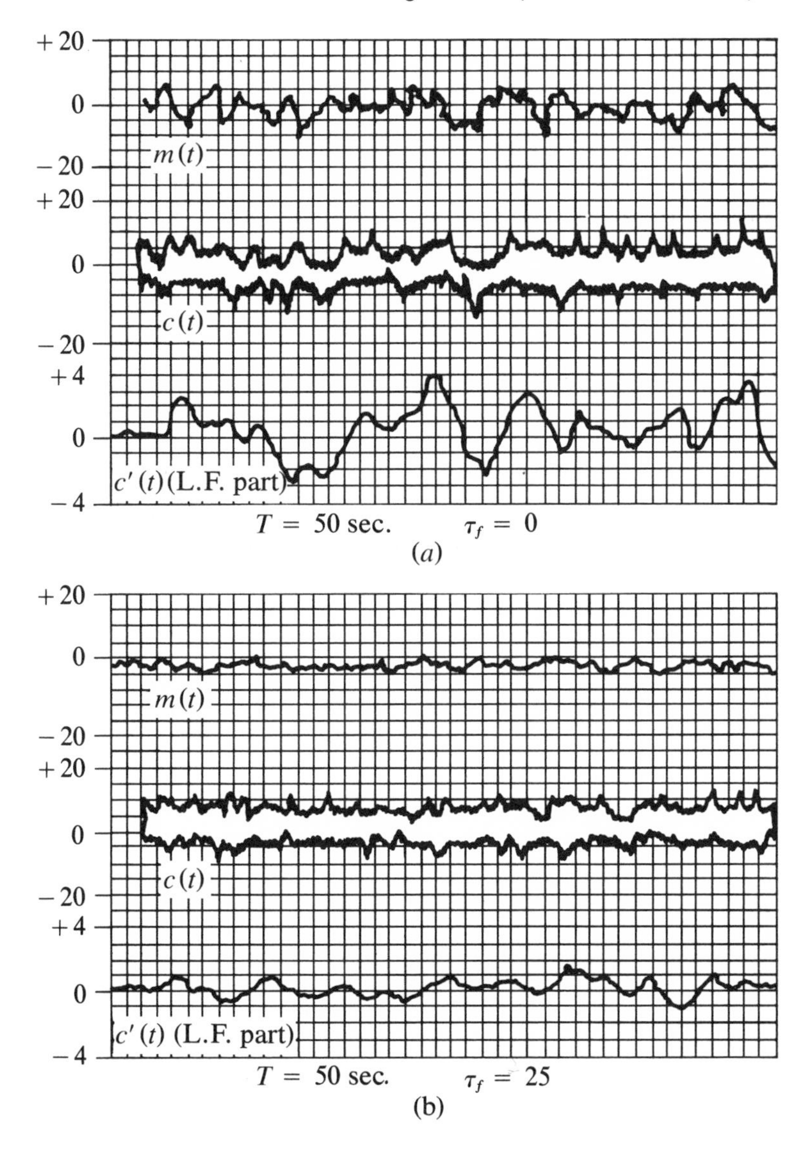

Figure 10.12
Effect of Noise on Process Variables for Control with and without a Filter. Each Time Division is 250 sec (Reprinted by Permission from Ref. 7)

Authors' Update on Chapter 10

A new method called *Model Algorithmic Control* for control systems design has been published recently. The method is based on impulse response of open-loop processes (impulse responses can be determined from step responses by simple mathematical manipulations) and therefore a process model, $G_p(s)$, is not needed for its implementation. The interested reader may consult the following references for details.

1. Tu, F. C. Y., Tsing, J. Y. H., Synthesizing a Digital Algorithm for Optimized Control, Instrumentation Technology, May 1979 pp. 52–55.
2. Morari, M., Garcia, C. E., Internal Model Control—A Flexible Algorithm for Computer Control of Industrial Processes, A.I.Ch.E. Meeting, New Orleans, LA, November 8–12, 1981. (Paper No. 57f)
3. Marchetti, J. L., Seborg, D. E. and Mellichamp, D. A., Predictive Control Based On Discrete Convolution Models, A.I.Ch.E. Meeting, New Orleans, LA, November 8–12, 1981. (Paper No. 57c)

CHAPTER 11

Stability of Sampled-Data Control Systems

In Chapter 1 we reviewed the stability aspects of conventional control systems. We noted that the stability of continuous-control systems was determined by the location of the roots of the characteristic equation of the system in the s plane. For a stable system all the roots had to be located in the left half of the s plane. Let us now derive the stability criteria for the sampled-data control systems with the aid of Z transforms. Consider the sampled-data control system shown in Figure 11.1. In accordance with our discussion in Chapter 9, the closed-loop pulse transfer function of this system to changes in set point is

$$\frac{C(Z)}{R(Z)} = \frac{D(Z)\, G_{h_0} G_p\,(Z)}{1 + D(Z)\, G_{h_0} G_p\,(Z)} \tag{11.1}$$

where

$$G_{h_0} G_p\,(Z) = Z\left[G_{h_0}(s)\, G_p\,(s)\right]$$

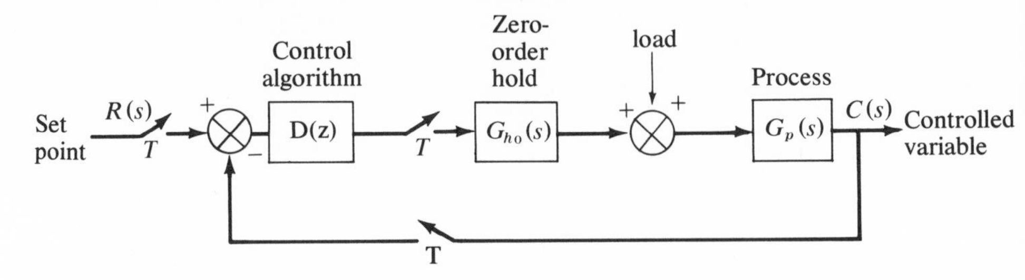

Figure 11.1
Typical Sampled-Data Control System

so that the open-loop pulse-transfer function of the system is

$$A(Z) = D(Z)G_{h_0}G_p(Z) = Z\left[G_{h_0}(s)\,G_p(s)\right]D(Z) \tag{11.2}$$

The characteristic equation of the sampled-data system is, therefore,

$$1 + A(Z) = 0 \tag{11.3}$$

The nature of the roots of Equation (11.3) determines the stability and transient behavior of the sampled-data control system. Recall that a stable region in the s domain is the region to the left of the imaginary axis, as shown in Figure 11.2 and that the relationship between Z and s is

$$Z = e^{Ts} \tag{11.4}$$

We would like to find the corresponding stable region in the Z plane. To do this we would vary s and examine its effect on Z, as described in the following paragraphs.

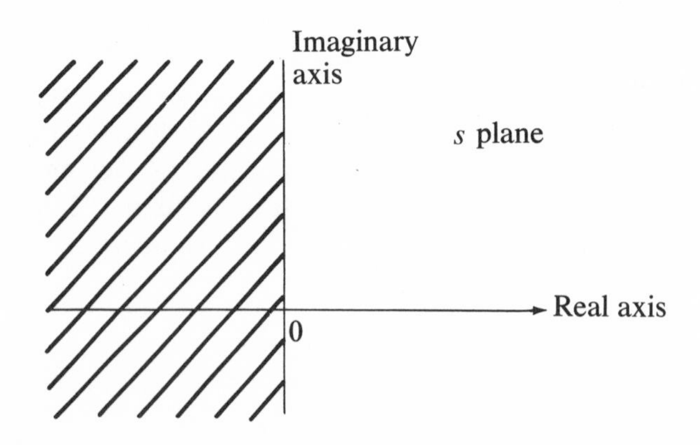

Figure 11.2
Stable Region in s plane

For values of s on the imaginary axis, $s = j\omega$. Therefore, in accordance with Equation 11.4

$$Z = e^{j\omega T} \tag{11.5}$$

As ω is varied from 0 to $\omega_s/4$ $(= 2\pi/4T)$, Z varies from $e^{jOT} = 1$ to $e^{j(2\pi/4T)T} = e^{j\pi/2} = \cos\pi/2 + j\sin\pi/2 = j$ that is, along the unit circle in the first quadrant of the Z plane, as shown in Figure 11.3*a* As ω increases from $\omega_s/4$ to $\omega_s/2$, Z varies from j to -1 along the unit circle in the second quadrant. As ω increases from $\omega_s/2$ to $3\omega_s/4$, Z moves from -1 to $-j$ along the unit circle in the third quadrant. Finally, as ω moves from $3\omega_s/4$ to ω_s, Z traverses from $-$j to $+1$ along the unit circle in the fourth quadrant. These movements are shown in Figure 11.3*b* and *c*. As ω is increased from ω_s to $2\omega_s$, the values of Z trace the unit circle once more. The process is repeated when ω increases or decreases through a range of ω_s. Now let us consider how the region to the left of the imaginary axis in the s plane maps on to the

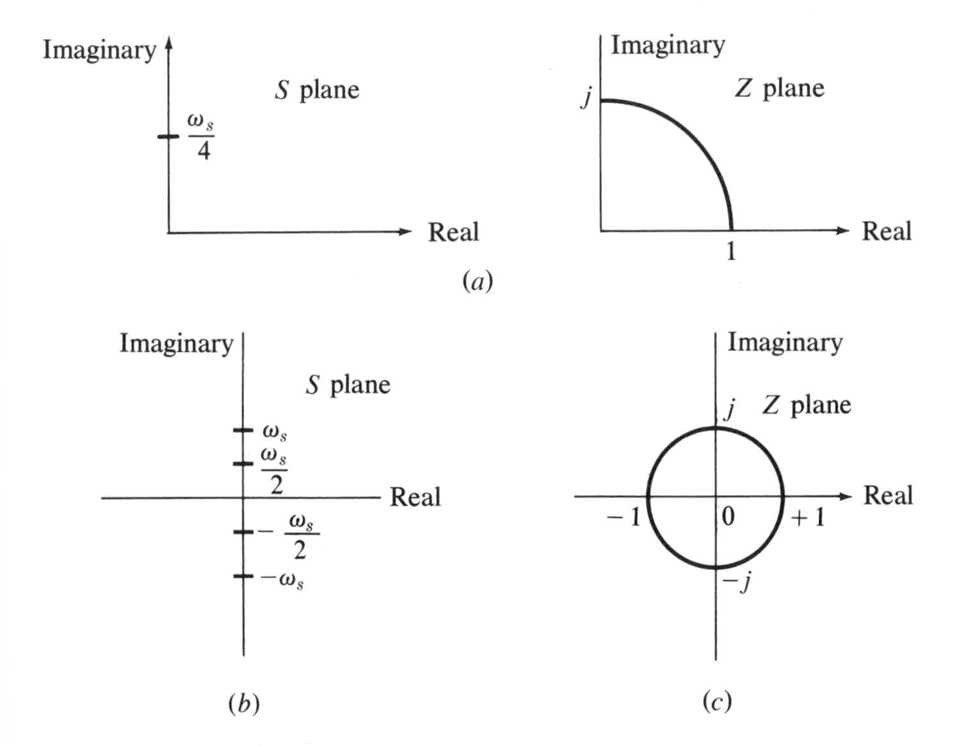

Figure 11.3
Transformations between s and Z Planes

Z plane. Consider a general point in the s plane whose coordinates are

$$s = \sigma + j\omega \tag{11.6}$$

Therefore,

$$Z = e^{sT} = e^{\sigma T}\,(e^{j\omega T})$$

When $\sigma = -\infty$ and $\omega = 0$ in the s plane, the value of Z is equal to zero. This implies that a point at infinity on the negative real axis of the s plane is mapped into the origin of the Z plane. Also, for $\sigma \leqslant 0$ and $\omega = 0$, $Z = e^{\sigma t}$ describes a line segment between 0 and 1. Thus the negative real axis of the s plane is mapped into that section of the positive real axis of the Z plane that falls inside the unit circle.

Finally, the magnitude of Z is given by

$$|Z| = |e^{sT}| \tag{11.7}$$

For a general point $s = \sigma + j\omega$, Equation (11.7) gives

$$|Z| = |e^{\sigma T}\, e^{j\omega T}| = e^{\sigma T} \tag{11.8}$$

since

$$|e^{j\omega T}| = |\cos \omega T + j \sin \omega T| = \sqrt{\cos^2 \omega T + \sin^2 \omega T} = 1$$

For $\sigma < 0$ (i.e., any point in the left half of the s plane), $e^{\sigma T}$ is less than one. Therefore, $|Z| < 1$, which represents the region covering the interior of the unit circle. Hence, the left half of the S plane is mapped into the area inside the unit circle of the Z plane. For $\sigma > 0$ on the other hand, $|Z|$ is greater than 1, and therefore, the right half of s plane is mapped into the area outside the unit circle. Thus we can

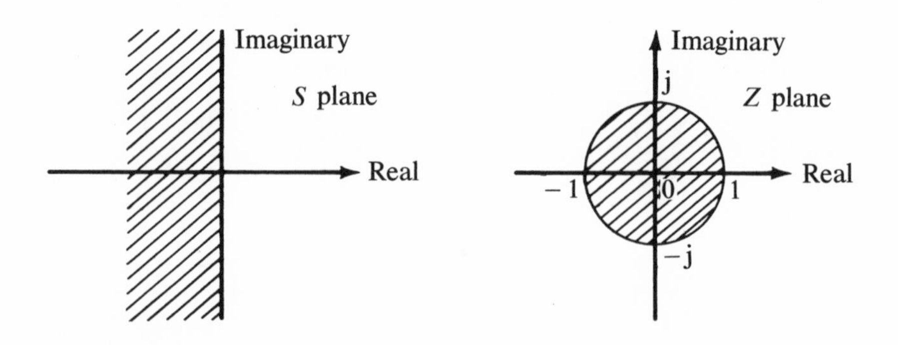

Figure 11.4
Stable Regions in S and Z Plane

state that a *sampled-data control system is stable if all the roots of the characteristic equation $1 + A(Z) = 0$ lie inside the unit circle about the origin of the Z plane*. The stable regions are sketched in Figure 11.4.

11.1 Schur–Cohn Stability Criterion[1]

This method provides an analytical method for determining the absolute stability of sampled-data control systems. The characteristic equation for the sampled-data system is

$$1 + A(Z) = 0 \tag{11.9}$$

Generally, $A(Z)$ will be in the form of a ratio of two polynomials. If we put the left-hand side of Equation (11.9) over a common denominator and denote the resulting numerator as $F(Z)$, then we can write

$$F(Z) = a_0 + a_1 Z + a_2 Z^2 + \cdots + a_n Z^n = 0$$

The first step in determining stability is to write the coefficients of a_k in determinant form as

$$\Delta_k = \begin{vmatrix} a_0 & 0 & 0 & \cdot\cdot & 0 & a_n & a_{n-1} & a_{n-k+1} \\ a_1 & a_0 & 0 & \cdot\cdot & 0 & 0 & a_n & a_{n-k+2} \\ \cdot\cdot & \cdot\cdot & a_0 & \cdot\cdot & \cdot\cdot & \cdot\cdot & \cdot\cdot & \cdot\cdot \\ \cdot\cdot & \cdot\cdot & \cdot\cdot & \cdot\cdot & & & & \\ a_{k-1} & a_{k-2} & a_{k-3} & \cdot\cdot & a_0 & 0 & 0 & a_n \\ \bar{a}_n & 0 & 0 & \cdot\cdot & 0 & \bar{a}_0 & \bar{a}_1 & \bar{a}_{k-1} \\ \bar{a}_{n-1} & \bar{a}_n & 0 & \cdot\cdot & 0 & 0 & \bar{a}_0 & \bar{a}_{k-2} \\ \cdot\cdot & \cdot\cdot & \bar{a}_n & \cdot\cdot & \cdot\cdot & \cdot\cdot & \cdot\cdot & \cdot\cdot \\ \cdot\cdot & \cdot\cdot & \cdot\cdot & \cdot\cdot & \cdot\cdot & \cdot\cdot & \cdot\cdot & \cdot\cdot \\ \bar{a}_{n-k+1} & \bar{a}_{n-k+2} & \bar{a}_{n-k+3} & \bar{a}_n & 0 & \cdot\cdot & 0 & \bar{a}_0 \end{vmatrix} \tag{11.10}$$

where $\bar{a}_n$ is the conjugate of a_n. Δ_k $(k = 1, 2, 3, \cdots, n)$ is a determinant that has $2k$ rows and $2k$ columns. The Schur–Cohn criteria states that all the roots of the characteristic equation lie inside the unit circle (i.e., the system is stable) if the following conditions are met:

$$\Delta_k < 0 \qquad \text{for } k \text{ odd}$$
$$\Delta_k > 0 \qquad \text{for } k \text{ even} \tag{11.11}$$

Now, as an illustration, let us develop the determinant for a few values of k.

k = 1: Δ_1 is as 2×2 determinant

$$\Delta_1 = \begin{vmatrix} a_0 & a_n \\ \bar{a}_n & \bar{a}_0 \end{vmatrix} \tag{11.12}$$

k = 2:

$$\Delta_2 = \begin{vmatrix} a_0 & 0 & a_n & a_{n-1} \\ a_1 & a_0 & 0 & a_n \\ \bar{a}_n & 0 & \bar{a}_0 & \bar{a}_1 \\ \bar{a}_{n-1} & \bar{a}_n & 0 & \bar{a}_0 \end{vmatrix} \tag{11.13}$$

Let us take an example to illustrate the method.

Example 1. Determine the stability of the sampled-data control system whose open-loop pulse-transfer function is given by

$$A(Z) = \frac{Z}{(2.45Z + 1)(2.45Z - 1)}$$

The characteristic equation is given by

$$1 + A(Z) = 0$$

or

$$F(Z) = (2.45Z)^2 - 1 + Z = 0$$
$$= 6Z^2 + Z - 1 = 0$$

Here n, the order of the characteristic equation, is 2. Therefore, two determinants Δ_1 and Δ_2 must be evaluated to determine stability. From Equations (11.12) and (11.13) we will get

$$\Delta_1 = \begin{vmatrix} -1 & 6 \\ 6 & -1 \end{vmatrix} = -35$$

$$\Delta_2 = \begin{vmatrix} -1 & 0 & 6 & 1 \\ 1 & -1 & 0 & 6 \\ 6 & 0 & -1 & 1 \\ 1 & 6 & 0 & -1 \end{vmatrix} = 1176$$

since

$$\Delta_1 < 0 \text{ and } \Delta_2 > 0$$

according to Equation (11.11) the system is stable.

If $F(Z)$ is a quadratic polynomial with real coefficients and the coefficient of Z^2 is unity, the Schur–Cohn criterion can be simplified. The necessary and sufficient conditions that the roots of the characteristic equation lie inside the unit circle in the Z plane are

$$\begin{aligned} |F(0)| &< 1 \\ F(1) &> 0 \\ F(-1) &> 0 \end{aligned} \tag{11.14}$$

Reference

1. Tou, J. T., *Digital and Sampled-data Control Systems,* McGraw-Hill, New York, 1959, p. 238.

CHAPTER 12

Modified Z Transforms

The Z-transform method enables us to determine the transient response of sampled-data control systems only at the sampling instants. To obtain the values of the response between sampling instants, modified Z transforms are useful. They are also helpful in analyzing sampled-data control systems containing transportation lag (i.e., dead time).

12.1. Definition and Evaluation of Modified Z Transforms[1,2,3]

Suppose that the transfer function of a process with dead time is represented by the following expression:

$$G_p(s) = G(S)e^{-\theta_d s} \tag{12.1}$$

where $G(s)$ contains no dead time and θ_d = dead time. If we substitute for θ_d the quantity $NT + \theta$, where N is the largest integer number of sampling intervals in θ_d and T is the sampling period, Equation 12.1 becomes

$$G_p(s) = G(s)\, e^{-(NT+\theta)s} \tag{12.2}$$

For example, if $\theta_d = 0.5$ and $T = 0.11$, then, $N = 4$ and $\theta = 0.50 - (4)(0.11) = 0.06$. It can be easily verified that for a given θ_d and T, θ lies between 0 and T.

Now, let us take the Z transform of Equation (12.2).

$$\begin{aligned} G_p(Z) &= Z\{G(s)\, e^{-(NT+\theta)s}\} \\ &= Z^{-N} Z\{G(s)\, e^{-\theta s}\} \end{aligned} \tag{12.3}$$

The quantity $Z\{G(s)\, e^{-\theta s}\}$ is defined as the modified Z-transform of $G(s)$ and is denoted by $Z_m\{G(s)\}$ or $G(Z, m)$. Thus

$$G(Z, m) = Z_m\,[G(s)] = Z\{G(s)\, e^{-\theta s}\}$$

Let us evaluate the modified Z transform of some simple functions.

1. Unit Step Function

$$f(t) = \begin{cases} u(t) & t \geqslant 0 \\ 0 & t < 0 \end{cases}$$

$$\begin{aligned} Z_m\{F(s)\} &= Z_m\left\{\frac{1}{s}\right\} = Z\left\{\frac{e^{-\theta s}}{s}\right\} \\ &= Z\{u(t-\theta)\} \\ &= \sum_{n=0}^{\infty} u(nT-\theta)\, Z^{-n} \\ &= 0 + u(T-\theta)\, Z^{-1} + u(2T-\theta)\, Z^{-2} \\ &\quad + u(3T-\theta)\, Z^{-3} \\ &= Z^{-1}\{1 + Z^{-1} + Z^{-2}\} + \cdots \end{aligned}$$

Therefore

$$Z_m[u(t)] = \frac{Z^{-1}}{1 - Z^{-1}} = \frac{1}{Z - 1}$$

2. Evaluate $z_m\{e^{-at}\}$

$$Z_m\{e^{-at}\} = Z_m\left\{\frac{1}{s+a}\right\} = Z\left\{\frac{e^{-\theta s}}{s+a}\right\}$$
$$= Z\{u(t-\theta)\, e^{-a(t-\theta)}\}$$
$$= \sum_{n=0}^{\infty} u(nT-\theta)\, e^{-a(nT-\theta)}\, Z^{-n}$$
$$= 0 + e^{-a(T-\theta)} Z^{-1} + e^{-a(2T-\theta)} Z^{-2} + e^{a(3T-\theta)} Z^{-3} + \cdots\cdot$$

Let

$$mT = T - \theta \text{ or } m = 1 - \frac{\theta}{T}.$$

Then,

$$Z_m\{e^{-at}\} = e^{-amT} Z^{-1} + e^{-amT} e^{-aT} Z^{-2} + e^{-amT} e^{-2aT} Z^{-3}$$
$$Z^{-1} e^{-amT}\{1 + Z^{-1} e^{-aT} + e^{-2aT} Z^{-2} + \cdots\cdot\}$$

Therefore,

$$Z_m\{e^{-at}\} = \frac{Z^{-1} e^{-amT}}{1 - Z^{-1} e^{-aT}}$$

This procedure can be applied to functions whose modified Z transforms are desired. A table of modified Z transforms is included in Appendix A.

12.2 Application of Modified *Z* Transforms to Systems with Dead Time

The use of modified Z transforms simplifies the analysis of systems containing a dead-time element. The procedure is best illustrated by an example.

Example 1. Determine the response of the system shown in Figure 12.1 to a unit step change in set point. Assume $T = 0.5$ and $D(Z)$ is a PI control algorithm with $K_c = 0.43$, $\tau_I = 1.57$. The closed-loop pulse transfer function of this system is

$$\frac{C(Z)}{R(Z)} = \frac{D(Z)\, G_{h_0} G_p(Z)}{1 + D(Z)\, G_{h_0} G_p(Z)} \tag{12.4}$$

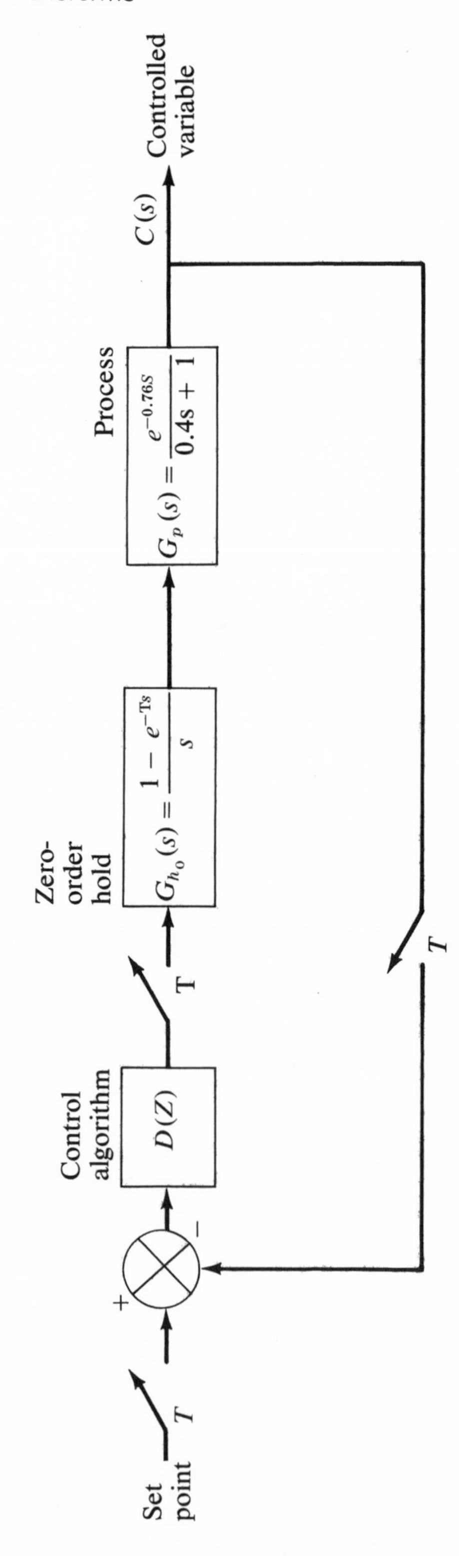

Figure 12.1
Sampled-Data Control System for a Process with Dead Time

where

$$G_{h_0} G_p(Z) = Z\{G_{h_0}(s)\, G_p(s)\}$$

Let us evaluate $G_{h_0} G_p(Z)$ by modified Z transforms.

$$G_{h_0} G_p(Z) = Z\left\{\frac{1 - e^{-sT}}{s} \cdot \frac{e^{-0.76s}}{0.4s + 1}\right\}$$

$$= Z\left\{\frac{e^{-0.76s}}{s(0.4s + 1)}\right\} - Z\left\{\frac{e^{-0.76s}\, e^{-Ts}}{s(0.4s + 1)}\right\} \tag{12.5}$$

Recall the theorem on translation of the function, which states that if the Z transform of $f(s)$ is $F(Z)$, then the Z transform of $e^{-Ts} f(s)$ is $Z^{-1} F(Z)$. Therefore,

$$G_{h_0} G_p(Z) = Z\left\{\frac{e^{-0.76s}}{s(0.4s + 1)}\right\} - Z^{-1} Z\left\{\frac{e^{-0.76s}}{s(0.4s + 1)}\right\}$$

$$= (1 - Z^{-1}) Z\left\{\frac{e^{-0.76s}}{s(0.4s + 1)}\right\} \tag{12.6}$$

Now, let us introduce modified Z transforms to simplify the calculation of $G_{h_0} G_p(Z)$.

In this example $\theta_d = 0.76$ and $T = 0.5$.

Therefore, N, the integral number of sampling intervals in θ_d, equals 1 and

$$\theta = \theta_d - NT = 0.76 - (1)(0.5) = 0.26 \tag{12.7}$$

Thus

$$G_{h_0} G_p(Z) = (1 - Z^{-1}) Z\left\{\frac{e^{-\theta_d s}}{s(0.4s + 1)}\right\} \tag{12.8}$$

$$= (1 - Z^{-1}) Z\left\{\frac{e^{-\theta s}\, e^{-NTs}}{s(0.4s + 1)}\right\}$$

$$= (1 - Z^{-1}) Z^{-N} Z\left\{\frac{e^{-\theta s}}{s(0.4s + 1)}\right\}$$

$$= (1 - Z^{-1}) Z^{-1} Z\left\{\frac{e^{-0.26s}}{s(0.4s + 1)}\right\}$$

$$= (1 - Z^{-1}) Z^{-1} Z_m\left\{\frac{1}{s(0.4s + 1)}\right\}$$

From the table of modified Z transforms,

$$Z_m \left\{\frac{a}{s(s+a)}\right\} = Z^{-1}\left(\frac{1}{1-Z^{-1}} - \frac{e^{-amT}}{1-e^{-aT}Z^{-1}}\right) \tag{12.9}$$

where

$$a = \frac{1}{0.4} = 2.5$$

$$m = 1 - \frac{\theta}{T} = 1 - \frac{0.26}{0.5} = 0.48$$

Substitution of these values in Equation (12.9) gives

$$Z_m \left\{\frac{1}{s(0.4s+1)}\right\} = Z_m \left\{\frac{2.5}{s(s+2.5)}\right\} \tag{12.10}$$

$$= Z^{-1}\left(\frac{1}{1-Z^{-1}} - \frac{0.5488}{1-0.2865\,Z^{-1}}\right\}$$

$$= \frac{Z^{-1}\left[1 - 0.2865\,Z^{-1} - 0.5488 + 0.5488\,Z^{-1}\right]}{(1-Z^{-1})(1-0.2865\,Z^{-1})}$$

$$= \frac{Z^{-1}(0.4512 + 0.2623\,Z^{-1})}{(1-Z^{-1})(1-0.2865\,Z^{-1})}$$

The expression for $G_{h_0}\,G_p(Z)$ thus becomes

$$G_{h_0}\,G_p(Z) = \frac{(1-Z^{-1})Z^{-1}Z^{-1}(0.4512 + 0.2623\,Z^{-1})}{(1-Z^{-1})(1-0.2865\,Z^{-1})} \tag{12.11}$$

$$G_{h_0}\,G_p(Z) = \frac{0.4512\,Z^{-2} + 0.2623\,Z^{-3}}{1-0.2865\,Z^{-1}}$$

Now let us return to Equation (12.4) and evaluate $D(Z)$. For a *PI* controller the output is related to the error by the equation

$$v_n = v_{n-1} + K_c\,(e_n - e_{n-1}) + \frac{K_c}{\tau_I}\,T\,e_n \tag{12.12}$$

Taking the Z transform of Equation 12.12 gives

$$V(Z) = Z^{-1}\,V(Z) + K_c\,E(Z) - Z^{-1}\,K_c\,E(Z) + \frac{K_c}{\tau_I}\,T\,E(Z) \tag{12.13}$$

Therefore,

$$D(Z) = \frac{V(Z)}{E(Z)} = \frac{\left(K_c + \frac{K_c}{\tau_I} T\right) - K_c Z^{-1}}{1 - Z^{-1}} \tag{12.14}$$

Substituting the given values for the controller constants

$$K_c = 0.43$$
$$\tau_I = 1.57 \tag{12.15}$$

into Equation (12.14), we get

$$D(Z) = \frac{0.57 - 0.43 Z^{-1}}{1 - Z^{-1}} \tag{12.16}$$

Now, for a unit step change in set point

$$R(t) = u(t)$$

and (12.17)

$$R(Z) = \frac{1}{1 - Z^{-1}}$$

Thus Equation (12.4) becomes

$$C(Z) = \frac{1}{1 - Z^{-1}} \times \left[\frac{\left(\frac{0.57 - 0.43Z^{-1}}{1 - Z^{-1}}\right)\left(\frac{0.4512Z^{-2} + 0.2623Z^{-3}}{1 - 0.2865 Z^{-1}}\right)}{1 + \left(\frac{0.57 - 0.43Z^{-1}}{1 - Z^{-1}}\right)\left(\frac{0.4512Z^{-2} + 0.2623Z^{-3}}{1 - 0.2865 Z^{-1}}\right)} \right] \tag{12.18}$$

The right side can be simplified and then expressed in power series of Z^{-1} by long division to give

$$C(Z) = 0 + 0Z^{-1} + 0.2565 Z^{-2} + 0.542 Z^{-3} + 0.658 Z^{-4} + 0.664 Z^{-5} + 0.649 Z^{-6} + \cdots \tag{12.19}$$

A plot of $C^*(t)$ versus time is shown in Figure 12.2

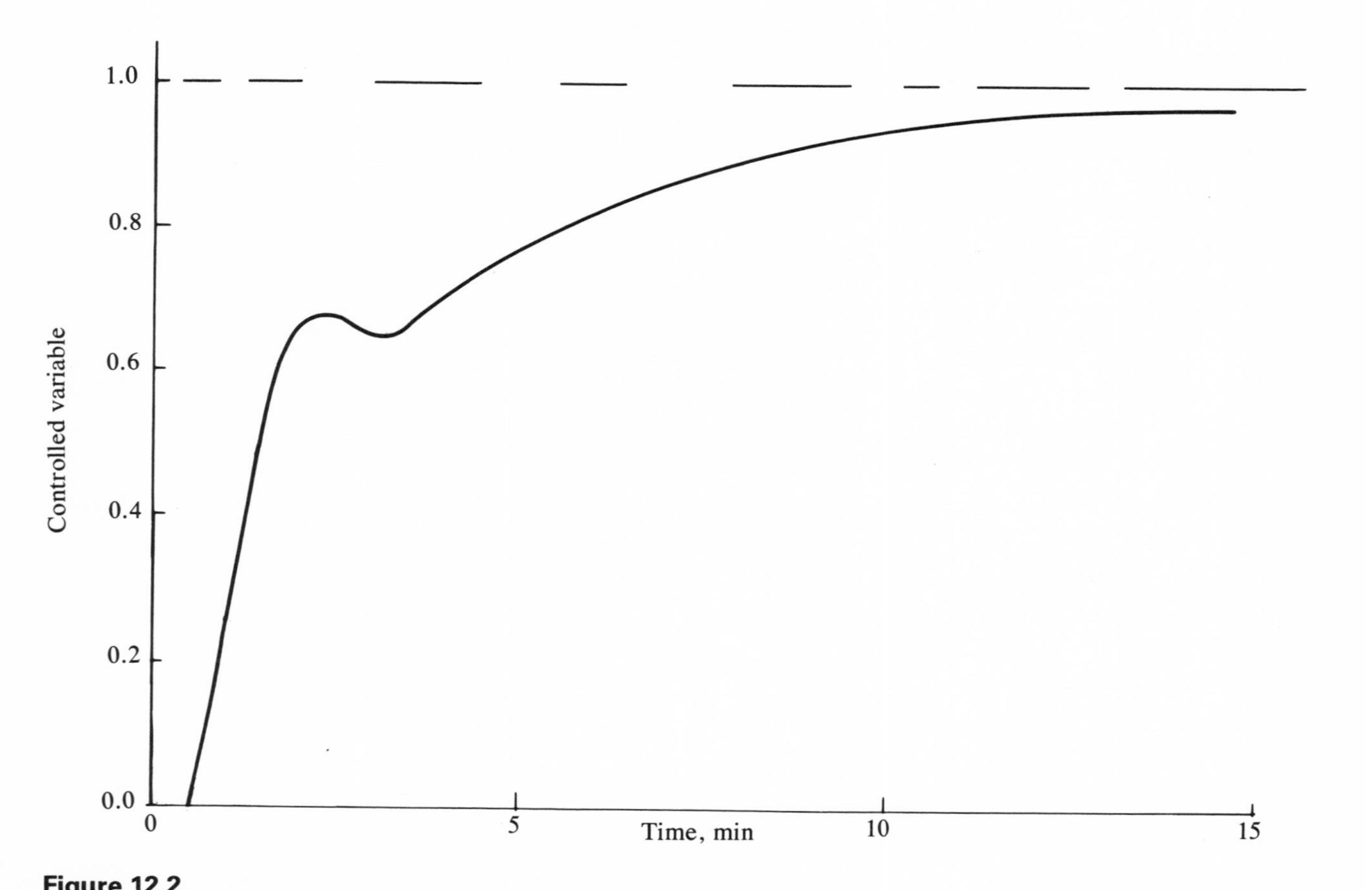

Figure 12.2
Response of System to Unit Step Change in Set Point

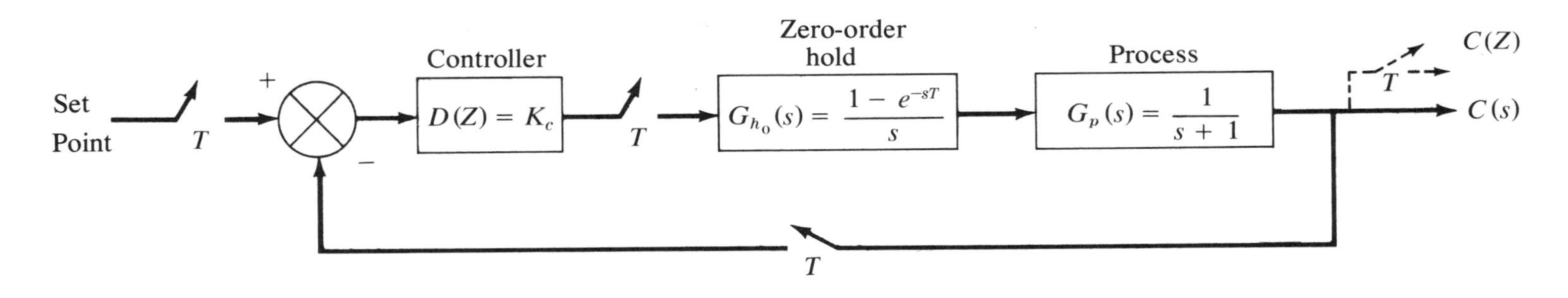

Figure 12.3
Proportional Sampled-Data Control System for a First-Order Process

12.3 Application of Modified Z Transforms to Determine Output Between Sampling Instants

Modified Z transforms offer an extremely valuable method for evaluating the response of sampled-data control systems between sampling instants. This is important, because a stable system and an unstable system can yield identical responses at the sampling instants. Since the Z-transform technique provides information only at the sampling instants, it would be impossible to tell which of the two systems is unstable. The procedure describing the use of modified Z transforms for determining the output between sampling instants is outlined in the following paragraphs.

Consider the block diagram of a sampled-data control system shown in Figure 12.3. Suppose the continuous response of the controlled variable to a step change in set point is as shown in Figure 12.4*a*. The Z-transform technique will give us the response of $C^*(t)$ as shown in Figure 12.4*b*. We would like to generate additional information about $C^*(t)$ between sampling instants so that the complete time response of C may be developed. To begin with, let us define a quantity m as

$$m = 1 - \frac{\theta}{T} \tag{12.20}$$

or

$$T = mT + \theta$$

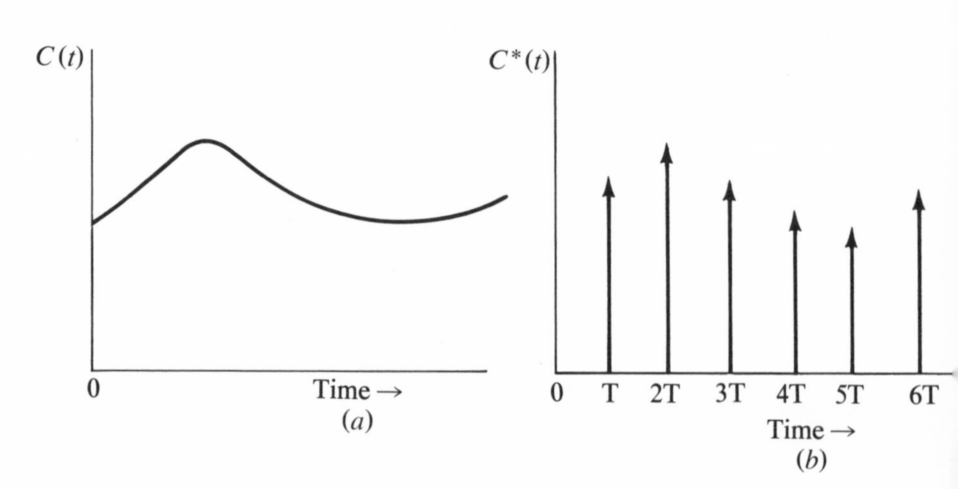

Figure 12.4
Response of System Shown in Figure 12.3

Equation (12.20) shows that when θ equals 0, m will equal 1, and when $\theta = T$, m will equal 0.

Now, let us add a fictitious dead-time element to the block diagram of Figure 12.3, as shown in Figure 12.5. Since the delay element is outside the block diagram, the response of $C(t)$ is unaffected. The fictitious delay element will enable us to delay the controlled variable so that we may obtain the inter-sampling information. For example, if the signal $C(t)$ at $t = 0$ is delayed θ time units and then sampled, we will obtain a different value of C^* as shown in Figure 12.6. Advancing the output of the sampler some more gives another value of C^*. In fact, as θ is varied between O and T (i.e., m between 0 and 1 as per Equation 12.20), the entire signal between these sampling instants can be reproduced. The repeated application of this procedure then will enable us to obtain the complete time response of the controlled variable. Let us illustrate the technique by applying it to an example problem.

Example 2[4]. Determine the response $C(t)$ of the system shown in Figure 12.3 for all values of time t, for a unit step change in set point. The closed-loop pulse transfer function of this system is

$$\frac{C(Z, m)}{R(Z)} = \frac{D(Z)\, G_2(Z)}{1 + D(Z)\, G_{h_0} G_p(Z)} \tag{12.21}$$

where

$$C(Z, m) = Z\{e^{-\theta s}\, C(s)\}$$

$$G_2(Z) = Z\{G_{h_0}(s)\, G_p(s)\, e^{-\theta s}\}$$

$$= Z_m\{G_{h_0}(s)\, G_p(s)\}$$

$$= G_{h_0}\, G_p(Z, m)$$

and

$$G_{h_0}G_p(Z) = Z\{G_{h_0}(s)\, G_p(s)\}$$

Thus Equation (12.21) can be written as

$$\frac{C(Z, m)}{R(Z)} = \frac{D(Z)\, G_{h_0}\, G_p\,(Z, m)}{1 + D(Z)\, G_{h_0} G_p(Z)}$$

Let us evaluate the Z transforms: For this illustration assume that

$$D(Z) = 1.5 \tag{12.22}$$

$$G_2(Z) = G_{h_0}\, G_p(Z, m) = Z_m\left[\frac{1 - e^{-sT}}{s} \cdot \frac{1}{s + 1}\right] \tag{12.23}$$

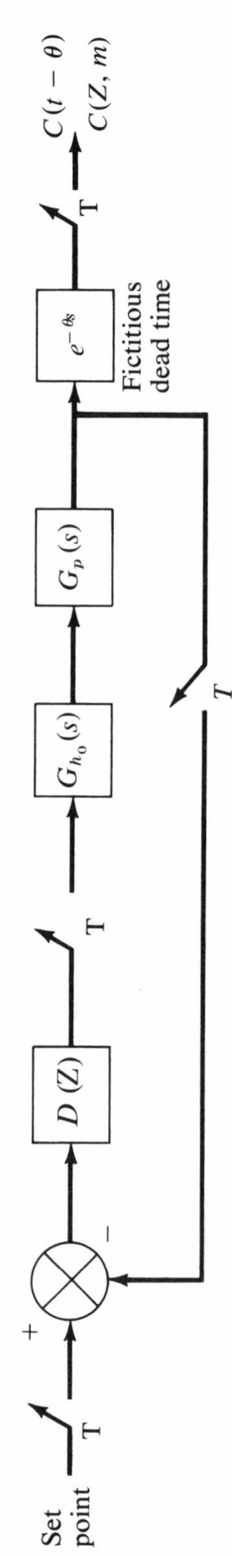

Figure 12.5
Sampled-Data System of Figure 12.3 with a Fictitious Dead-Time

$$= Z\left[\frac{(1 - e^{-sT})}{s(s + 1)} e^{-\theta s}\right]$$

$$= \frac{Z^{-1}\left[(1 - e^{-mT}) + (e^{-mT} - e^{-T})Z^{-1}\right]}{1 - e^{-T} Z^{-1}}$$

and

$$G_{h_0}G_p(Z) = Z\{G_{h_0}(s)\, G_p(s)\}$$

$$= Z\left\{\frac{1 - e^{-sT}}{s(s + 1)}\right\} \tag{12.24}$$

$$= \frac{(1 - e^{-T})Z^{-1}}{1 - e^{-T} Z^{-1}}$$

For a step change in $R(t)$

$$R(Z) = \frac{1}{1 - Z^{-1}}$$

Substituting in Equation 12.21 for $R(Z)$ gives

$$C(Z, m) = R(Z)\frac{D(Z)\, G_{h_0}G_p\,(Z, m)}{1 + D(Z)\, G_{h_0}G_p(Z)} \tag{12.25}$$

$$= \frac{1}{1 - Z^{-1}} \cdot \frac{1.5Z^{-1}\left[(1 - e^{-mT}) + (e^{-mT} - e^{-T})Z^{-1}\right]}{\left[1 - e^{-T} Z^{-1} + 1.5(1 - e^{-T})Z^{-1}\right]}$$

For $T = 1$, this equation gives, after some simplification,

$$C(Z, m) = \frac{1.5Z^{-1}\left[(1 - e^{-mT}) + (e^{-mT} - e^{-T})Z^{-1}\right]}{1 - 0.42Z^{-1} - 0.58Z^{-2}} \tag{12.26}$$

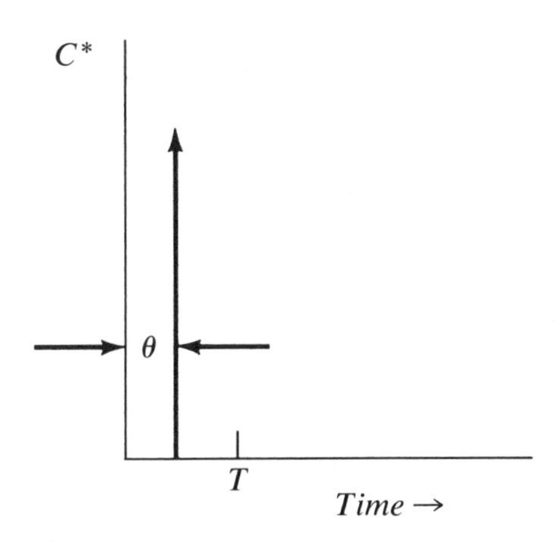

Figure 12.6 Response Delayed by θ time Units and then Sampled

Equation (12.26) can be expressed as a power series by long division. Thus

$$C(Z, m) = 1.5\,(1 - e^{-mT})Z^{-1} + 1.5(0.58\,e^{-mT} + 0.05)Z^{-2} + 1.5\,(0.602 - 0.341\,e^{-mT})Z^{-3} + \cdots \quad (12.27)$$

The coefficient in front of Z^{-n} gives the response C^* between $(n - 1)$th and nth sampling instant. By varying m between 0 and 1, the complete response between any two sampling instants can be evaluated. Thus between 0 and T the response is given by

$$C^* = 1.5\,(1 - e^{-mT}) \quad (12.28)$$

By varying m between 0 and 1, the complete response between O and T can be evaluated. The complete response thus obtained for the problem is sketched in Figure 12.7.

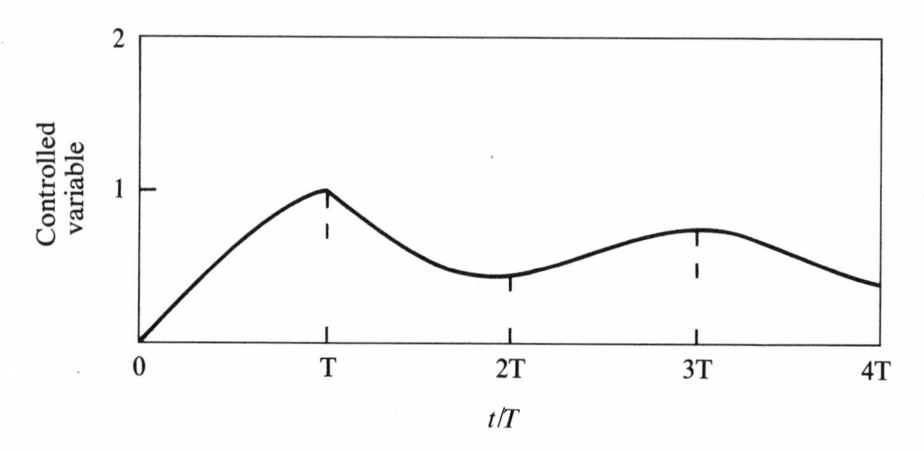

Figure 12.7 Complete Response of System of Figure 12.3

References

1. Barker, R. H., The Pulse Transfer Function and Its Application to Sampling Servo Systems, *Proc. IEE, London*, **99**, Part 4, 1952, 302–317.
2. Jury, E. I., Syntheses and Critical Study of Sampled-data Control Systems, *Trans. A.I.E.E.*, **75**, Part 2, 1956, 141–151.
3. Jury, E. I., Additions to the Modified Z-Transform Method, *IRE WESCON Conventional Record*, Part 4, August 21, 1957. pp. 136–156.
4. Smith, C. L., *Digital Computer Process Control*, Intext Publishers, Scranton, Pa., 1972.

PART II

Design and Application of Advanced Control Concepts

CHAPTER 13

Process Modeling and Identification

From our study of Chapter 10 it should be clear that the design of Z transform based computer-control algorithms require that the process transfer function be known. The knowledge of the process transfer function is also required for the implementation of advanced control strategies. A process transfer function may be developed from a theoretical analysis of the process or from experimental tests in the plant. In the former approach, we write appropriate unsteady-state balances (e.g., those involving conservation of mass, energy, and momentum), linearize the resulting differential equations if the differential equations turn out to be nonlinear, express the variables in deviation form, and finally take the Laplace transform so as to obtain the process transfer function. In the experimental approach we introduce a suitable change in the input and record the output response. These data are subsequently analyzed to determine an approximate process transfer function.

The development of dynamic mathematical models for specific

process systems (e.g., distillation columns, heat exchangers, chemical reactors) is beyond the scope of this text. For the interested reader we have included at the end of this chapter a bibliography on mathematical modeling of chemical processes.

The dynamic mathematical models are extremely useful in the design and analysis of process-control systems. A common application of the models is in the design phase of the project. In this instance, the process-control system based on the model may be simulated on a digital, analog, or hybrid computer so as to assess the relative benefits of different control strategies. Of course, this approach suffers somewhat, because some factors that can affect the operation of the real process may not be included in the model.

In dealing with existing plants that are in operation, a more direct approach would be to employ experimental tests to develop a process model or a process transfer function. This is the subject of this chapter.

Chemical engineering processes may be described either as being open-loop stable or open-loop unstable. Most processes that we as control engineers must work with are open-loop stable and are therefore self-regulating. This means that in manual control if a step change in the process input is made, the process output will ultimately reach a new steady-state level. On the other hand, some commercially important processes exhibit open-loop unstable behavior and are termed nonself-regulating processes. Examples of this class of processes include some level control systems and exothermic chemical reactors. In an exothermic chemical reactor a change in the coolant flow causes a temperature upset, which affects reaction rate and in turn temperature and so on; thus, in theory at least, the temperature can rise without a bound. In some level control systems a step change in the manipulated variable causes the level to change indefinitely. The experimental identification of process parameters of open-loop unstable systems is generally difficult. Deshpande[1] has presented a technique for obtaining process models of open-loop unstable systems from closed-loop tests, and the reader interested in the modeling of such systems is referred to that article for details. In this chapter we describe three experimental techniques, in order of increasing complexity, for identifying the dynamics of open-loop stable processes. The first is based on step response and gives the parameters of a first- or second-order model having a dead-time element. The second, called pulse testing, is a frequency-domain method, which yields the frequency response diagram of the open-loop process. The third technique is a time-domain method which yields the gain and time constants of the process model.

13.1 Process Models from Step-Test Data

This technique consists of subjecting the process, while it is operating under steady-state conditions with the feedback controller in manual, to a step change in input and recording the resulting transient response. The notion of a transfer function is involved in the development of the model parameters, and therefore, some care should be exercised in selecting the size of the step input. The step size should be sufficiently large so that the output data is distinguishable from the process noise but not so large as to drive the system out of the linear range. One should also make certain that the process is free of load upsets throughout the duration of the test. Generally, at least a couple of tests should be conducted, one involving a positive step change and another involving a negative step change, to ascertain that the process loads are absent and that the range of linearity is not exceeded.

The plot of step response of an open-loop process versus time is called a process reaction curve. Numerous methods are available for the determination of model parameters from the process reaction curve. Ziegler and Nichols[2] and Miller[3] have described a simple procedure for finding the parameters of a first-order plus dead-time model. Oldenbourg and Sartorius[4] were probably the first investigators to propose a procedure for determining the parameters of a second-order model, which was later extended by Sten[5] to include the estimation of dead time. Other approaches to model parameter evaluation have been proposed by Smith[6], Cox[7], Smith and Murrill[8], Meyer et al.[9], Csaki and Kis[10], Naslin and Miossec,[11] and Sundaresan et al.[12]. Most of the available methods for parameter estimation of second-order models depend on the accurate location of the inflection point, which is prone to error. Sundaresan et al.[12] appear to have overcome this difficulty. We describe their procedure, with the permission of the publisher, in the following paragraphs. The interested reader may wish to consult the earlier references to learn about the previous methods.

The transfer function of an overdamped second-order plus dead-time model is given by

$$G_p(s) = \frac{e^{-\theta_d s}}{(\tau_1 s + 1)(\tau_2 s + 1)} \tag{13.1}$$

Whereas the transfer function of an underdamped second-order plus dead-time model is given by

$$G_p(s) = \frac{e^{-\theta_d s}}{\frac{1}{\omega_n^2} s^2 + \frac{2\zeta}{\omega_n} s + 1} \tag{13.2}$$

The objectives are to determine the three parameters θ_d, τ_1, and τ_2 or θ_d, ω_n, ζ of the model from the process reaction curve. The steady-state gain K_p is easily determined from the plot as the ultimate change in the controlled variable per unit change in the process input and therefore is not one of the unknown parameters in this analysis. The plot of a typical process response to which we wish to fit the second-order plus dead-time model is shown in Figure 13.1. The method of Sundaresan, which is based on the work of Lees[13] and Gibilaro and Waldram,[14] recognizes the fact that the process reaction curve resembles certain statistical distribution functions and therefore can be described by its first moment, which in turn could be utilized for parameter evaluation. First, let us consider the evaluation of the overdamped system parameters K_p, τ_1, and τ_2.

As pointed out by Sundaresan et al., the first moment of the response function $c(t)$ is

$$m_1 = \int_0^\infty (1 - c(t))dt \tag{13.3}$$

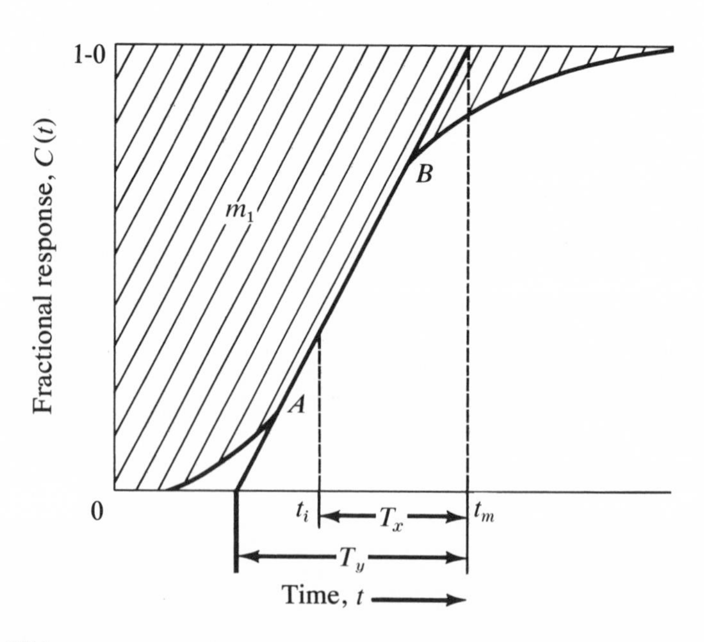

Figure 13.1
Evaluation of Second-Order Dead-Time Parameters from an Overdamped Curve

Thus m_1 is simply the shaded area in Figure 13.1.

The authors also point out that the Laplace transform is related to the moment-generating function. The relationship between the first moment m_1 and the transfer function $G_p(s)$ is

$$m_1 = -\left.\frac{d\,G_p(s)}{d\,s}\right|_{s=0} \qquad (13.4)$$

$$= \theta_d + \tau_1 + \tau_2$$

Now, consider the time-domain solution of Equation (13.1) for a step change in input, which is given by the equation

$$C(t) = \left[1 - \frac{\tau_1}{(\tau_1 - \tau_2)} e^{-\frac{t - \theta_d}{\tau_1}} + \frac{\tau_2}{(\tau_1 - \tau_2)} e^{-\frac{t - \theta_d}{\tau_2}}\right] u(t - \theta_d) \qquad (13.5)$$

By differentiating Equation (13.5) twice and setting the resulting second derivative equal to zero, we may find the expression for point of inflection

$$t_i = \theta_d + \alpha \ln \eta \qquad (13.6)$$

where

$$\eta = \frac{\tau_1}{\tau_2} \qquad (13.7)$$

and

$$\alpha = \frac{\tau_1 \tau_2}{\tau_1 - \tau_2} \qquad (13.8)$$

The slope M_i at the point of inflection is

$$M_i = \frac{(\eta)^{1/(1-\eta)}}{(\eta - 1)\alpha} \qquad (13.9)$$

This slope is that of a tangent that passes through t_i and intersects the final value of $c(t)$ at time t_m, as shown in Figure 13.1. This value of time is given by

$$t_m = \theta_d + \alpha \left[\ln \eta + \frac{\eta^2 - 1}{\eta} \right] \tag{13.10}$$

Equations (13.4), (13.9), and (13.10) may be combined so as to get

$$(t_m - m_1)M_i = \frac{\eta^{1/(1-\eta)}}{(\eta - 1)} \ln \eta \tag{13.11}$$

Note that if in this last equation η is changed to $1/\eta$, the right side remains unchanged. Therefore, in the solution of Equation (13.11) we need to consider only the range of 0 to 1 for η. Let us now write Equation (13.11) in the form

$$\lambda = \chi e^{-\chi} \tag{13.12}$$

where

$$\lambda = (t_m - m_1)M_i \tag{13.13}$$

and

$$\chi = \ln \eta/(\eta - 1) \tag{13.14}$$

It is clear from Equation (13.12) that the maximum value of λ is e^{-1} which occurs when the system is critically damped, that is, $\eta = 1$ or $\chi = 1$. For the overdamped case, η is less than 1, and therefore, χ lies between 0 and e^{-1}. At the extreme, where η equals zero and, therefore, $\lambda = 0$, the model reduces to a first-order system. Thus for approximating the true response by a second-order model with or without dead time the values of λ lie between 0 and e^{-1}. A plot of Equation (13.12) is shown in Figure 13.2. The procedure for parameter evaluation is as follows:

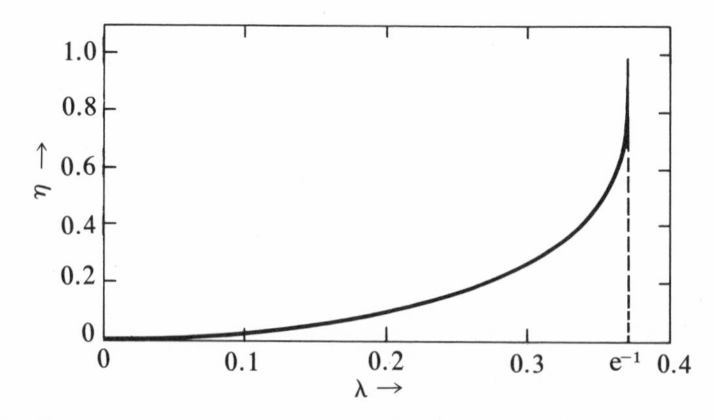

Figure 13.2
Plot of λ for Overdamped Second-Order Approximations

1. Determine the shaded area of Figure 13.1 by numerical integration.
2. Draw a tangent that passes through the straight line portion of the response curve.
3. Denote the slope of this line as M_i and its intersection with the final value of $C(t)$ as t_m.
4. Compute λ as per Equation (13.13).
5. From Figure 13.2 determine η.
6. Solve Equation (13.9) for α.
7. Solve Equations (13.4), (13.7), and (13.8) for θ_d, τ_1, and τ_2.

The equations resulting from the application of this procedure are

$$\tau_1 = \frac{\eta^{\eta/(1-\eta)}}{M_i}$$

$$\tau_2 = \frac{\eta^{1/(1-\eta)}}{M_i} \tag{13.15}$$

$$\theta_d = m_1 - \frac{\eta^{1/(1-\eta)}}{M_i}\left(\frac{\eta + 1}{\eta}\right)$$

For critically damped systems $\eta = 1$ and these equations reduce to

$$\tau_1 = \tau_2 = \frac{1}{M_i e} \tag{13.16}$$

$$\theta_d = m_1 - \frac{2}{M_i e}$$

For underdamped second-order approximations, the following procedure has been recommended. Note that in this case λ is greater than e^{-1}. Figure 13.3 shows a typical response for an underdamped system. The time domain solution of the system of Equation (13.2) to a step change in input is

$$c(t) = u(t - \theta_d)\left[1 - e^{-(t-\theta_d)\omega_n \xi}\left\{\frac{\xi}{\sqrt{1-\xi^2}} \sin\sqrt{1-\xi^2}\,\omega_n \times (t - \theta_d)) + \cos(\sqrt{1-\xi^2}\,\omega_n (t - \theta_d))\right\}\right] \tag{13.17}$$

Using this equation and the application of a procedure similar to that employed in the overdamped case, the authors have derived the following equations for the system parameters

$$\lambda = (t_m - m_1)M_i = \frac{\cos^{-1}\xi}{\sqrt{1-\xi^2}} \exp\left(\frac{-\xi}{\sqrt{1-\xi^2}} \cos^{-1}\xi\right) \tag{13.18}$$

$$\omega_n = \frac{\cos^{-1}\xi}{\sqrt{1-\xi^2}} \frac{1}{(t_m - m_1)} \tag{13.19}$$

$$\theta_d = m_1 - \frac{2\xi}{\omega_n} \tag{13.20}$$

It may be noted that in the computation of m_1, the areas below the ordinate of 1 are taken as positive and those above 1 are taken as negative. Also, although Equation (13.18) provides for an infinite number of roots for ζ for a given value of λ, in the region of our interest $(0 \leq \zeta \leq 1)$ λ is a monotonic function of ζ, decreasing from $\pi/2$ at $\zeta = 0$ to e^{-1} at $\zeta = 1$. A plot of ζ versus λ is shown in Figure 13.4. The other system parameters are then determined from Equations (13.19) and (13.20).

The method described in this chapter is simple and easy to use, but its application would require a digital computer program for numerical integration of Equation (13.3). Let us apply the method to a couple of example problems.

Example 1. The process transfer function of a third-order system is given by

$$G(s) = \frac{C(s)}{X(s)} = \frac{1}{(s+1)(0.5s+1)(2s+1)} \tag{13.21}$$

where the time constants are expressed in minutes. The true response

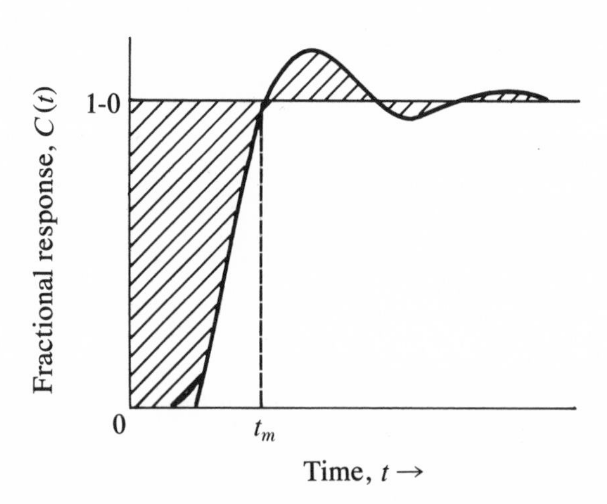

Figure 13.3
Response of Second-Order Underdamped System with Dead Time

of $C(t)$ to a step change in input $X(t)$ is shown in Figure 13.5. Approximate this process by a second-order plus dead-time model.

From Figure 13.5 the shaded area m_1 is computed to be 3.5 min. The tangent line drawn along the straight-line portion of the plot has a

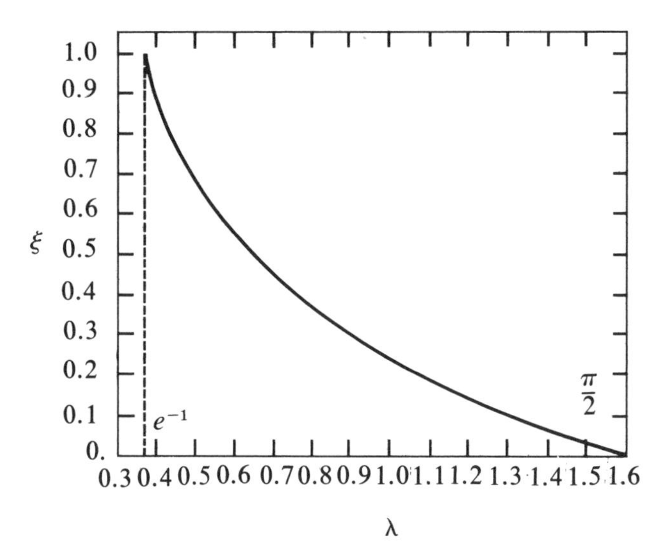

Figure 13.4
Plot of λ versus ξ for Second-Order Underdamped Approximations

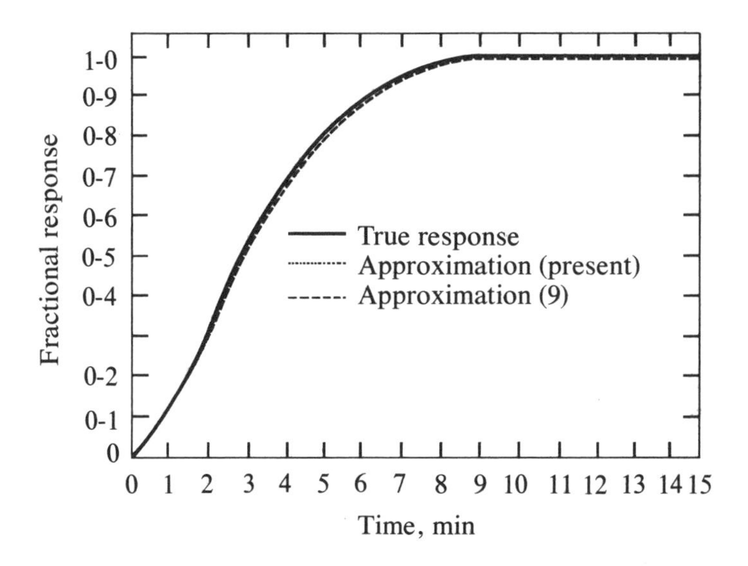

Figure 13.5
Comparison of True Response with Approximate Response of an Overdamped Second-Order System

slope $M_i = 0.23\ \text{min}^{-1}$, and it intersects $C(t) = 1$ line at $t_m = 5.1$ min. For $\lambda = (t_m - m_1)M_i = 0.368$, Figure 13.2 gives $\eta = 0.9$. Substituting these values into Equation (13.15) gives $\theta_d = 0.33$, $\tau_1 = 1.5$, and $\tau_2 = 1.67$. Thus the approximating transfer function is

$$G_p(s) = \frac{e^{-0.33s}}{(1.67s + 1)(1.5s + 1)} \tag{13.22}$$

Example 2. Consider the block diagram of an open-loop cascade control system having an inner closed-loop shown in Figure 13.6. The response of $C(t)$ to a step change in $R(t)$ is shown in Figure 13.7. From this figure $M_i = 0.417\ \text{min}^{-1}$, $t_m = 3.4$ min, and $m_1 = 1.99$ min, which gives a value of λ equal to 0.588. From Figure 13.4 at this value of λ the damping parameter ζ equals 0.588. These values together with Equations (13.19) and (13.20) give $\theta_d = 0.566$ min and $\omega_n = 0.826\ \text{min}^{-1}$ (or $\tau = 1/0.826 = 1.21$ min). Thus the system of Figure 13.6 may be approximated as

$$R(s) \rightarrow \boxed{\frac{e^{-0.566s}}{1.465s^2 + 1.42s + 1}} \rightarrow C(s)$$

The step response of the model is compared with the true response of the process in Figure 13.7. The results show good agreement.

13.2. Pulse Testing for Process Identification

Unlike the reaction curve method, this method is a frequency-domain method yielding a frequency-response diagram of the open-

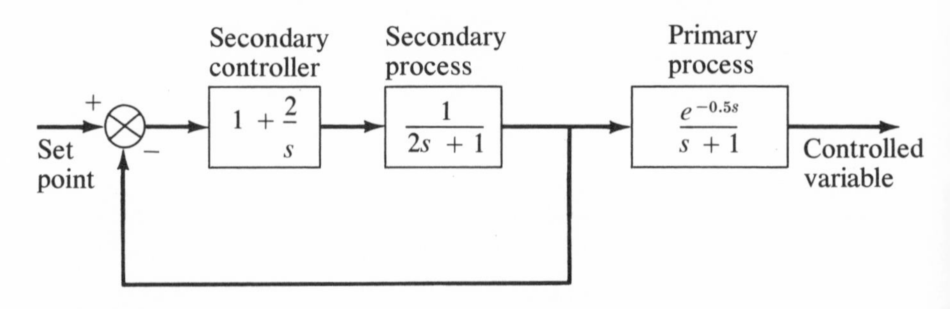

Figure 13.6
Open-loop Cascade System with a Closed Inner Loop

loop process. As in the case of the reaction curve method, one must ascertain that load upsets are absent during the test. Also, since the notion of Laplace transforms is involved, the range process linearity should not be exceeded throughout the duration of the test.

In the pulse-testing method, a pulse of arbitrary shape is applied at the input of the process [$X(s)$ in Figure 13.8], while it is operating under steady-state conditions with the controller in manual, and the transient response of the process is recorded. These records (see Figure 13.9) assist us in the development of a frequency-response diagram, as described in the following paragraphs. For a historical discussion and examples of early applications of pulse testing, the interested reader is referred to Hougen[15a] and Clements and Schnelle.[15b] The Laplace transform of the input variable $X(s)$ is related to the Laplace transform of the output $Y(s)$ via the transfer function $G(s)$, which is the product of $G_u(s)$, $G_p(s)$, and $G_m(s)$. Thus

$$G(s) = \frac{Y(s)}{X(s)} \tag{13.23}$$

If the definition of Laplace transforms is substituted in Equation (13.23), it can be written as

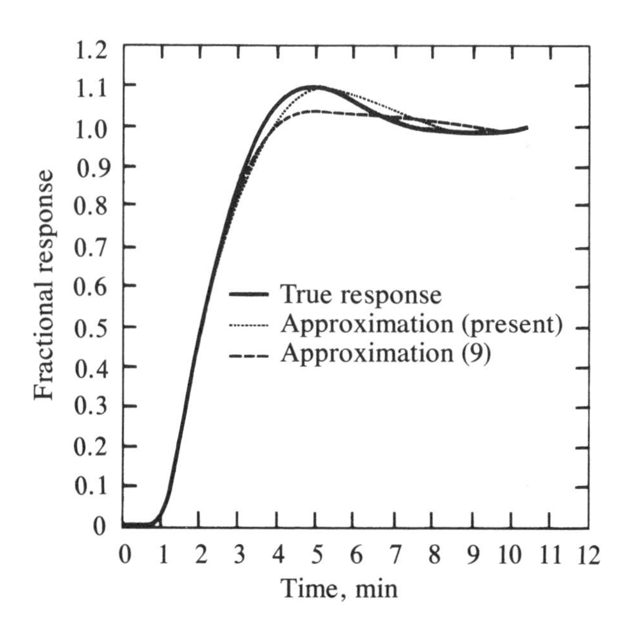

Figure 13.7
Comparison of Time Response with Approximate Response for an Underdamped Second-Order Process

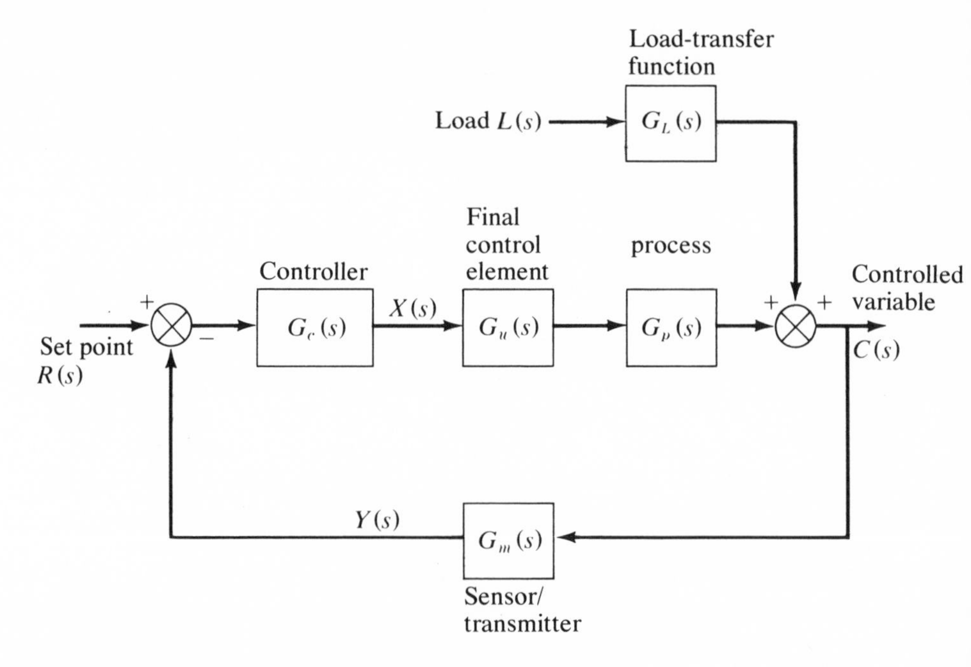

Figure 13.8
Typical Closed-Loop Control System

$$G(s) = \frac{\int_0^\infty Y(t)\, e^{-st}\, dt}{\int_0^\infty X(t)\, e^{-st}\, dt} \tag{13.24}$$

From Figure 13.9 it is clear that the upper limits on the two integrals can be replaced by t_y and t_x, the time periods during which changes in Y and X occur. If, in addition, $j\omega$ is substituted for s, Equation (13.24) becomes

$$G(j\omega) = \frac{\int_0^{t_y} Y(t)\, e^{-j\omega t}\, dt}{\int_0^{t_x} X(t)\, e^{-j\omega t}\, dt} \tag{13.25}$$

The numerator in Equation (13.25) is the Fourier transformation of the time function $Y(t)$ and the denominator is the Fourier transformation of the time function $X(t)$. Since $e^{-j\omega t}$ can be replaced by $\cos \omega t - j \sin \omega t$, Equation (13.25) can be written as

$$G(j\omega) = \frac{\int_0^{t_y} Y(t) \cos(\omega t)\, dt - j \int_0^{t_y} Y(t) \sin(\omega t)\, dt}{\int_0^{t_x} X(t) \cos(\omega t)\, dt - j \int_0^{t_x} X(t) \sin(\omega t)\, dt} \tag{13.26}$$

$$= \frac{A - jB}{C - jD} \tag{13.27}$$

where A, B, C, and D represent the four integrals in Equation (13.26). If the right-hand side is multiplied and divided by the complex conjugate of the denominator, Equation (13.27) becomes

$$\begin{aligned} G(j\omega) &= \frac{A - jB}{C - jD} \cdot \frac{C + jD}{C + jD} \\ &= \frac{AC + BD + j(AD - BC)}{C^2 + D^2} \\ &= \frac{(AC + BD)}{C^2 + D^2} + j\,\frac{(AD - BC)}{C^2 + D^2} \\ &= \text{Re}\ G(j\omega) + j\ \text{Im}\ G(j\omega) \end{aligned} \tag{13.28}$$

From this equation the amplitude ratio AR and the phase angle ϕ are determined by taking the magnitude and the angle of the complex number, respectively.

Thus

$$\begin{aligned} AR = |G(j\omega)| &= \sqrt{\text{Re}\ G(j\omega)^2 + \text{Im}\ G(j\omega)^2} \\ &= \sqrt{\left(\frac{AC + BD}{C^2 + D^2}\right)^2 + \left(\frac{AD - BC}{C^2 + D^2}\right)^2} \end{aligned} \tag{13.29}$$

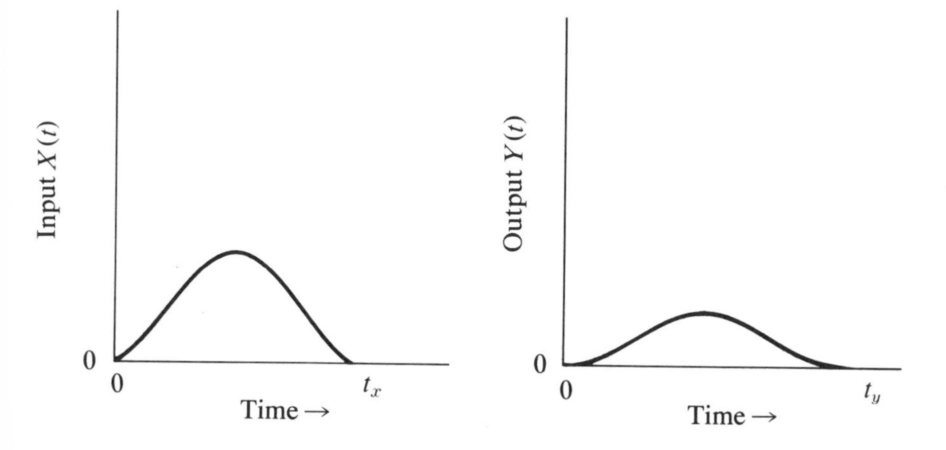

Figure 13.9
Typical Input-Output Plots from Pulse Tests

and

$$\phi = \arctan\left(\frac{AD - BC}{AC + BD}\right) \tag{13.30}$$

To use these equations, a specific value of frequency ω is chosen. The integrations are performed, yielding a single value of A, B, C, and D. Substituting these values in Equations (13.29) and (13.30) then gives a point on the frequency-response diagram, corresponding to the chosen frequency. The procedure is repeated for numerous values of ω, and the complete frequency response diagram is prepared. Since the time functions $Y(t)$ and $X(t)$ would be available in graphical form, the integrals in Equation (13.26) are most conveniently evaluated using numerical integration techniques on the digital computer. The listing of a digital computer program[16] written in Fortran IV for analyzing pulse test data is included in Appendix D. The program deck and the accompanying user's manual are available for purchase at ISA headquarters.

The pulse-testing technique gives us a frequency-response diagram of the open-loop process. This is as far as we need go if our purpose is to determine the tuning constants of conventional controllers. If, on the other hand, we require a process transfer function for the purpose of implementing computer control or advanced control applications, it may be fitted to the frequency response diagram as described in the following examples.

Example 3. The results from a pulse test on an industrial open-loop cascade control system with a closed inner loop are shown in Figure 13.10. The use of these data in conjunction with the pulse analysis program results in frequency-response data which are plotted in Figure 13.11. We wish to fit a transfer function of the form

$$G_p(s) = \frac{K_p\, e^{-\theta_d s}}{\tau_p s + 1} \tag{13.31}$$

to the data shown in Figure 13.11. The following step-by-step procedure may be employed for this purpose:

1. From the pulse analysis program the steady-state gain K_p, which is the ratio of the areas under the input and output curves in Figure 13.10, equals 1.

2. The corner frequency ω_c for the first-order process occurs at an AR equal to 0.707 K_p. Thus, from Figure 13.11*a* at $AR = 0.707$, the corner frequency ω_c equals 2.5 radians/min. Therefore, the time constant of the first-order process is

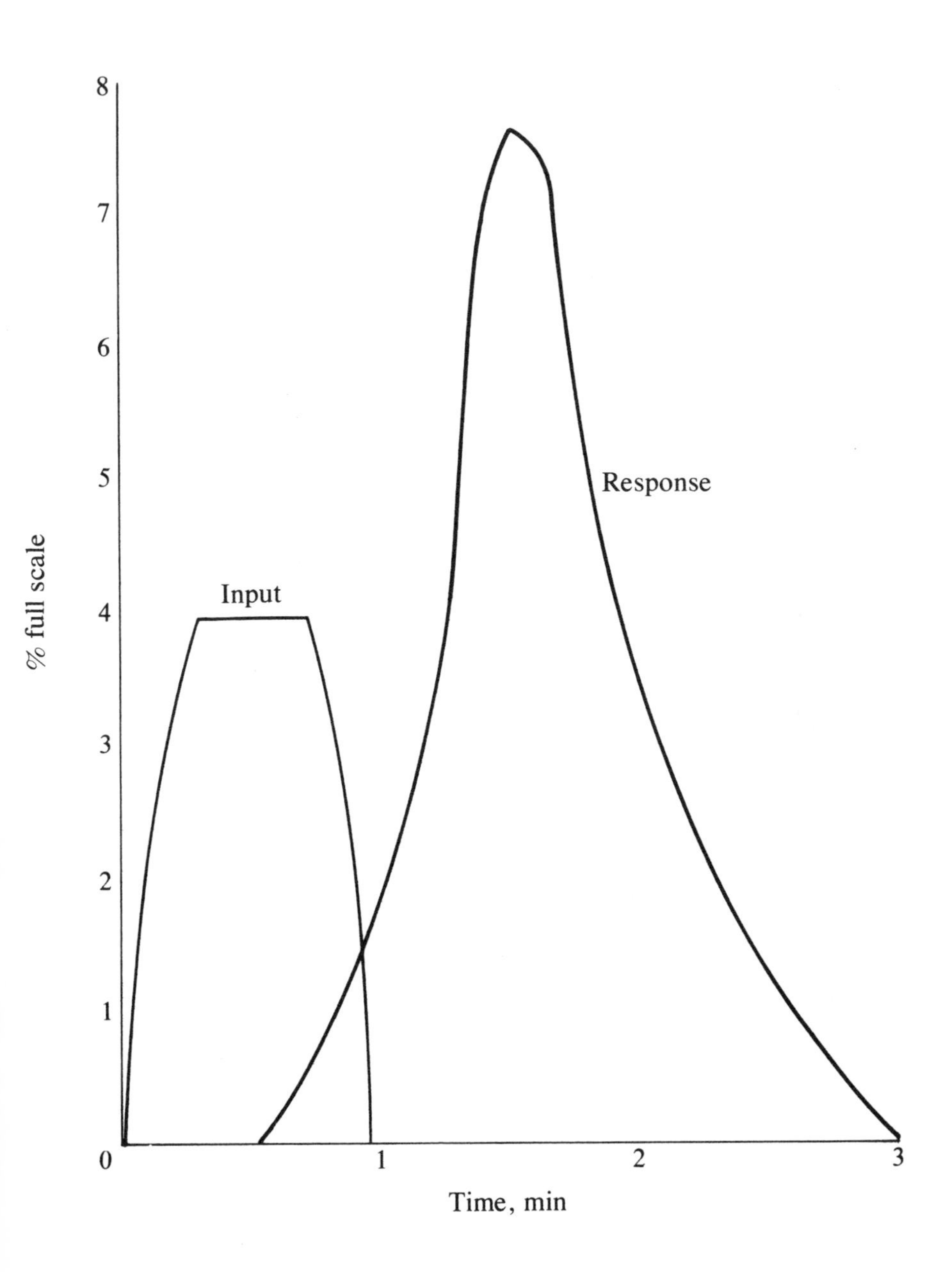

Figure 13.10
Input Pulse and Transient Response

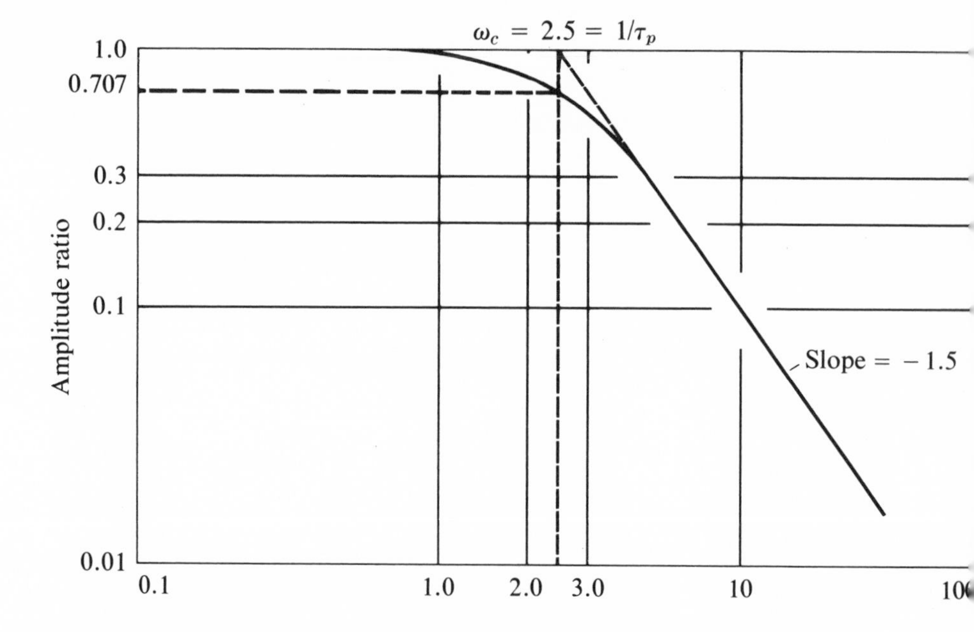

Figure 13.11a
Frequency-Response Diagram—Amplitude Ratio versus Frequency

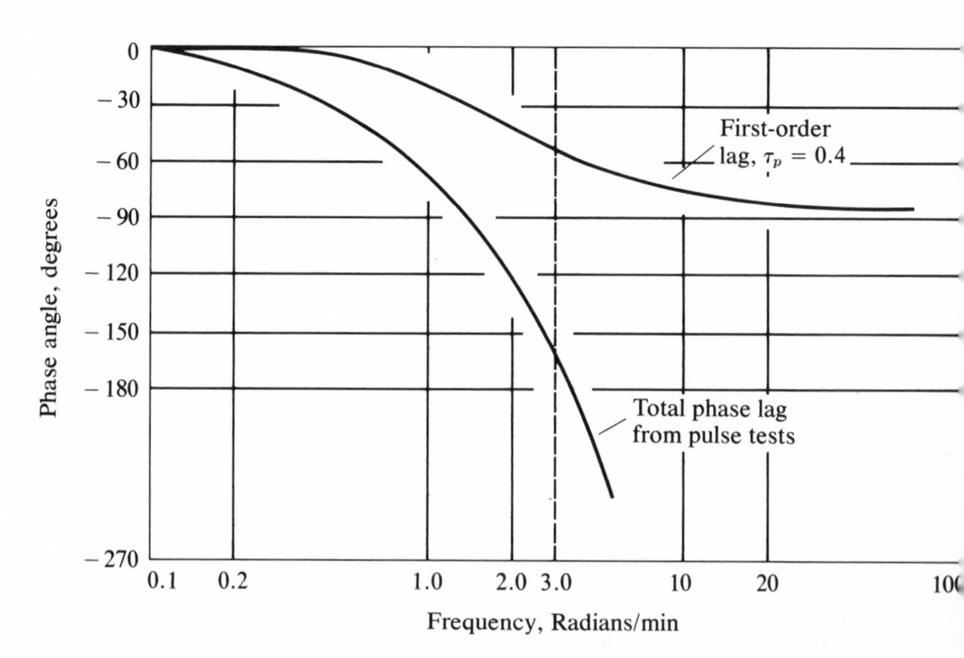

Figure 13.11b
Frequency-Response Diagram—Phase Angle versus Frequency

$$\tau_p = \frac{1}{\omega_c} = \frac{1}{2.5} = 0.4$$

3. Now we plot the phase angles for this first-order lag with τ_p = 0.4, as shown in Figure 13.11*b*.

4. For a first-order process the phase angle at the corner frequency equals 45°. Therefore, at this frequency the total phase angle is the sum of the phase angles of first-order lag and that of dead time. Thus, from Figure 13.11*b*

$$\phi_{\text{total}} = -154 = \text{phase angle due to dead time} \Big|_{\omega_c} - 45°$$

or

$$-154 = -\omega_c\, \theta_d \left(\frac{180}{\pi}\right) - 45$$

$$= -(2.5)\left(\frac{180}{\pi}\right)\theta_d - 45$$

Therefore,

$$\theta_d = \frac{(109)\,(\pi)}{(180)\,(2.5)} = 0.76 \text{ min}$$

and the approximate model is

$$G_p = \frac{1e^{-0.76s}}{(0.4s + 1)}$$

In the previous example we considered the fitting of frequency-response curves to a first-order lag with dead time. To fit second-order models we note the following characteristics of the second-order systems:

1. As in first-order systems the steady-state gain is the ratio of areas under the input and output pulse curves. The zero-frequency asymptote on the amplitude-ratio portion of the Bode plot also gives the steady-state gain of the process.

2. For overdamped second-order approximations, the two time constants are found by noting the corner frequencies (also called breakpoint frequencies) on the amplitude curve of the Bode plot. To do this, we fit the amplitude data as far out in frequency as possible to a first-order curve by trial and error. We then note the corner frequency of the first-order curve, which is the reciprocal of the time constant τ_1. Then we graphically subtract the first-order

curve from the amplitude data which gives the amplitude data for another first-order lag to which a first-order lag is fitted so as to obtain τ_2.

3. For underdamped second-order systems the time constant is found from the breakpoint frequency, and the damping factor is found by comparing the resonant peak with the amplitude data of known second-order systems. The Bode plot of an underdamped second-order system for several values of the damping factor ζ is shown in Figure 13.12.

4. Once the amplitude data are fitted to a known transfer function, its phase angles are plotted and compared with the experimen-

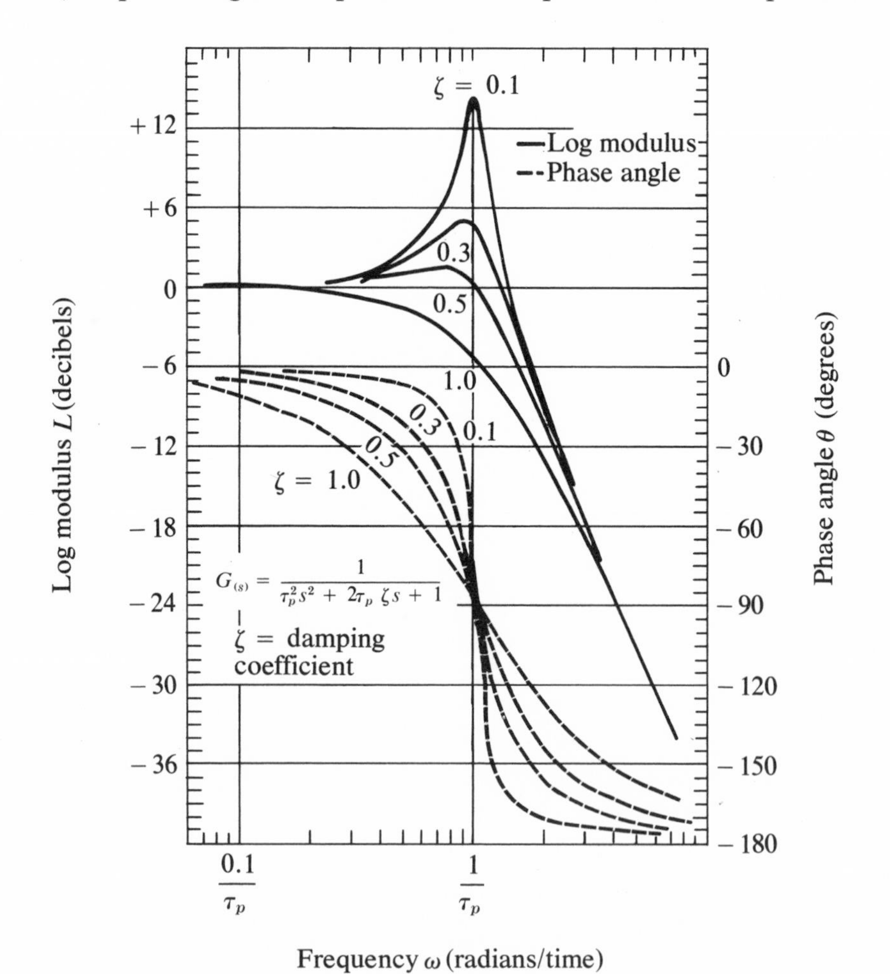

Figure 13.12
Second-Order System Bode Plots [From Ref. 17 with Permission]

tal phase angle data. The difference, if any, is the contribution of dead time. Since for a dead-time element the phase angle equals $\omega\theta_d$, θ_d can be easily evaluated.

In Figures 13.13*a* and 13.13*b*, taken from Luyben,[17] we show two examples of fitting higher-order transfer functions. The details of the fitting procedure have been left out as an exercise for the reader.

The test input during a pulse test begins and ends at the same value. Generally, the output will also return to the initial value. However, under certain circumstances the output will not return to the initial steady state. This occurs when the open-loop process contains a pure integrator.[17] For example, if the inflow to a tank is

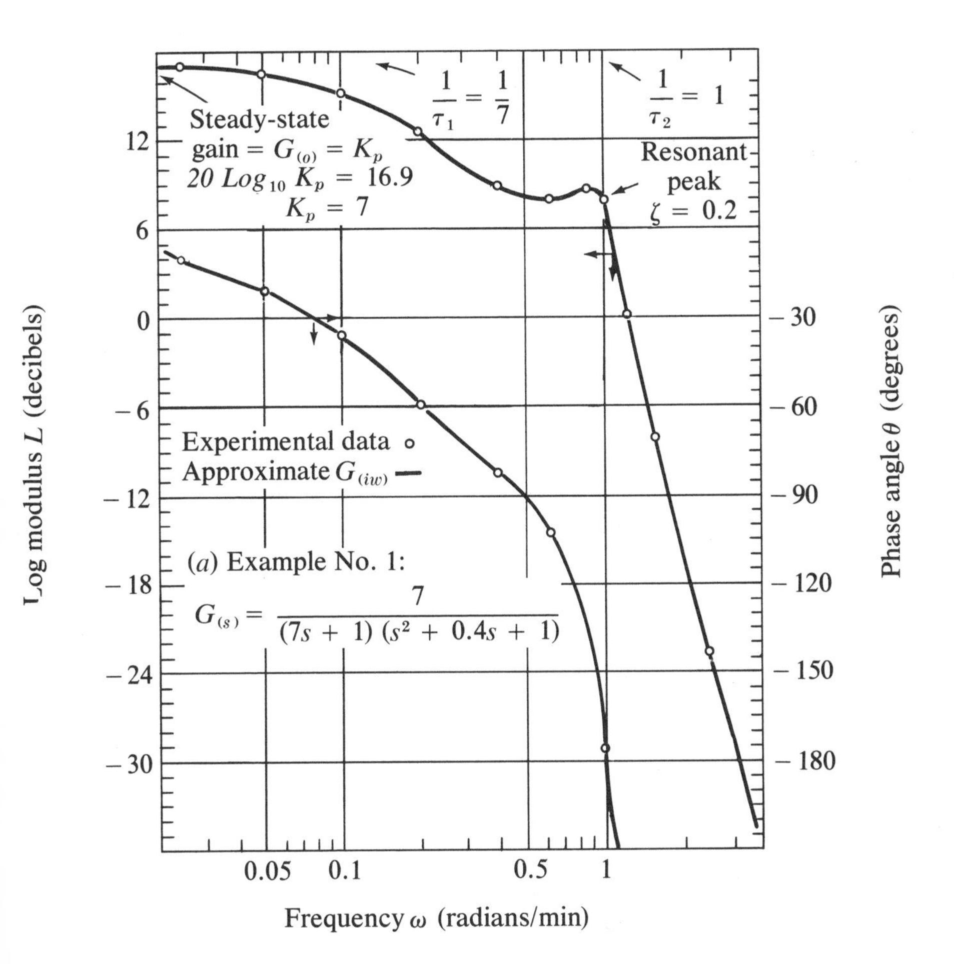

Figure 13.13*a*
Fitting Approximate Transfer Function to Experimental Frequency-Response Data (From Ref. 17 with Permission)

pulsed while the flow out of the tank is held constant, the tank level will rise to a new level and will not return to its initial value on conclusion of the test. The computer program listing of Appendix D is capable of handling nonclosing pulses that arise when working with processes containing an integrating element.

Let us now discuss the criteria for pulse selection and for the selection of the step size in the numerical integration of Equation (13.26).

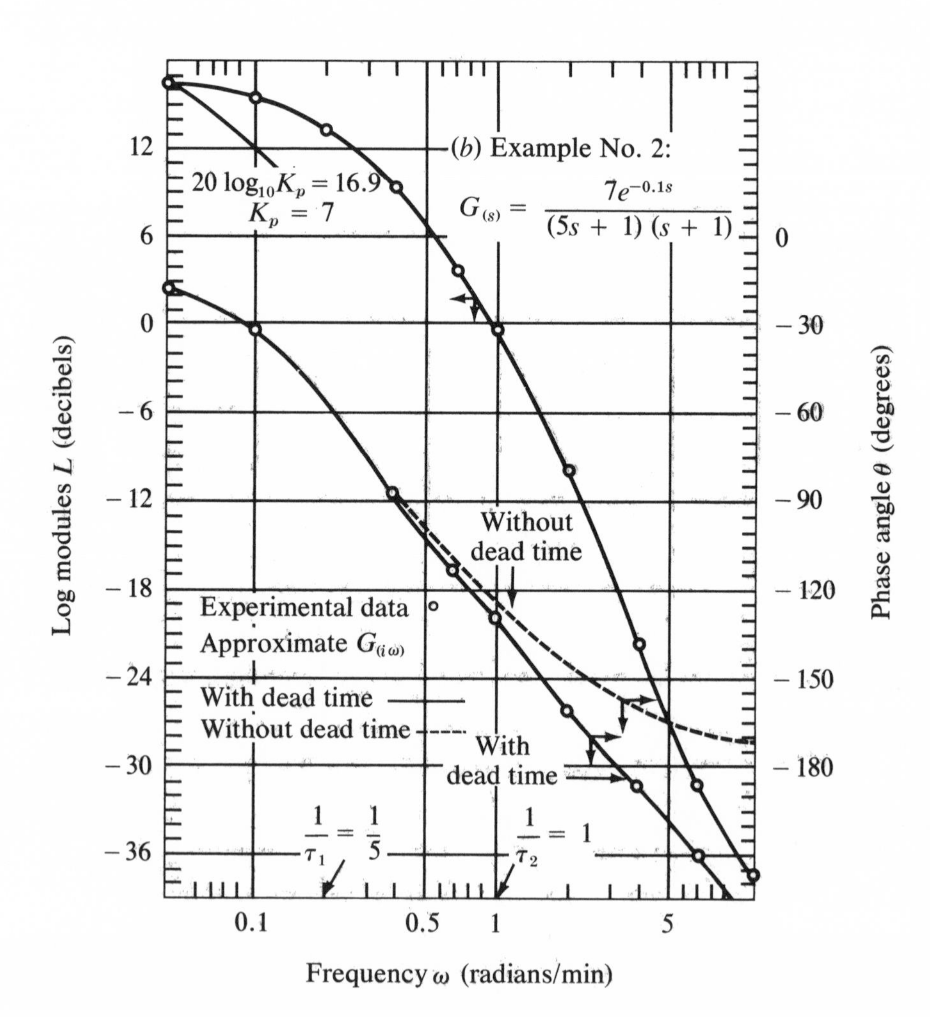

Figure 13.13b
Fitting Approximate Transfer Function to Experimental Frequency-Response Data (From Ref. 17 with Permission)

Shape of the Input Pulse. Consider the definition of $G(j\omega)$

$$G(j\omega) = \frac{\int_0^{t_y} Y(t)\, e^{-j\omega t}\, dt}{\int_0^{t_x} X(t)\, e^{-j\omega t}\, dt} \tag{13.31}$$

The shape of the pulse should be such that its frequency content (i.e., amplitude of FIT, the abbreviation for Fourier integral transforms) should be finite over the frequency range of interest. For a rectangular pulse, FIT goes to zero at $\omega = 2\pi/\mathrm{T}$, $4\pi/\mathrm{T}$, and so on. For a displaced cosine* pulse FIT goes to zero at $4\pi/T$, $6\pi/T$, and so on. Therefore, the latter may be a more suitable input. However, a near rectangular pulse is so much easier to generate and therefore is used in most applications.

Magnitude of the Pulse. Recall that we are after a linear dynamic model in the form of $G(j\omega)$. It must be a linear model, because the notion of a transfer function applies only to a linear system. In general, the process is nonlinear, and we are obtaining a model that is linearized around the steady-state operating level. If the height of the pulse is too high, we may drive the process out of the linear range. Therefore, pulses of various heights should be tried. ISA[16] recommends 10% of span as a starting point for magnitude of the input pulse. It is a good idea to test with positive and negative pulses. The computed $G(j\omega)$ should be identical if the region of linearity is not exceeded.

Duration of the Input Pulse. The FIT of a rectangular pulse of width D is:

$$\mathrm{FIT} = h\,\frac{\sin \omega D}{\omega} - i\,\frac{h}{\omega}(1 - \cos \omega D) \tag{13.32}$$

when frequency $\omega = 2\pi/D$, FIT goes to zero, and the calculation of the transfer function is meaningless. Therefore, the smaller D can be made, the higher is the frequency to which $G(j\omega)$ can be found.

According to Luyben,[17] a good rule of thumb is to keep the width of the pulse less than about half the smallest time constant of interest.

*For a Displaced cosine function

$$f(t) = 1 - \cos \frac{2\pi}{T} t \qquad 0 \leq t \leq T$$

If the dynamics of the process are completely unknown, it may take a few trials to establish a reasonable pulse width.

If D is too small for a given pulse height the system is disturbed very little, and it becomes difficult to separate the real output signal from noise and experimental error.

Size of ΔT in Numerical Integration

According to Messa et al.,[18] the criterion for picking the number of data points is that enough data points must be used so that the approximating function matches the real function over all intervals. Based on the data from several investigators, they suggest the size of Δt should be such that $\omega_0(\Delta t) \leq \pi/2$ where ω_0 is the upper limit on the desired frequency and Δt is the time interval between points. The upper limit of frequency is determined by the frequency content of the input signal. Thus, for a rectangular pulse of width D, we have

$$\mathrm{FIT} = h\,\frac{\sin \omega D}{\omega} - i\,\frac{h}{\omega}\,(1 - \cos \omega D) \tag{13.33}$$

FIT goes to zero when $\omega = 2\pi/D$. Therefore, calculation of FIT and the transfer function near this frequency is meaningless. Also, Murill et al.[19] have shown that FIT amplitude beyond $\omega = 2\pi/D$ is very small. Therefore, the upper limit of frequency for a rectangular pulse is $\omega_0 = 2\pi/D$. As an example, if D = 5 min, the upper limit of frequency equals (2) (3.14)/5 = 1.26 rad/min. Now if we select Δt of 1 min, $\omega_0\,\Delta t$ for this case will be (1.26) rad/min × (1.0) min = 1.26 rad. This is well below the suggested value of $\pi/2$ (equals 1.56 rad). Therefore, 1 min for the size of Δt is entirely satisfactory.

13.3. Time Domain Process Identification[20]

In many industrial applications the parameters of the process model vary as a function of time, because of changing operating conditions, feed stocks, and other reasons. In such cases, it is advisable to identify the process parameters from time to time and update the controller settings if necessary. The pulse-testing procedure is not especially convenient in such on-line process identification applications. This section describes a time-domain procedure for determining the dynamics of open-loop processes.

As we pointed out in Chapter 1, many chemical engineering processes can be described by an overdamped second-order plus dead-time process model of the form

$$G(s) = \frac{X(s)}{U(s)} = \frac{K\,e^{-\theta_d s}}{(T_1 s + 1)\,(T_2 s + 1)} \tag{13.34}$$

where

X = process output

U = process input

K = process steady-state gain

θ_d = dead time

T_1, T_2 = process time constants

The objectives are to determine K, T_1, and T_2. It is assumed that an estimate of dead time is available from previous analysis. The procedure for determining these parameters consists of the following steps:

1. Apply a regulated disturbance (e.g., step or pulse) to the input of the process while it is operating at steady state in manual control.
2. Record the output of the process as well as the regulated disturbance as a function of time.
3. Assume trial values of K, T_1, and T_2. If the results of a step test are available, initial estimates of these parameters may be obtained quickly by one of techniques cited in a previous section (for examples see Refs. 4, 5, 6, or 7).
4. Using these trial values and the input data of step 1, predict the output via the time domain equation corresponding to Equation (13.34).
5. The difference between the experimental output and the predicted output gives an indication of error in the trial parameters.
6. Update the parameters and repeat steps 4 and 5 until the error is minimized to an acceptable level.

The time-domain equation corresponding to Equation (13.34) can be obtained in one of two ways; in the first approach, we begin with the following differential equation which represents the overdamped second-order process with dead time:

$$\frac{d^2X}{dt^2} + a\,\frac{dX}{dt} + bX = c\,u(t - \theta_d) \tag{13.35}$$

This equation is the inverse of Equation (13.34). Then we would express the derivatives in finite difference form so as to obtain an algebraic equation for output X_i in terms of input u_i. The second approach is to include a zero-order hold ahead of $G(s)$, evaluate G_{h_0} $G(Z)$ and invert it to obtain an equation for X_i in terms of u_i. The latter approach gives, as will be shown below,

$$X_i = C_1 X_{i-1} + C_2 X_{i-2} + C_3 u_{j-1} + C_4 u_{j-2} \tag{13.36}$$

where

$j = i - k$ (the dead time is assumed to be an integral multiple of sampling periods T, i.e., $\theta_d = kT$).

X_i = deviation of the process output from the original steady-state value at time i.

u_{j-1} = deviation of the process input from the original steady-state value at time $j - 1$.

c_1, c_2, c_3, c_4 = constants which are functions of K, T_1, and T_2, and T.

When a process does not have dead time, $j = i$. But when the process does have dead time, θ_d must be evaluated by one of the methods cited in a previous section of this chapter. In the absence of appreciable process noise, a crude estimate of dead time may be obtained by noting the time between the introduction of the input disturbance and the first deviation of the process output from its initial steady-state value. In sampled data systems the sampling period T must be short enough so that θ_d can be closely approximated by an integral multiple of T. In conventional control systems, equally spaced data points from the u and X plots must be taken frequently enough to ensure that an integral multiple of the sampling interval is a good approximation for dead time, θ_d. Moore[21] presents an analysis when θ_d is not an integral multiple of the sampling period. This will not be considered in this section.

To determine the values of the constants c_1, c_2, c_3, and c_4, a Z transform derivation is necessary. In the following analysis the dead time θ_d is taken as zero and incorporated in the final equation at the end. Thus the process transfer function whose parameters are to be determined is given by:

$$G(s) = \frac{X(s)}{U(s)} = \frac{K}{(T_1 s + 1)(T_2 s + 1)} \tag{13.37}$$

Assuming a zero-order hold, the Z transform of the hold-process combination is given by

$$G_{h_0}G(Z) = Z\{G_{h_0}(s)\, G(s)\} = \frac{X(Z)}{U(Z)}$$

$$= Z\left\{\frac{1-e^{-sT}}{s}\cdot\frac{K}{(T_1s+1)(T_2s+1)}\right\} \quad (13.38)$$

Therefore,

$$\frac{X(Z)}{U(Z)} = K(1-Z^{-1})Z\left\{\frac{1}{s(T_1s+1)(T_2s+1)}\right\} \quad (13.39)$$

Expand by partial fractions. Then

$$\frac{X(Z)}{U(Z)} = K(1-Z^{-1})Z\left\{\frac{1}{s}+\frac{T_1/(T_2/T_1-1)}{T_1s+1}+\frac{T_2/(T_1/T_2-1)}{T_2s+1}\right\} \quad (13.40)$$

taking Z transform, we get

$$\frac{X(Z)}{U(Z)} = K(1-Z^{-1})\left[\frac{1}{1-Z^{-1}}+\frac{1/(T_2/T_1-1)}{1-e^{-T/T_1}Z^{-1}}+\frac{1/(T_1/T_2-1)}{1-e^{-T/T_2}Z^{-1}}\right] \quad (13.41)$$

$$= \frac{b_1Z^{-1}+b_2Z^{-2}}{1-a_1Z^{-1}+a_2Z^{-2}}$$

with

$$a_1 = e^{-T/T_1}+e^{-T/T_2};\ a_2 = e^{-T/T_1}\cdot e^{-T/T_2}$$

$$b_1 = K\left(1-a_1+\frac{T_1\,e^{-T/T_2}-T_2\,e^{-T/T_1}}{T_1-T_2}\right) \quad (13.42)$$

and

$$b_2 = K\left(a_2-\frac{T_1\,e^{-T/T_2}-T_2\,e^{-T/T_1}}{T_1-T_2}\right)$$

Cross multiplying the terms of Equation (13.41) and inverting gives

$$X_i = a_1X_{i-1}-a_2X_{i-2}+b_1u_{i-1}+b_2u_{i-2} \quad (13.43)$$

substituting j to account for dead time, we get

$$X_i = a_1X_{i-1}-a_2X_{i-2}+b_1u_{j-1}+b_2u_{j-2} \quad (13.44)$$

The output X_i at the ith sampling instant is calculated using Equation (13.44), where a_1, a_2, b_1, and b_2 are defined in Equation (13.42) as functions of K, T_1, and T_2. There is an option on how to set up the discrete model. One choice is to use actual data for the values of X_{i-1} and X_{i-2} in the calculation of X_i. The other choice is to use the calculated values for X_{i-1} and X_{i-2}. Since the former approach may yield sporadic results because of noise, we have selected the latter approach. The latter approach tends to smooth out the effect of noise.

The problem of modeling the process is thus reduced to finding the values of K, T_1, and T_2 that best fit the data. An obvious approach would be to determine the constants a_1, a_2, b_1, and b_2 using a straightforward linear regression. Then one could solve for K, T_1, and T_2 with Equation (13.42). However, this procedure would yield four equations and three unknowns, an overdetermined set. Therefore, a nonlinear regression approach is necessary in which K, T_1, and T_2 are used as the regression variables. The approach is outlined in the next few paragraphs.

For a given set of values of K, T, T_1, T_2, the values of a_1, a_2, b_1, and b_2 are calculated, using Equation (13.42). Then each output data point X_i is predicted, using Equation (13.44). If the difference between the experimental value and the predicted value for each data point is squared and summed over the entire set of data, a measure of fit is obtained. This measure is expressed as:

$$\text{ERROR} = \sum_{i=0}^{m} (X_{i,a} - X_{i,c})^2 \tag{13.45}$$

where

ERROR = the measure of fit

m = number of data points

$X_{i,a}$ = actual (experimental) value of X_i

$X_{i,c}$ = calculated value of X_i

The lower the value of ERROR, the more accurately the regression variables fit the data. Thus to find the optimum set, the regression variables K, T_1, and T_2 must be adjusted until ERROR is minimized. This involves numerous interations.

Finding optimum values of the regression variables can be accomplished by adjusting all the variables simultaneously or adjusting each variable one at a time. The first method usually involves first and second partial derivatives of ERROR with respect to K, T_1, and T_2. The calculation of these partial derivations involves very bulky equations, which are prone to significant roundoff error. The single-variable approach will be used here. The method is easy to use and is shown to function well, given the constraints that K, T_1, and T_2 are positive real numbers.

Since the regression operation is unidimensional in nature, only one variable is operated on at a time, with the other two variables held constant. Two different search techniques are applied to each variable. The first of these was developed by Davies, Swann, and

Campey[22] (the DSC search). This method is excellent in that it quickly finds the approximate location of the optimum value of a variable. In other words, the DSC search "brackets" the optimum value. The second search, developed by Powell[22] rapidly zeroes in on the optimum once the variable has been bracketed.

The method is cyclic in nature. First K is optimized. Next T_1 is optimized using the new value of K and the old value of T_2. Then T_2 is optimized using the new values of K and T_1. At that point K is reoptimized, because there are new values for T_1 and T_2. Then T_1 and T_2 are reoptimized. This cycle is repeated until a prescribed accuracy is obtained, as will be explained. A step-by-step procedure for the optimization of K is presented here. Exactly the same operations are required for optimizing T_1 and T_2, except that a different variable is searched while the other two variables are held constant.

Let $F(K)$ = ERROR, evaluated at K with T_1 and T_2 held constant. Let DK = an increment size.

1. Calculate $F(K)$ with the initial values of K, T_1, and T_2.
2. Calculate $F(K + DK)$. If $F(K + DK) \leqslant F(K)$, go to step 3. If $F(K + DK) > F(K)$, let $DK = -DK$ and recalculate $F(K + DK)$.
3. Let the index $j = 1$.
4. Let $K_{j+1} = K_j + DK$.
5. Calculate $F(K_{j+1})$.
6. If $F(K_{j+1}) \leqslant F(K_j)$, double DK, increase index j by 1, and return to step 4. If $F(K_{j+1}) > F(K_j)$, denote K_{j+1} by K_m, K_j by K_{m-1}, and K_{j-1} by K_{m-2}. Let $DK = -0.5\,DK$ and return to steps 4 and 5 for only one more calculation.
7. Denote the last value for K by K_{m+1}.

These first seven steps complete Part I of the DSC search. A graphical representation of this initial procedure is shown in Figure 13.14.

8. In the four equally spaced values of K in the set (K_{m-2}, K_{m-1}, K_{m+1}, K_m) discard either K_m or K_{m-2}, whichever is the farthest from the K corresponding to the smallest value of ERROR in the set. Let the remaining three values of K be denoted by K_a, K_b, and K_c, where K_b is the center point and $K_a = K_b - DK$ and $K_c = K_b + DK$.
9. Carry out a quadratic interpolation to estimate XSOPT, a pseudo-optimum value of K, where:

$$\text{XSOPT} = K_b + \frac{DK\left[F(K_a) - F(K_c)\right]}{2\left[F(K_a) - 2F(K_b) + F(K_c)\right]} \qquad (13.46)$$

10. Calculate F (XSOPT) and denote it as FSOPT.

This concludes Part II of the DSC search. The Powell search is now applied.

11. Let

$$\begin{aligned} &\text{XSTAR} = \text{XSOPT} \qquad && FA = F(K_a) \\ &\text{FSTAR} = \text{FSOPT} \qquad && FB = F(K_b) \\ & && FC = F(K_c) \end{aligned} \tag{13.47}$$

12. If XSTAR and whichever of the set (K_a, K_b, K_c) corresponding to the smallest ERROR, differ by less than the prescribed accuracy in K, XACCUR, terminate the K search. Otherwise, discard from the set (K_a, K_b, K_c) the one that corresponds to the greatest

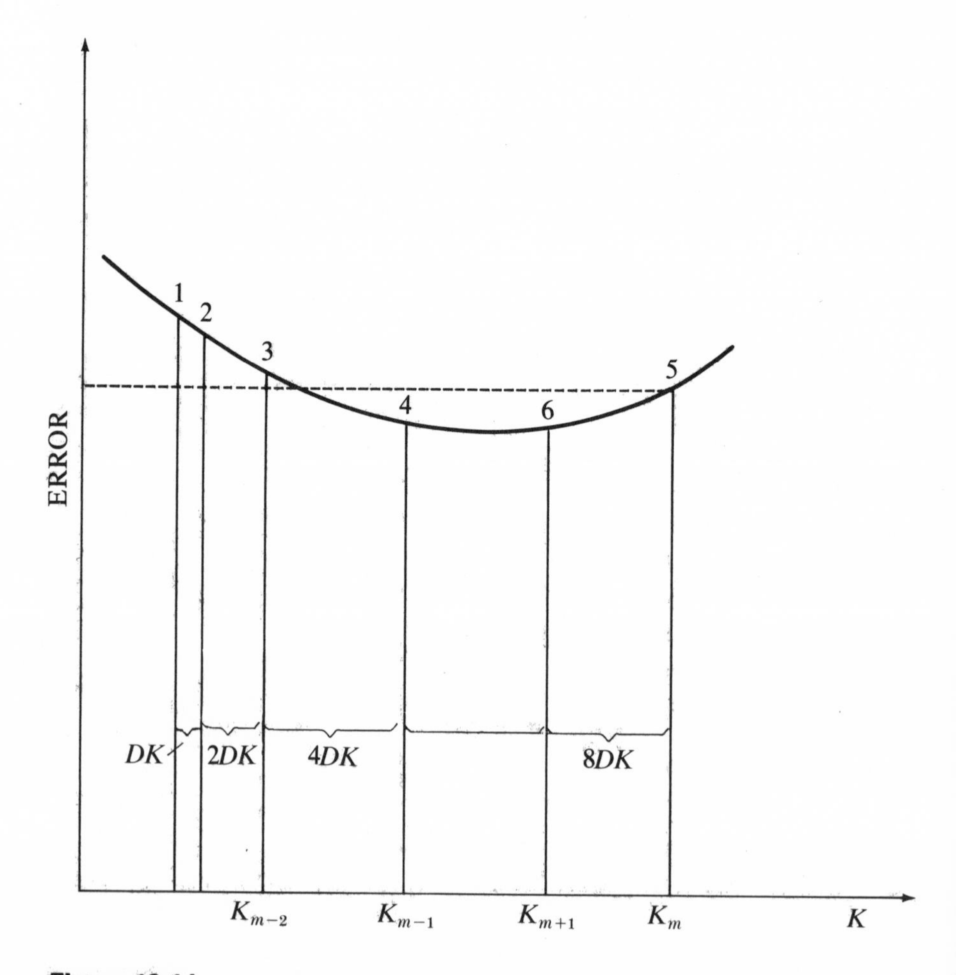

Figure 13.14
DSC Search part 1

value of ERROR [unless the bracket on the minimum of ERROR will be lost by doing so, in which case discard instead the K so as to maintain the bracket].

13. Let XSTAR replace that K which was discarded from the set (K_a, K_b, K_c). Let FSTAR be the corresponding value of ERROR. Thus we will again have three elements in the set.

14. Calculate a new XSTAR with the following equations:

$$\text{ANUM} = (K_b^2 - K_c^2)FA + (K_c^2 - K_a^2)FB + (K_a^2 - K_b^2)FC$$

$$\text{DENOM} = (K_b - K_c)FA + (K_c - K_a)FB + (K_a - K_b)FC$$

$$\text{XSTAR} = \frac{1}{2}\,\frac{\text{ANUM}}{\text{DENOM}}$$

15. Calculate F(XSTAR) and denote it by FSTAR.

16. Go to step 12.

Thus the variable K has been optimized. The next step is to optimize T_1 using the new K value, XSTAR and the original T_2. Then T_2 is optimized and the cycle is repeated, as explained earlier. At the end of each cycle the values of K, T_1, and T_2 are subtracted from those of the previous cycle. If these differences between subsequent cycles are within a prescribed accuracy, YACCUR, the procedure is terminated. At this point, then, we will have the values of K, T_1, and T_2. Thus the second-order process model is defined, and the entire modeling procedure is complete.

Modeling Program Development and Testing. A Fortran program, called MODEL, has been written to fit process input-output data to a second-order process model. This program is listed in Appendix E. The inputs to the program are the process input-output data, the number of data points, the sampling period (T), and initial "guesses" for K, T_1, and T_2. The outputs of the program are values of K, T_1, and T_2 at the end of each cycle and, of course, the optimum process parameters. The general flow chart for the program MODEL is displayed in Figure 13.15.

There are two accuracy specifications utilized in the body of the program. The first one, XACCUR, is the Powell search termination criterion used in comparing subsequent values of XSTAR. The other specification, YACCUR, is the minimum difference requirement for subsequent cycle values of K, T_1, and T_2. If YACCUR is exceeded in any one of the three variable comparisons, another cycle is initiated. The value of YACCUR, 0.001, is five times smaller than

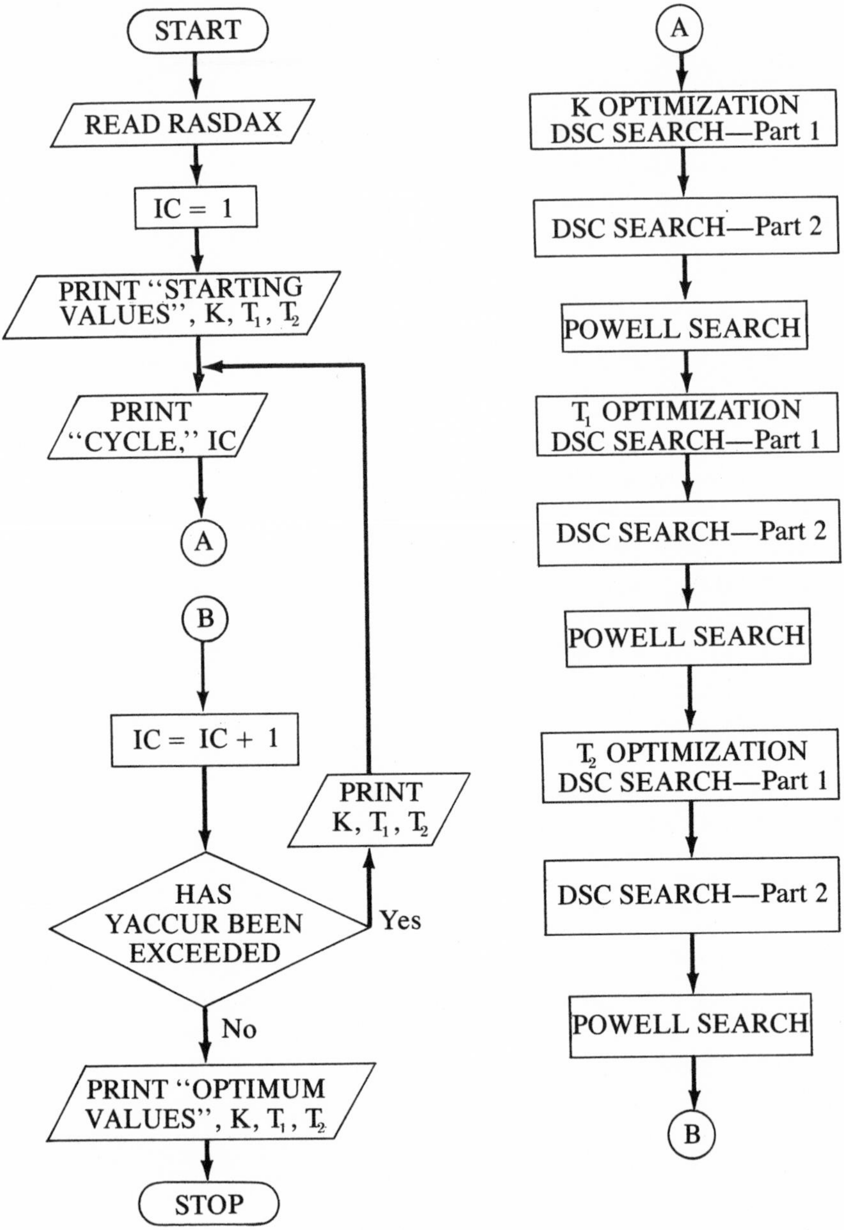

START

READ RASDAX

IC = 1

PRINT "STARTING VALUES", K, T_1, T_2

PRINT "CYCLE," IC

A

B

IC = IC + 1

HAS YACCUR BEEN EXCEEDED

Yes

PRINT K, T_1, T_2

No

PRINT "OPTIMUM VALUES", K, T_1, T_2

STOP

A

K OPTIMIZATION DSC SEARCH—Part 1

DSC SEARCH—Part 2

POWELL SEARCH

T_1 OPTIMIZATION DSC SEARCH—Part 1

DSC SEARCH—Part 2

POWELL SEARCH

T_2 OPTIMIZATION DSC SEARCH—Part 1

DSC SEARCH—Part 2

POWELL SEARCH

B

Nomenclature:

YACCUR = Termination criteria; minimum allowable difference of regression variables of subsequent cycles

IC = Cycle Number

Figure 13.15
MODEL—General Flow Chart

the value chosen for XACCUR, 0.005. The ratio of 1/5 was selected so that unnecessary computer time would not be spent in an attempt to optimize any one variable when the other two variables were not at their respective optimum values.

To be used in conjunction with MODEL, five process data files are presented for illustration of the method. These files are named RASDAX, where X is a number from 1 to 5. These files are listed in Appendix E1. The format for all of the files is the same. The first three entries are the initial guesses of K, T_1, and T_2 in that order. The fourth entry is the number of sampling periods for which input-output data pairs are recorded. Next, the process input data are listed. Finally, the process output data are listed. The sampling period for all of the sets is 1 min. The criteria for the selection of the sampling time is that a sufficient number of sample values would result so as to obtain a good representation of the continuous input-output data. It is important to note that both the input and output data begin when time equals zero, the point at which the input disturbance is first introduced.

The first data set, named RASDA1, was taken from the work done by Patke.[23] This work was based on actual data from a real process obtained by Patke. The input data was a 5-min positive rectangular pulse to the air-top pressure of a steam valve which regulated the steam pressure to a U tube heater in a watertank. There was a steady flow of water in and out of the tank. The output data was the transient response of water temperature in the tank. From Patke's work the dimensionless gain of the process system was calculated to be 0.308:

The second data set, named RASDA2, was exactly the same as RASDA1, except the input pulse was in the negative direction. Because of the nonlinearities in the system, the process gain was calculated to be 0.441.

The third data set, named RASDA3, was taken from an experiment performed by G. R. Ranade.[24] He used the same equipment that Patke used, but with a step-testing procedure. He introduced a positive step change in air-top pressure to the steam valve and graphically fitted the output temperature response to a first-order plus dead-time model. His values for process gain and time constant were 6.67°F/psi (corresponding to 0.3 dimensionless gain) and 7.17 min, respectively.

To further test the accuracy of MODEL, two additional data sets, RASDA4 and RASDA5, were generated. For both of these sets, an exact process model was chosen arbitrarily. For RASDA4 the transfer function of the chosen model was:

$$G(s) = \frac{X(s)}{U(s)} = \frac{5.5432}{(5.111s + 1)\,(2.222s + 1)} \tag{13.48}$$

The input, U, was a step change of magnitude 1.5. For RASDA5, the input was a rectangular pulse of magnitude 1.5 and duration 5 min. In each case, the output function $X(s)$ was solved and inverted into the time domain using standard Laplace transform techniques. Finally, for numerous values of time, the output $X(t)$ was calculated and entered into the data files. Thus two data files from known, exact, process models could be used as input to MODEL to test its convergence accuracy.

To evaluate the convergence of MODEL to the correct process constants, the program was run with the five different data sets to determine if it converged on the known process model parameters. The initial values of K, T_1, and T_2 were chosen arbitrarily. A summary of the first test results is shown in Table 13.1. For the experimental data sets RASDA1, RASDA2, and RASDA3, the maximum deviation between the previously calculated experimental values and the computer program values was less than 7%. For the exact model data sets RASDA4 and RASDA5, the maximum deviation between the actual model values and the calculated values was less than 1.2%. These accuracy levels are quite adequate in process control work.

A second test was conducted in which the starting values of K, T_1, and T_2 were varied so as to observe their effect on the number of cycles and accuracy. It was found that the accuracy of the optimum values was not significantly different, even when the starting values were far from the final values.

In this section we have presented an extremely useful technique for process identification in the time domain. It must be pointed out that the discussion on starting values and levels of convergence are based on limited experience; therefore, the designer should exercise caution when using the material in this section.

The application of this identification procedure in computer control is straightforward. From time to time the control loop is bypassed and a pulse test is conducted. The input and output data are entered into the computer program described earlier. This program would probably reside in "background" memory. The program would output process parameters which can then be used to update the controller tuning constants. Fertik's[25] control parameter charts may be used for determining the optimum tuning constants for digital PI or PID controllers.

Table 13-1
Comparison of Calculated Model Parameters with Actual Model Parameters

Set Name	*Description of Input Disturbance*	*Initial Guesses*	*Actual Parameters (if known)*	*Calculated Model Parameters*
RASDA1	Pulse length: 5 min Pulse height: 1.5 (Patke's work)	$K = 0.5$ $T_1 = 0.5$ $T_2 = 3.0$	$K = 0.308$	$K = 0.313$ $T_1 = 0.960$ $T_2 = 2.781$
RASDA2	Pulse length: 5 min. Pulse height: 1.5 (Patke's work)	$K = 0.5$ $T_1 = 0.5$ $T_2 = 3.0$	$K = 0.441$	$K = 0.469$ $T_1 = 0.957$ $T_2 = 5.394$
RASDA3	Step change height: 1.5 (Ranade's work)	$K = 5.0$ $T_1 = 4.0$ $T_2 = 2.0$	$K = 6.667$ $T = 7.17$ (assumed first order)	$K = 6.672$ $T_1 = 6.873$ $T_2 = 0.303$
RASDA4	Step change height: 1.5 (Exact Model)	$K = 6.0$ $T_1 = 5.0$ $T_2 = 1.0$	$K = 5.5432$ $T_1 = 5.111$ $T_2 = 2.222$	$K = 5.544$ $T_1 = 5.141$ $T_2 = 2.197$
RASDA5	Pulse length: 5 min Pulse height: 1.5 (Exact Model)	$K = 6.0$ $T_1 = 5.0$ $T_2 = 1.0$	$K = 5.5432$ $T_1 = 5.111$ $T_2 = 2.222$	$K = 5.546$ $T_1 = 5.133$ $T_2 = 2.208$

References

1. Deshpande, P. B., Process Identification of Open-loop Unstable Systems, *A. I. Ch. E. J.*, **26**, 2, 1980.
2. Ziegler, J. G., Nichols, N. B., Optimum Settings for Automatic Controllers, *Trans. ASME*, **64**, 11, November 1942, 759.
3. Miller, J. A., et al., A Comparison of Controller Tuning Techniques, Control Engineering, **14**, 2, December 1967, 72.
4. Oldenbourg, R. C., Sartorius, H., *The Dynamics of Automatic Controls*, American Society of Mechanical Engineers, New York, 1948, p. 276.
5. Sten, J. W., Evaluating Second-Order Parameters, *Instrumentation Tech.* **17**, 9, September 1970, 39–41.
6. Smith, O. J. M., A Controller to Overcome Dead-time, *ISA J.*, **6**, 2, February 1959, 28–33.
7. Cox, J. B., et al., A Practical Spectrum of DDC Chemical-Process Control Algorithms, *ISA J.*, **13**, 10, October 1966, 65–72.
8. Smith, C. L., Murrill, P. W., *ISA J.*, **13**, 9, 1966, 48.
9. Meyer, J. R., et al., Simplifying Process Response Approximations, *Instruments and Control Systems*, **40**, 12, December 1967, 76–79.
10. Csaki, F., Kis, P., *Period. Polytech.*, **13**, 1969, 73.
11. Naslin, P., Miossec, C., Automatisme, **16**, 1971, 513.
12. Sundaresan, K. R., et al., Evaluating Parameters from Process Transients, *Ind. Eng. Chem. Process Des. Dev.*, **17**, 3, 1978, 237–241.
13. Lees, F. P., *Chemical Engineering Science,* **26**, 1971, 1179.
14. Gibilaro, L. G., Waldram, S. P., *Chemical Engineering Science*, **4**, 1972, 197.

15a. Hougen, J. O., *Experiences and Experiments with Process Dynamics*, CEP Monograph Series, Vol. 60, No. 4, 1964.

15b. Clements, W. C., Schnelle, K. B., Ind. Eng. Chem. Process Des. Dev. 2, 1963, 94.

16. Dynamic Response Testing of Process Control Instrumentation, ISA-S26 Standard, Instrument Society of America, 400 Stanwix Street, Pittsburgh, Pa., October 1968.
17. Luyben, W. L., *Process Modeling, Simulation, and Control for Chemical Engineers*, McGraw-Hill, New York, 1973.
18. Messa, C. J., et al., Criteria for Determining the Computational Error in Numerically Calculated Fourier Integral Transforms, *I & EC Fundamentals*, **8**, 4, 1969.
19. Murrill, P. W., Pulse Testing Methods, *Chemical Engineering*, February 24, 1969.
20. Schaefer, R. A., On-Line Process Identification and Control, M. Eng. Thesis, University of Louisville, 1978.
21. Moore, C. F., et al., Formulating the Nonlinear, Least-Square Regres-

sion Model for Fast On-line Analysis, Paper No. 22D, A. I. Ch. E. National Meeting, New Orleans, La., 1969.

22. Himmelblau, D. M., *Applied Nonlinear Programming*, McGraw-Hill, New York, 1972.
23. Patke, N. G., Deshpande, P. B., Pulse Testing for Process Identification and Control, Paper No. NG77, ISA Conference, Niagara Falls, NY, November 1977.
24. Ranade, G. R., Seminar Report in Advanced Control Topics, unpublished report, Department of Chemical Engineering, University of Louisville, 1977.
25. Fertik, H. A., Tuning Controllers for Noisy Processes, *ISA Trans.* **14**, 4, 1975.

SHORT BIBLIOGRAPHY ON MODELING OF CHEMICAL ENGINEERING SYSTEMS

1. Luyben, W. L., *Modeling, Simulation, and Control for Chemical Engineers*, McGraw-Hill, New York, 1973.
2. Franks, R. G. E., *Modeling and Simulation for Chemical Engineers*, Wiley, New York, 1972.

Authors' Update on Section 13.3, Time Domain Process Identification.

The program listed in this section has been extended to include determination of dead-time. It has also been applied to include determination of optimal tuning constants of digital P, PI, or PID controllers. The interested reader may consult the following references for details.

1. Brantley, R. O., Schaefer, R. A., Deshpande, P. B., On-Line Process Identification, I & E C Proc. Des. Dev. *21*, 2, April 1982.
2. Stark, P. A., Ralston, D. L., Deshpande, P. B., A Comparative Assessment of Two Recent On-Line Identification Process Identification Techniques, American Control Conference, Arlington, Va. June 1982.
3. Watson, K. R., Deshpande, P. B., Optimal Tuning Constants for Computer Control Algorithms, 75th A.I.Ch.E. Meeting, Los Angeles, CA. November 14–18, 1982.

CHAPTER 14

Algorithms for Processes with Dead Time

In Chapter 1 we examined the detrimental effects of dead time on the response of first-order systems. The presence of dead time necessitates lowering of controller gain to maintain stability. We observed that when the apparent dead time θ_d exceeded the dominant time constant of the system τ, the peak offsets, following a step change in load, could approach those of the uncontrolled situation, even with best PID controller tuning. Under these conditions, the settling time approaches 9 θ_d. Since many chemical engineering processes exhibit apparent dead-time characteristics and since dead time is detrimental to control, there is considerable incentive to develop advanced control algorithms that can compensate for such time delays. In this chapter we shall describe the design and applications of three advanced control algorithms which are very effective in improving the control of such processes having large dead times. They are (1) Smith predictor algorithm, (2) analytical predictor algorithm, and (3) general-purpose algorithm.

14.1. Smith Predictor Algorithm

This is perhaps the best known of the dead-time compensation techniques in use today. It was developed by the late O. J. M. Smith[1] in 1957. The technique is a model-based approach to better control of systems with long dead times and has come to be known as the Smith predictor. This was one of the advanced control strategies developed years ago which was shelved because of a lack of practical hardware to implement it (it requires a pure dead-time element having a delay equal to that of the process—difficult if not impossible with analog hardware). When digital computers for online control applications appeared on the market, the Smith predictor was "rediscovered" and tried in many applications. Let us develop the Smith predictor algorithm for a process that can be represented by a first-order lag plus dead-time model. The block diagram of a conventional control system for this process is shown in Figure 14.1.

As shown in Figure 14.1, the process is conceptually split into a pure lag and a pure dead time. If the ficticious variable B could be measured somehow, we could connect it to the controller, as shown in Figure 14.2. This would move the dead time outside the loop. The controlled variable C would repeat whatever B did after a delay of θ_d. Since there is no delay in the feedback signal B, the response of the system would be greatly improved. This scheme, of course, cannot be implemented, because B is an unmeasurable (fictitious) signal.

Now, suppose we develop a model of the process and apply the manipulated variable M to the model as shown in Figure 14.3. If the model were perfect and $L = 0$ (i.e., no load disturbances are present), then c will equal c_m and $E_m = c - c_m = 0$. The arrangement shown in Figure 14.3 reveals that although the fictitious process

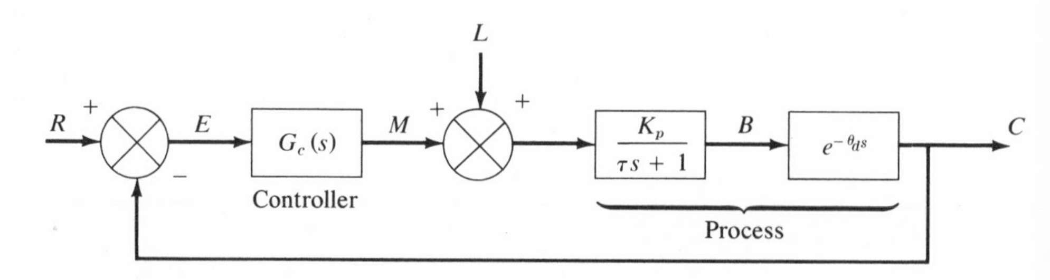

Figure 14.1
Conventional Feedback Loop Having Deadtime

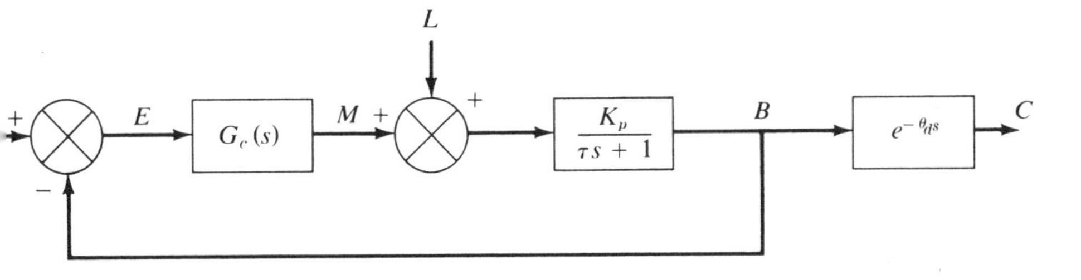

Figure 14.2
Desired Configuration of the Feedback Loop

variable B is unavailable, we can get at B_M in the model. The B_M will be equal to B unless modeling errors or load upsets are present. We use B_M as the feedback signal, as shown in Figure 14.4. The difference, $C - C_M$, is the error E_m, which arises because of modeling errors or load upsets.

The arrangement shown in Figure 14.4 will control the model well but perhaps not the process, if load upsets occur or if our model is inaccurate. To compensate for these errors, a second feedback loop is implemented using E_m, as shown in Figure 14.5. This is the Smith predictor control strategy. The $G_c(s)$ is a conventional PI or PID controller, which can be tuned much more tightly because of the elimination of dead time from the loop.

Some literature present the Smith predictor strategy, as shown in Figure 14.6. Some elementary manipulations will show that the two systems of Figures 14.5 and 14.6 are equivalent.

The closed-loop transfer function of the system shown in Figure 14.5 for $L = 0$ is

$$\frac{C(s)}{R(s)} = \frac{G_c(s)\,G_p'(s)\,e^{-\theta_d s}}{1 + G_c(s)G_M'(s) - G_c(s)G_M'(s)e^{-\theta_M s} + G_c(s)\,G_p'(s)\,e^{-\theta_d s}} \tag{14.1}$$

If $G_M'(s) = G_p'(s)$ and $\theta_M = \theta_d$, Equation (14.1) reduces to

$$\frac{C(s)}{R(s)} = \frac{G_c(s)\,G_p'(s)}{1 + G_c(s)\,G_p'(s)}\,e^{-\theta_d s} \tag{14.2}$$

Equation 14.2 is the closed-loop transfer function of the desired configuration, based on the fictitious signal B, which was shown earlier in Figure 14.2. To be assured of success, the model parameters must be known to a high degree of accuracy.

As an example, let us consider the design of Smith predictor

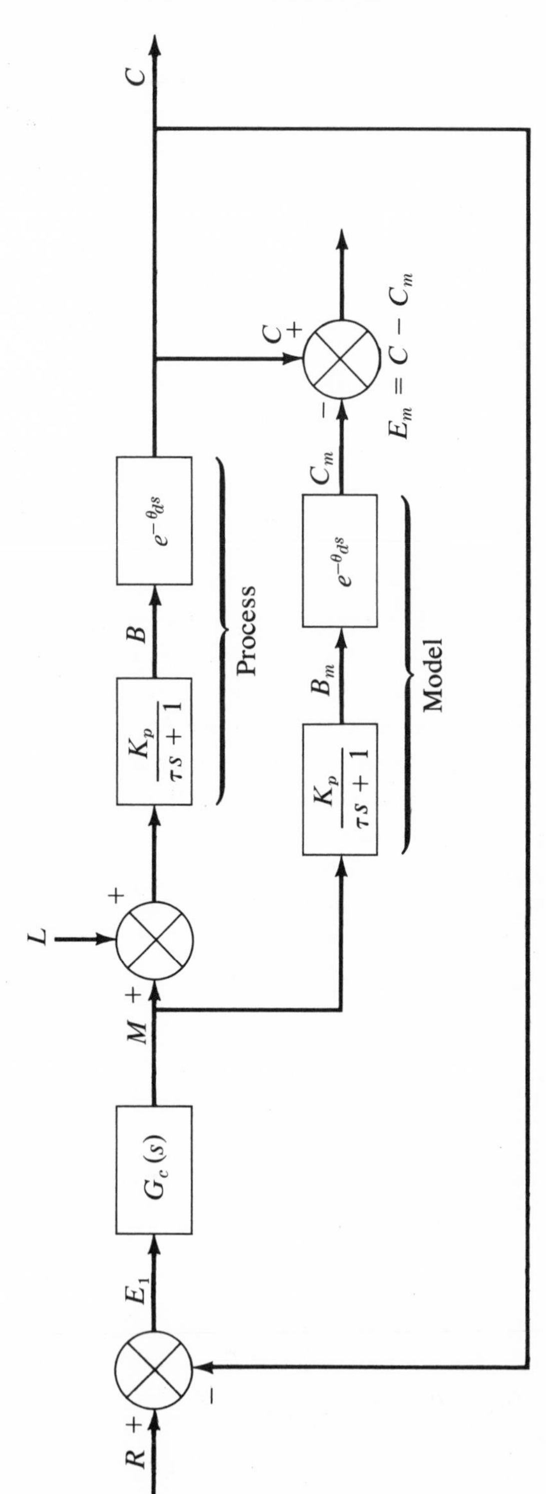

Figure 14.3
Feedback Arrangement Incorporating a Process Model

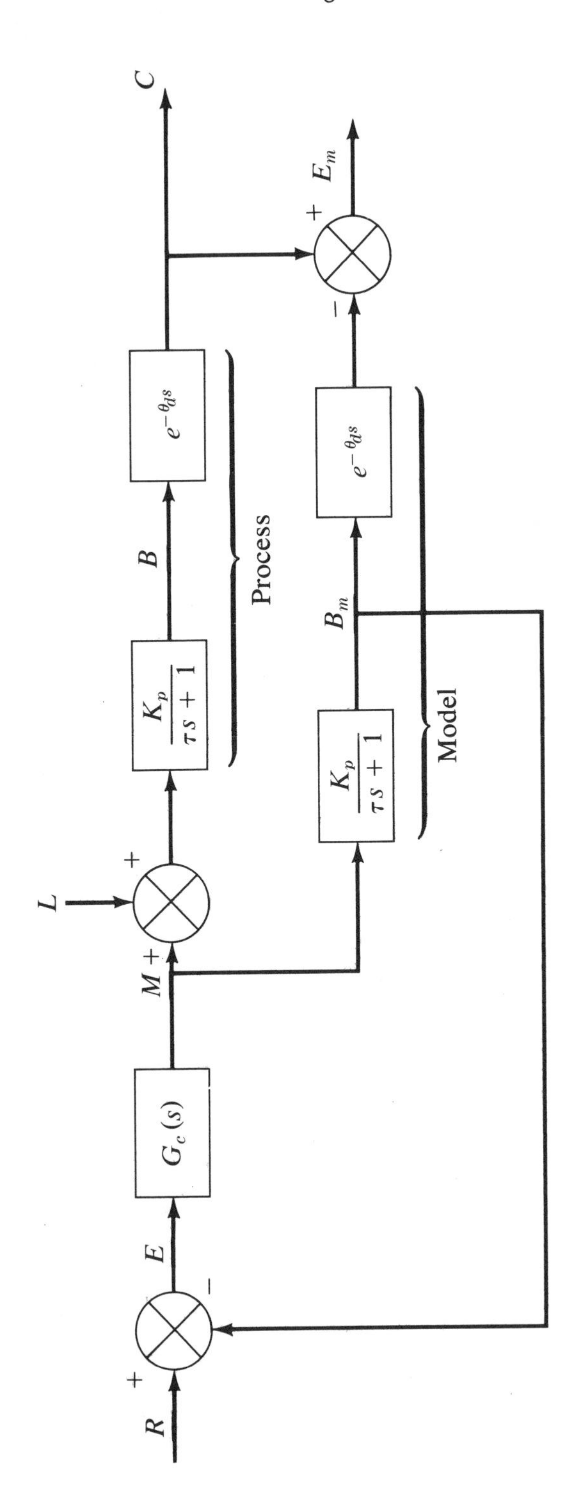

Figure 14.4
Preliminary Smith Predictor Scheme

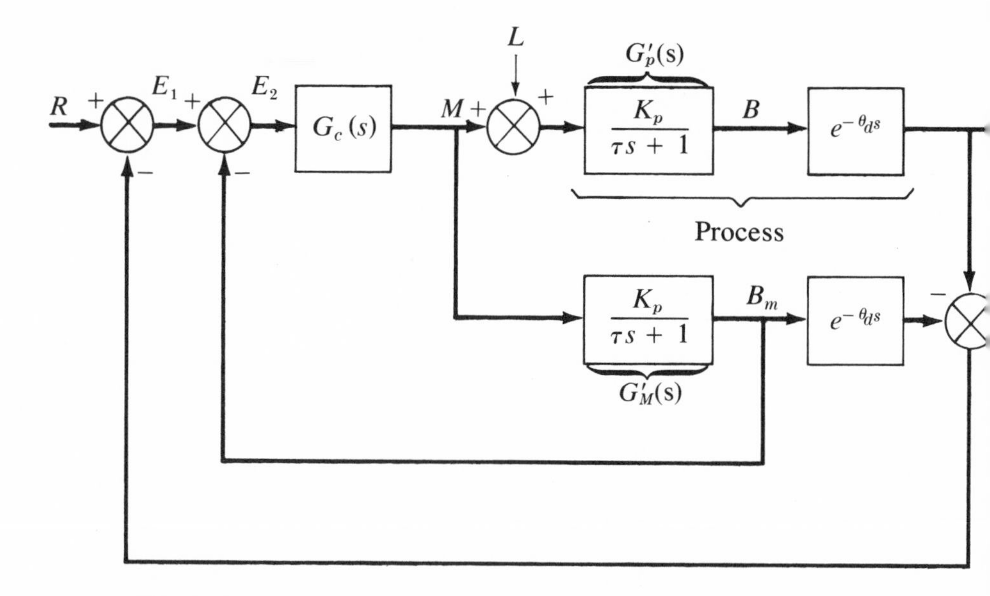

Figure 14.5
Final Smith Predictor Control System

algorithm for computer control of a first-order process having dead time whose transfer function is given by

$$G_p(s) = \frac{K_p e^{-\theta_d s}}{\tau s + 1} \tag{14.3}$$

where

K_p = process gain

θ_d = dead time

τ = process time constant

The Smith predictor scheme is redrawn for this computer control application as shown in Figure 14.6. The implementation of this scheme would require that we develop equations for B_M, C_M, and u_K.

From Figure 14.6 observe that the model output C_M is related to the input u by

$$\frac{C_M(Z)}{U(Z)} = Z\{G_{h_0}(s)\, G_p(s)\} \tag{14.4}$$

where

$C_M(Z) = Z$ transform of model output

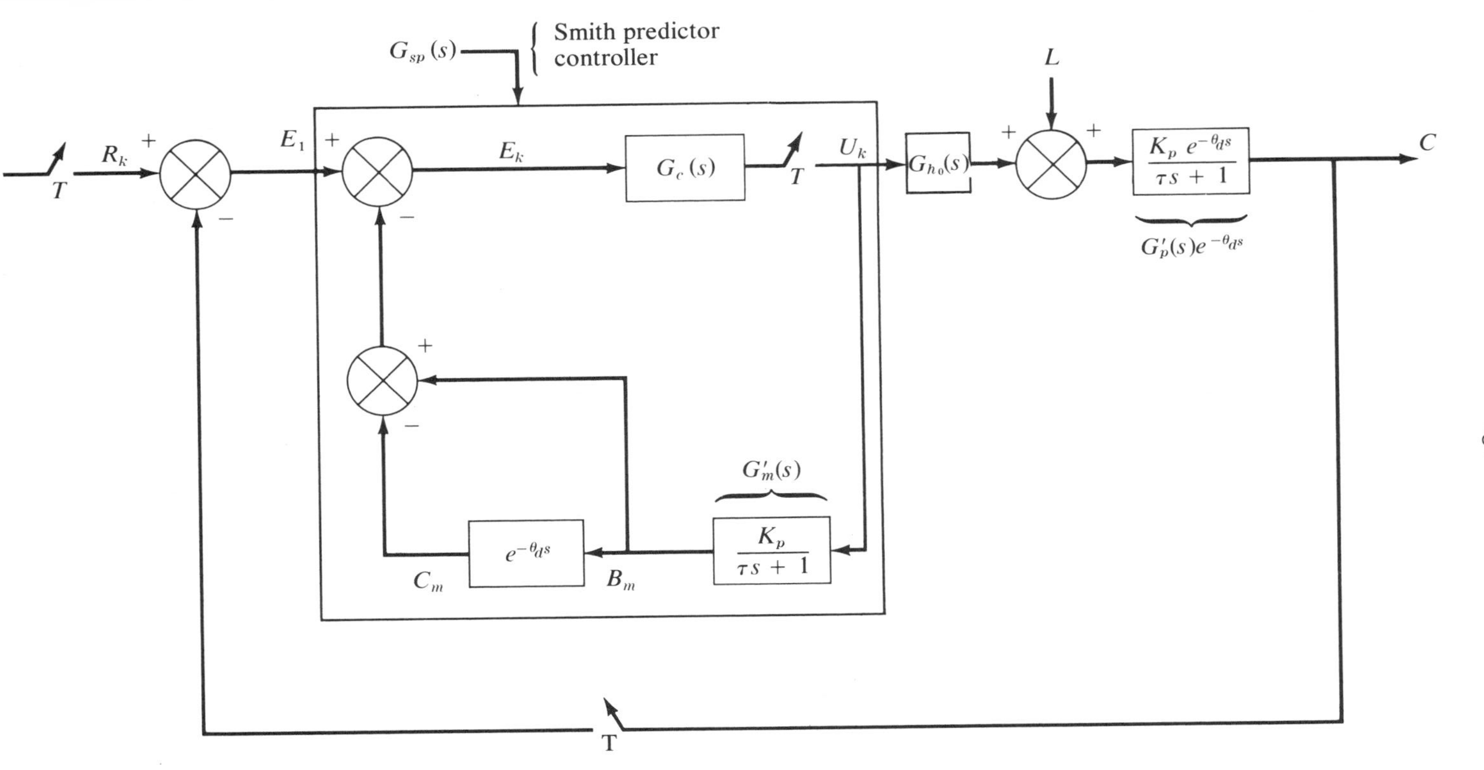

Figure 14.6
Smith Predictor System Redrawn for Computer Control Application (If Samplers Are Removed, This Block Diagram Reduces to the One in Figure 14.5)

$U(Z)$ = Z transform of input

$G_{h_0}(s)$ = transfer function of zero-order hold, $\dfrac{1 - e^{-sT}}{s}$

$G_p(s)$ = transfer function of process, $\dfrac{K_p e^{-\theta_d s}}{\tau s + 1}$

Substituting the transfer functions in Equation (14.4) we get

$$\frac{C_M(Z)}{U(Z)} = Z\left\{\frac{1 - e^{-sT}}{s} \cdot \frac{K_p e^{-\theta_d s}}{\tau s + 1}\right\} \tag{14.5}$$

If we denote the integral number of sampling periods in dead time as N, then the following equality holds:

$$\theta_d = (N + \beta)T \tag{14.6}$$

where β = a fraction between 0 and 1. With this expression for θ_d, Equation (14.5) can be written as

$$\frac{C_M(Z)}{U(Z)} = Z\left\{\frac{1 - e^{-sT}}{s} \cdot \frac{K_p\, e^{-(N + \beta)T/s}}{\tau s + 1}\right\} \tag{14.7}$$

Taking Z transform of Equation (14.7) gives

$$\frac{C_M(Z)}{U(Z)} = K_p\, Z^{-N}(1 - Z^{-1})\, Z^{-1}\left(\frac{1}{1 - Z^{-1}} - \frac{e^{-(1-\beta)T/\tau}}{1 - e^{-T/\tau}Z^{-1}}\right) \tag{14.8}$$

cross multiplying and inverting gives an equation for the model output containing time delay. Thus

$$C_{M,k} = A_2 C_{M,(k-1)} + A_2 k_p \left(\frac{1}{A_3} - 1\right) u_{k-(N+2)} + K_p\left(1 - \frac{A_2}{A_3}\right) u_{k-(N+1)} \tag{14.9}$$

where

$$A_2 = e^{-T/\tau}$$
$$A_3 = e^{-\beta T/\tau}$$

Now, we need to get the model output containing no dead time. For this case,

$$\frac{B_M(Z)}{U(Z)} = Z\left\{\frac{1 - e^{-sT}}{s} \cdot \frac{K_p}{\tau s + 1}\right\}$$
$$= K_p(1 - Z^{-1})\left[\frac{1}{1 - Z^{-1}} - \frac{1}{1 - e^{-T/\tau}Z^{-1}}\right] \tag{14.10}$$

cross multiplying and inverting gives

$$B_{M,k} = K_p(1 - A_2)u_{k-1} + A_2 B_{M,k-1} \tag{14.11}$$

From Figure 14.6 the error is computed from the equation

$$E_k = R_k - C_k - (B_{M,k} - C_{M,k}) \tag{14.12}$$

Note that if the model is perfect

$$C_k = C_{M,k} \tag{14.13}$$

and the input to the controller will be based on

$$E_k = R_k - B_{M,k} \tag{14.14}$$

Thus, dead time has been compensated. The digital controller must be a conventional control algorithm (i.e., P, PI, or PID). If the controller is based on Z transforms, it incorporates the effect of dead time, and the Smith Predictor algorithm is not applicable.

Experimental Applications. A few applications of the Smith predictor algorithm have appeared in the literature. Meyer, et al.[2] have listed these references in their paper. They have described an application of the Smith predictor algorithm for automatic control of top product composition in a distillation column. A schematic of the pilot-scale distillation control system is shown in Figure 14.7. The process transfer function relating the top product composition X_D to feed rate F and reflux R_e is

$$X_D(s) = \frac{e^{-60s}}{1002s + 1} R_e(s) + \frac{0.167\, e^{-486s}}{895s + 1} F(s) \tag{14.15}$$

The delay and time constants are given in seconds. The parameters of the digital PI controller were tuned according to the relations[3]

$$K_c = \frac{0.984}{K_p}\left(\frac{T}{\tau}\right)^{-0.986} \tag{14.16}$$

$$\tau_I = 1.644\tau\left(\frac{T}{\tau}\right)^{0.707}$$

where T is the sampling period. The authors found that the controlled variable exhibited oscillatory behavior, and hence a further trial-and-error tuning procedure was adopted. Table 14.1 shows the controller constants suggested by Equation (14.16) and those adopted in the experimental work.

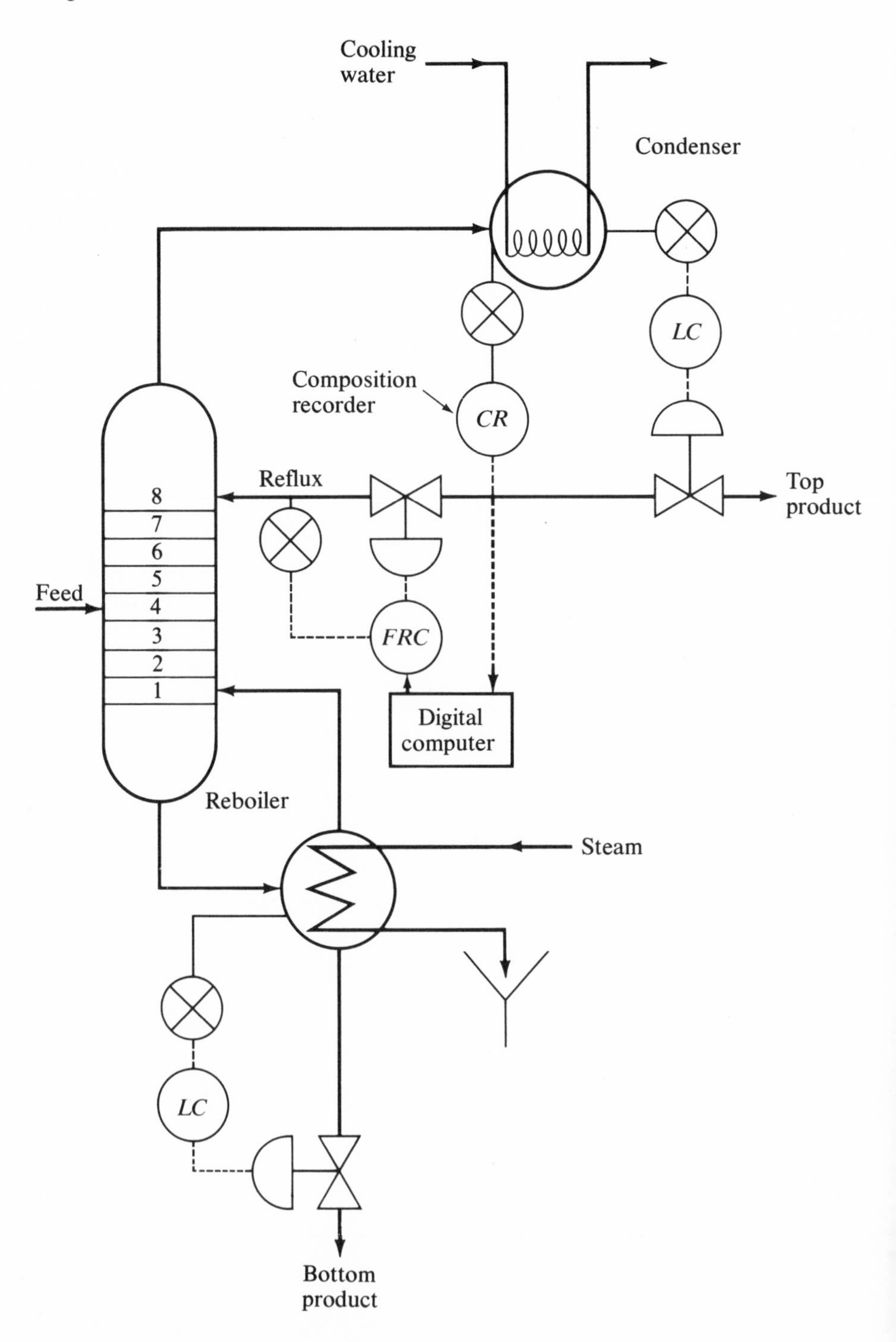

Figure 14.7
Schematic Diagram of the Distillation Column

Table 14.1.
Comparison of Controller Settings of Meyer et al.

	K_c, *grams/s/%*	τ_I, *sec*
Calculated according to Equation 14.16	15.8	225
Experimentally tuned	10.0	250

The differences between calculated and experimental tuning constants are indicative of modeling errors.

The experimental responses of the system to a step change in set point and load are sketched in Figures 14.8 and 14.9. These figures

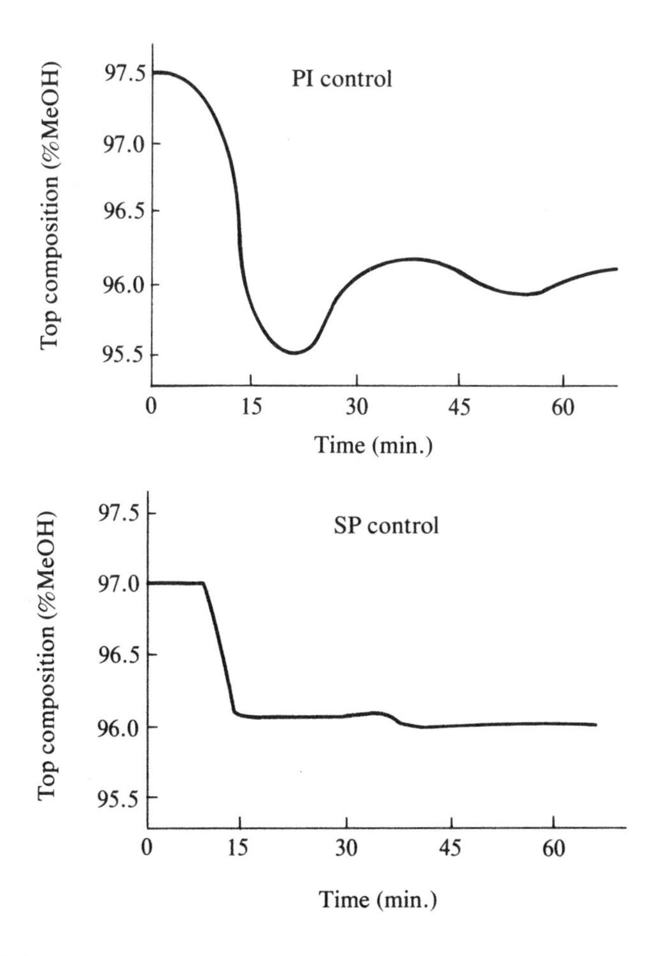

Figure 14.8
Experimental Response Using PI and SP Control System to 1% Decrease in Composition Set Point

clearly show the superior performance of the Smith predictor algorithm in comparison with the digital PI controller without dead-time compensation.

14.2 Analytical Predictor Algorithm[3]

In the Smith predictor algorithm the process model was used to compensate for the dead time prior to sending the signal to the controller. An alternate approach would be to use a process model to "predict" the future value of the controlled variable and use the predicted value as the input to the controller. This is the basic notion behind the analytical predictor (AP) algorithm originally proposed by Moore. A block diagram illustrating the analytical predictor concept is shown in Figure 14.10. The analytical predictor predicts the value of the controlled variable T' time units in future from current inputs where T' is the sum of the system dead time plus one-half of the sampling period, that is,

$$T' = \theta_d + \frac{1}{2} T$$

or

$$T' = (N + \beta) T + 0.5 T \tag{14.17}$$

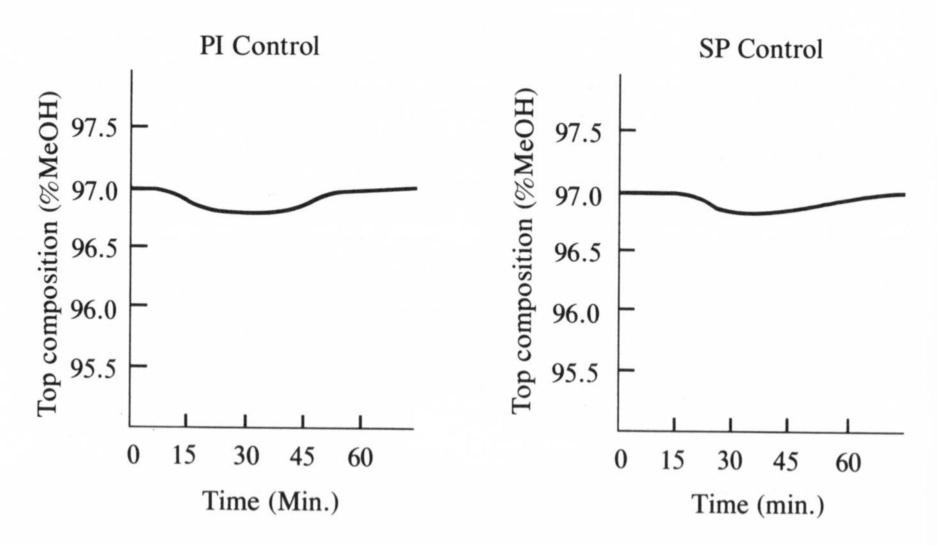

Figure 14.9
Experimental Response Using PI and SP Control Scheme to 22% Decrease in Feed Flow Rate

$$\frac{T'}{T} = N + \beta + 0.5$$

Control degradation of sampled-data systems occurs because of sampling. In many instances the dynamic effect of sampling is similar to that of pure dead time equal to one-half the sampling period. It is for this reason the analytical predictor predicts the value of the controlled variable over T' units of time which includes the process dead time and the effect of sampling.

Consider the design of the analytical predictor algorithm for a first-order process with dead time. The differential equation representing this class of models is

$$\tau \frac{dC}{dt} + C(t) = K_p\, u(t - \theta_d) \tag{14.18}$$

where

τ = process time constant

K_p = process gain

θ_d = total dead time (dead time in process + dead time in measurement line)

The analytical solution of Equation (14.18) gives, as shown by Moore, the following expression for the predicted output

$$\hat{C}_{k+T'} = A_1 \left\{A_3 A_2^N\ C_k + A_2^N\ K_p (1 - A_3)\, u_{k-(N+1)} + K_p (1 - A_2) \sum_{i=1}^{N} A_2^{i-1}\, u_{k-i}\right\} + K_p (1 - A_1)\, u_k\,. \tag{14.19}$$

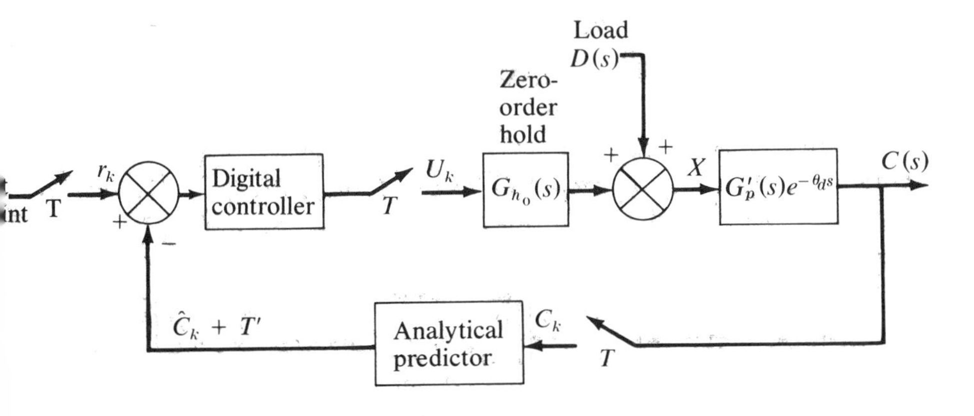

Figure 14.10
Analytical Predictor (AP) Control System

where A_2 and A_3 have been defined just below Equation (14.9) and $A_1 = e^{-T/2\tau}$. This predicted output is used to calculate the controller output according to

$$u_k = K_c\,(r_K - \hat{C}_{K+T'}) \tag{14.20}$$

Equation (14.20) represents a proportional control algorithm and, therefore, an offset

$$\lim_{t\to\infty} R(t) - C(t) = K_c\,K_p/(K_cK_p + 1)$$

will develop for our system whenever a step change in set point is made. This offset can be eliminated by "calibrating" the set point according to

$$\rho_K = \frac{K_c\,K_p + 1}{K_cK_p}\,r_K \tag{14.21}$$

The control algorithm then becomes

$$u_K = K_C\,(\rho_K - \hat{C}_{K+T'}) \tag{14.22}$$

Moore has also developed algorithms to compensate the analytical predictor algorithm for measurable and unmeasurable disturbances. If the disturbance d_K can be measured, the controller calculaions should be done according to

$$u_K = K_c\,(\rho_K - \hat{C}_{K+T'}) - d_K \tag{14.23}$$

Note from Figure 14.10 that the input to the process is

$$X_K = u_K + d_K \tag{14.24}$$

In the presence of disturbances, the input to the process is X_k. However, if the control calculations are based as shown in Equation (14.23), the effect d_K will be cancelled.

In the more usual situations where disturbances cannot be measured, Moore suggests the following procedure.

If a disturbance is present, it affects the controlled variable C_K. By comparing the predicted value of the controlled variable with C_K, an estimate of the disturbance can be obtained as

$$\hat{d}_K = \hat{d}_{K-1} + \omega\,T\,(C_K - \hat{C}_K) \tag{14.25}$$

The constant ω can be used as a tuning parameter to be chosen on line or through simulation. At the instant when the loop is switched to computer control, d_o is assumed to be zero, and d_1 is calculated as

$$\hat{d}_1 = \omega\,T\,(C_K - \hat{C}_K) \tag{14.26}$$

Thereafter, d_K is computed from Equation (14.25). The predicted value $\hat{C}_K$ is given by

$$\hat{C}_k = A_2 C_{k-1} + K_p \left(1 - \frac{A_2}{A_3}\right) (u_{k-(N+1)} + \hat{d}_{k-1}) + \qquad (14.27)$$

$$K_p \left(\frac{A_2}{A_3}\right) (1 - A_3)\,(u_{k-(N+2)} + \hat{d}_{k-1})$$

The estimated disturbance $\hat{d}_k$ can be incorporated into Equation (14.19) to provide a better prediction of the future output. Thus

$$\hat{C}_{k+T'} = A_1 A_3 A_2^N\, C_k + A_2^N\, K_p (1 - A_3)\,(u_{k-(N+1)} + \hat{d}_k) +$$

$$K_p (1 - A_2) \sum_{i=1}^{N} A_2^{i-1}\,(u_{k-i} + d_k) + K_p (1 - A_1)\,(u_k + \hat{d}_k) \qquad (14.28)$$

The controller equation for this case is

$$u_K = K_c (\rho_K - \hat{C}_{K+T'}) - \hat{d}_K \qquad (14.29)$$

Substituting from Equation (14.21) for ρ_K and from Equation (14.28) for $\hat{C}_{K+T'}$ we get

$$u_k = K_c \left\{\frac{K_c K_p + 1}{K_c K_p} r_k - A_1 A_3 A_2^N C_k - A_1 A_2^N K_p (1 - A_3)(u_{k-(N+1)} - \hat{d}_k)\right.$$

$$\left. - A_1 K_p (1 - A_2) \sum_{i=1}^{N} A_2^{i-1}\,(u_{K-i} + \hat{d}_k) - K_p (1 - A_1)\,(u_k - \hat{d}_k)\right\}$$

$$- \hat{d}_k. \qquad (14.30)$$

Simplifying this equation gives

$$u_k = C_1 r_k - C_2 C_k - C_3 (u_{k-(N+1)} + \hat{d}_k) -$$

$$C_4 \sum_{i=1}^{N} A_2^{i-1}\,(u_{k-i} + \hat{d}_k) - C_5 \hat{d}_k - \hat{d}_k \qquad (14.31)$$

where

$$C_1 = \frac{K_c K_p + 1}{K_p\left[1 + K_c K_p (1 - A_1)\right]}; \qquad C_2 = \frac{K_c A_1 A_3 A_2^N}{1 + K_c K_p (1 - A_1)}$$

$$C_3 = \frac{K_c K_p A_1 A_2^N (1 - A_3)}{1 + K_c K_p (1 - A_1)}; \qquad C_4 = \frac{K_c K_p A_1 (1 - A_2)}{1 + K_c K_p (1 - A_1)}$$

and

$$C_5 = \frac{K_c K_p (1 - A_1)}{1 + K_c K_p (1 - A_1)}$$

In this final form of the algorithm there are terms involving current measurements, past controller outputs, set point, and disturbance. The term $\hat{d}_K$ is computed via Equations (14.25) and (14.27). Note that in Equation (14.31) the presence of $\hat{d}_K$ provides a form of integral action, since the value of the estimated disturbance will continue to change with time until the measured value of the controlled variable C_K matches the predicted value $\hat{C}_K$.

Moore suggests the deadbeat tuning procedure for K_c and ω. The equations are

$$K_c = \frac{A_1}{K_p (1 - A_1)} \tag{14.32}$$

$$\omega = \frac{1}{T K_p (1 - A_2)}$$

Experimental Applications

Meyer et al.[2] have surveyed the experimental applications of the Analytical predictor algorithm. They have applied the analytical predictor technique to the automatic control of a pilot-scale distillation column. The transfer function of the process is given in Equation (14.15). In this case as well the authors found that further on-line tuning of K_c and ω was necessary. Table 14.2 shows the calculated and experimentally determined constants.

Table 14.2
Comparison of Calculated and Experimentally Determined Controller Constants

	K_c, *grams/sec/%*	ω, *grams/%/sec*^{2}
Calculated according to Moore[3]	32.9	0.29
Experimental values	10	0.06

Again, the differences in tuning constants are indicative of modeling errors. The response of the system to a step change in set point and load is shown in Figures 14.11 and 14.12. These Figures indicate that the AP control scheme performs better than the PI controller.

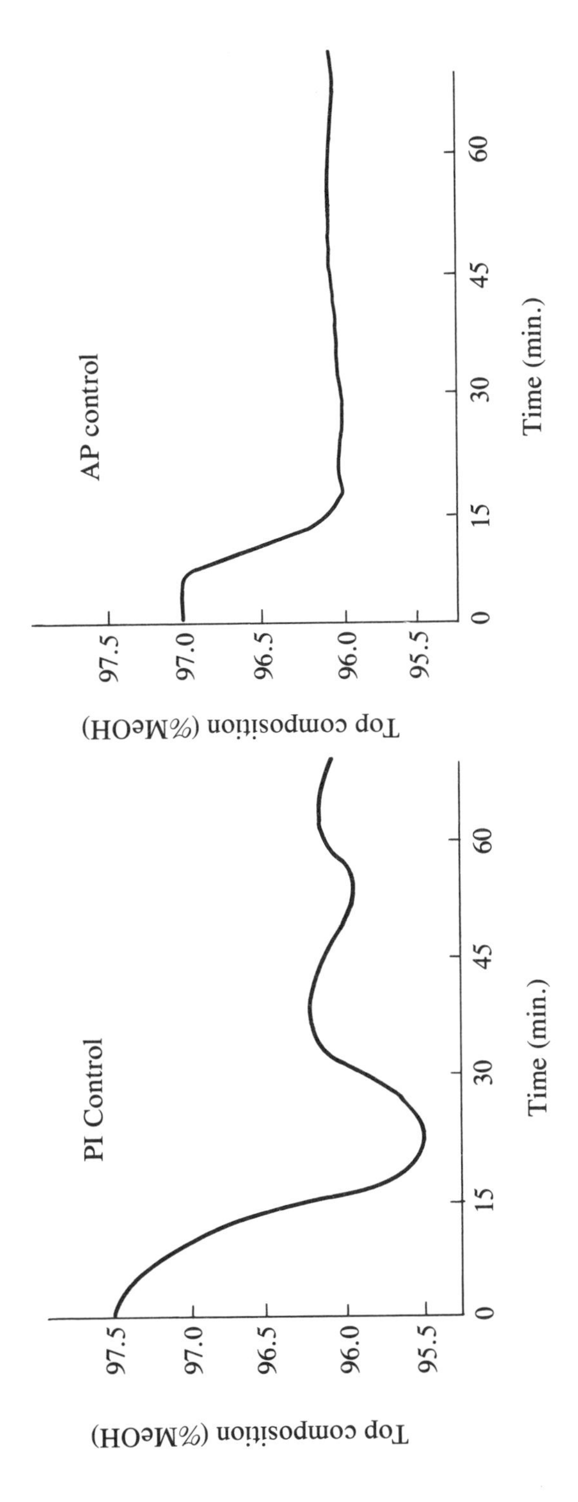

Figure 14.11
Experimental Response using PI and AP Control System to 1% Decrease in Composition Set Point

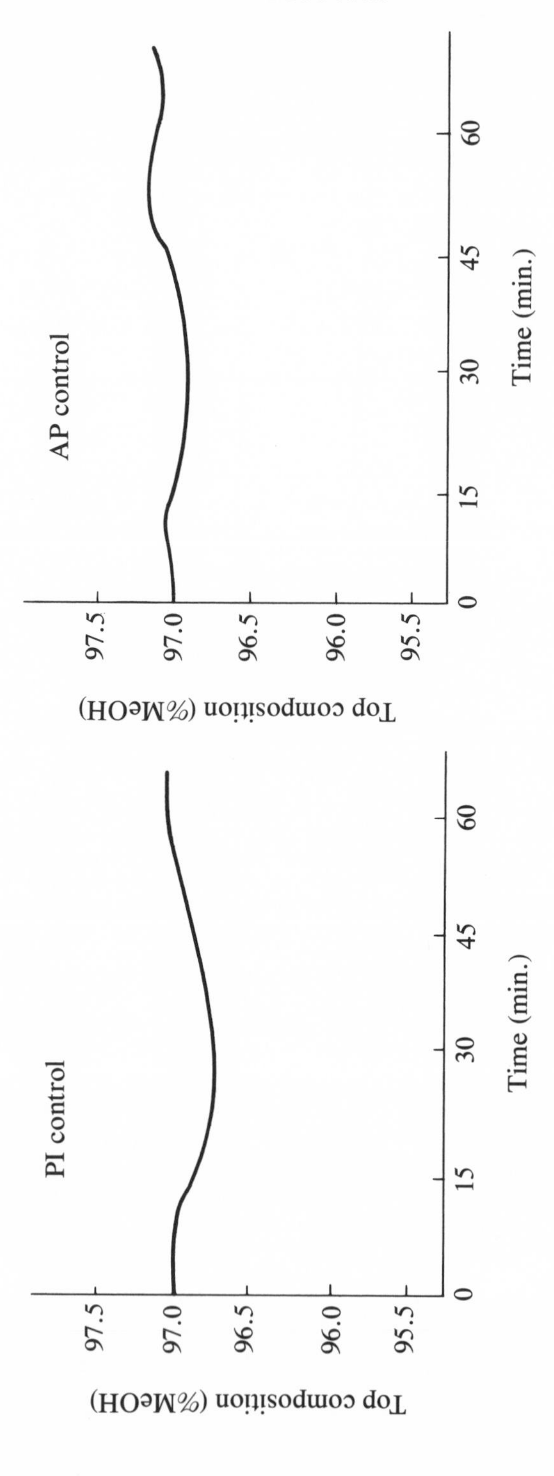

Figure 14.12
Experimental Response using PI and AP Control Scheme to 22% Decrease in Feed Flow Rate

14.3. Algorithm of Gautam and Mutharasan[4]

Computer-control loops frequently employ the digital equivalent of the conventional controller. The loops are often tuned for set point changes. The response of the system is not as good for load changes. Therefore, it is highly desirable to have an algorithm that performs equally well for set point and load changes.

Gautam and Mutharasan[4] have presented a general-purpose control algorithm for a first-order process with dead-time that appears to perform well in the presence of set point *and* load changes.

We shall present the algorithm for a general first-order lag plus dead-time model. Numerical values of the dead time, gain, and time constant are based on process identification studies on an industrial system. From Figure 14.13 we note that in the absence of load disturbances L equals zero and M equals X (or M^* equals X^*). The relationship between Y^* and X^* (or M^*) in the Z domain is

$$\frac{Y(Z)}{X(Z)} = Z\{G_{h_0}(s)G_p(s)\} \tag{14.33}$$

where

$$G_{h_0}(s) = \text{transfer function of the zero-order hold,}$$
$$= \frac{1 - e^{-Ts}}{s}$$

$$G_p(s) = \text{process transfer function, } \frac{K_p\, e^{-\theta_d s}}{\tau s + 1}$$

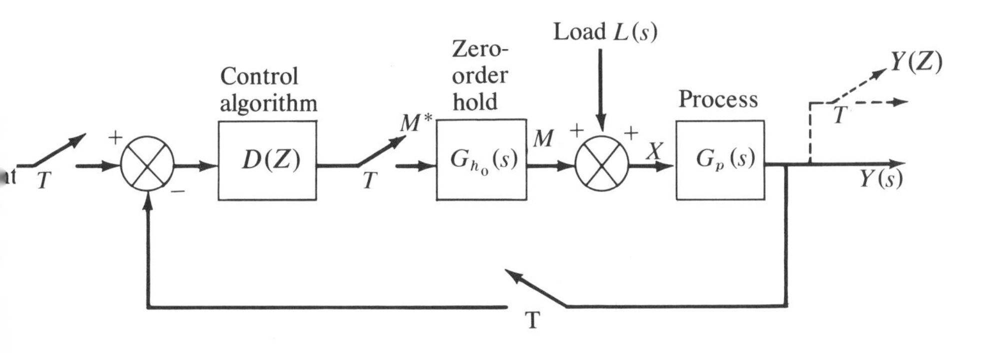

Figure 14.13
Computer-Control System

T = sampling period, 0.5 min (chosen arbitrarily)

θ_d = dead-time, 0.76 min

K_p = process gain, – 1

τ = time constant, 0.4 min

Thus

$$\frac{Y(Z)}{X(Z)} = Z\left\{\frac{1 - e^{-sT}}{s} \cdot \frac{K_p\, e^{-\theta_d s}}{\tau s + 1}\right\} \tag{14.34}$$

With modified Z transforms, Equation (14.34) gives

$$\frac{Y(Z)}{X(Z)} = \frac{K_p Z^{-(N+1)}\,(1 - e^{-mT/\tau}) + K_p Z^{-(N+2)}\,(e^{-mT/\tau} - e^{-T/\tau})}{1 - e^{-T/\tau} Z^{-1}} \tag{14.35}$$

This equation can be inverted to obtain the input-output relationship for the process. Thus

$$y_k = e^{-T/\tau}\, y_{k-1} + K_p\left\{(1 - e^{-mT/\tau})\, x_{k-(N+1)} + (e^{-mT/\tau} - e^{-T/\tau})\, x_{k-(n+2)}\right\} \tag{14.36}$$

where the subscript k refers to the kth sampling instant, N is the largest integral number of sampling periods in θ_d, and m, θ, are given by

$$m = 1 - \frac{\theta}{T}$$
$$\theta = \theta_d - NT \tag{14.37}$$

By introducing A_1, A_2, and A_3 to represent the constants defined earlier, Equation (14.36) can be solved for $x_{k-(n+1)}$. Thus

$$x_{K-(N+1)} = \frac{A_3}{K_p(A_3 - A_2)}\, y_K - \frac{A_2 A_3}{K_p(A_3 - A_2)}\, y_{K-1} - \frac{A_2(1 - A_3)}{(A_3 - A_2)}\, x_{k-(N+2)} \tag{14.38}$$

But in the presence of load disturbances Figure 14.13 indicates that the input to the process is given by

$$x_{K-(N+1)} = L_{K-(N+1)} + M_{K-(N+1)} \tag{14.39}$$

Therefore, an estimate of the load condition can be obtained by combining Equations (14.38) and (14.39). Thus

$$L_{K-(N+1)} = \frac{A_3}{K_p(A_3 - A_2)} y_K - \frac{A_2 A_3}{K_p(A_3 - A_2)} y_{K-1} - \frac{A_2(1 - A_3)}{(A_3 - A_2)} x_{K-(N+2)} - M_{K-(N+1)} \tag{14.40}$$

and therefore, the manipulated variable should be calculated from the equation

$$\begin{aligned} M_K &= \frac{y_{set}}{K_p} - L_{K-(N+1)} \\ &= \frac{y_{set}}{K_p} - \frac{A_3}{K_p(A_3 - A_2)} y_K + \frac{A_2 A_3}{K_p(A_3 - A_2)} y_{k-1} + \\ &\quad \frac{A_2(1 - A_3)}{(A_3 - A_2)} x_{K-(N+2)} + M_{K-(N+1)} \end{aligned} \tag{14.41}$$

The value of $x_{k-(N+2)}$ for use in this equation is obtained from the recursive equation, derived from Equation (14.38). Thus

$$x_{K-(N+2)} = \frac{A_3}{K_p(A_3 - A_2)} y_{K-1} - \frac{A_2 A_3}{K_p(A_3 - A_2)} y_{K-2} - \frac{A_2(1 - A_3)}{(A_3 - A_2)} x_{K-(N+3)} \tag{14.42}$$

In this development, since the sampling period is 0.5 min and dead time is 0.76 min, N equals 1. This makes $\theta = 0.26$ and $m = 0.48$. Substituting these values into Equations (14.41) and (14.42) gives the algorithm in a form that is suitable for implementation on the control computer. Thus

$$\begin{aligned} M_K &= -y_{set} + 2.216 y_K - 0.6349 y_{K-1} + 0.5813 X_{K-3} + M_{K-2} \\ &\quad + \omega(y_{set} - y_K) \end{aligned} \tag{14.43}$$

where ω = tuning parameter which is to be determined by trial and error and

$$x_{K-3} = -2.216 y_{K-1} + 0.6351 y_{K-2} - 0.5812 x_{K-4} \tag{14.44}$$

The last term in Equation (14.43) has been added to improve the speed of response of the system. If $K = 1$ denotes the time at which the loop is switched to computer control, M_1 is the first value of the controller output computed from Equation (14.43) and the values of x_{K-3} for the next three sampling periods, that is, x_{-2}, x_{-1}, x_0 are the steady-state values of the process output y_{set}. Thereafter, x_{K-3} is computed from Equation (14.44). Note that the contribution of the last term in Equation (14.43) is significant only when the process shows a deviation from set point.

Simulation Results. The process model and the control algorithm were implemented on a digital computer. The differential equation representing the process model was numerically solved by the fourth-order Runge–Kutta method. The integration step size was 0.02 min. The responses of the process to set point and load changes are shown in Figures 14.14 and 14.15, respectively. The best value of ω turned out to be -0.1 for both set point changes and load changes. Also shown in these figures is the response of the process using the PI control algorithm given in the following equation:

$$M_K = M_{K-1} - K_c \left[(e_K - e_{K-1}) + \frac{T}{\tau_I} e_K \right] \tag{14.45}$$

where

$$K_c = \text{gain, } 0.3$$

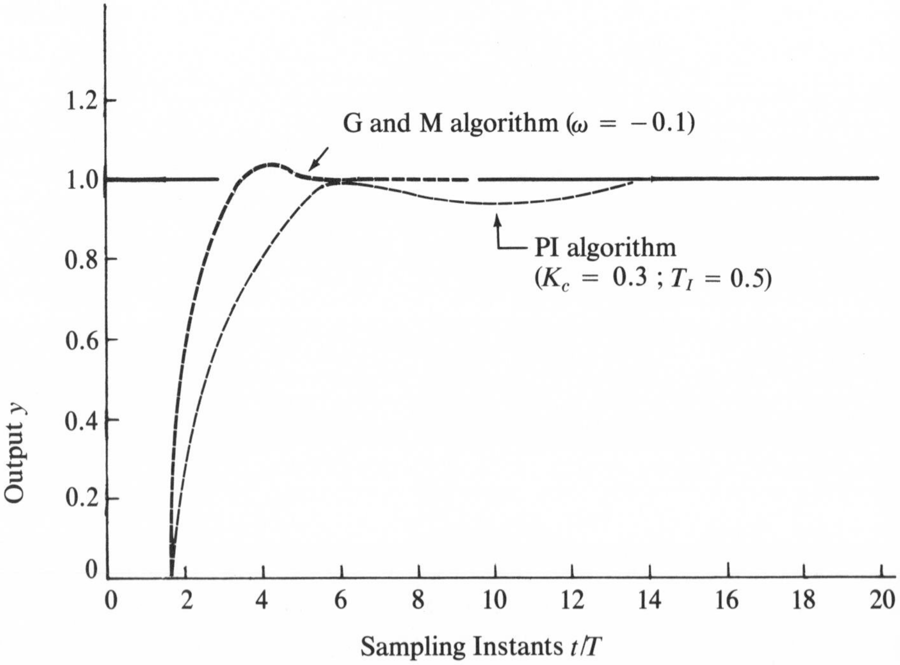

Figure 14.14
Set-Point Response of G and M Algorithm

$$\tau_I = \text{reset, 0.5 min}$$

The best values of proportional gain and reset were also found by trial and error. From Figures 14.14 and 14.15, it can be seen that the present algorithm performs well, for both set-point and load changes.

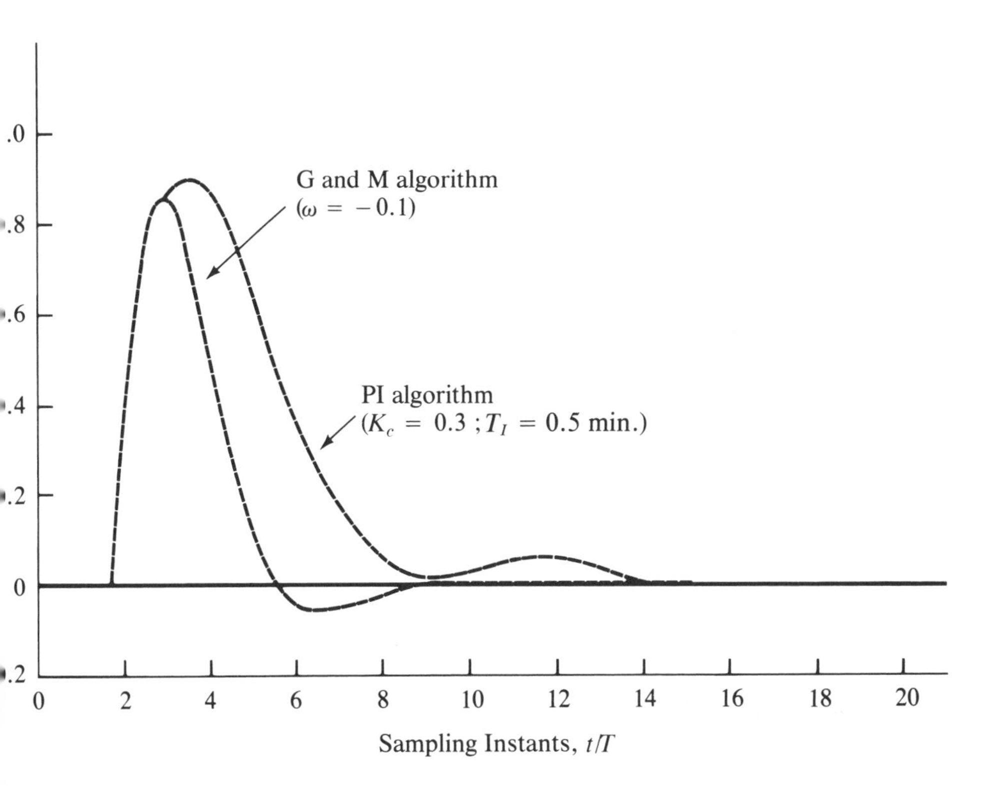

Figure 14.15
Response of G & M Algorithm to a Load Change

NOMENCLATURE

A_1	constant; exp $(-T/2\tau)$	s	Laplace transform variable
A_2	constant; exp $(-T/\tau)$	T'	total dead time; $\theta_d + T$ or $\theta_d + T/2$
A_3	constant; exp $(-\beta T/\tau)$	T	sampling time
B,C	controlled variables	u_K	input function
d_k	disturbance		
E	error		

$G_{h_0}(s)$	zero-order hold
$H(s)$	measuring element transfer function
K_c	controller gain
K_T	integral controller constant
K_L	gain relating C to L
K_p	process gain
K_r	reset gain
L	load disturbance
M	manipulated variable
m	constant; $1 - \theta/T$
N	integral part of θ_d/T
R, r_k	set point
R_e	reflux
x	input variable
x_D	top composition
y	output variable
Z	Z-transform variable

Greek letters

β	fractional part of θ_d/T
ω	tuning parameter
ρ_k	calibrated set point
θ	fractional dead time; $\beta \times T$
θ_d	process dead time
τ	time constant of the process

References

1. Smith, O. J. M., Close Control of Loops with Dead Time, *Chemical Engineering Progress*, **53**, 5, 1957, 217–219.
2. Meyer, C., et al., An Experimental Application of Time Delay Compensation Techniques to Distillation Column Control, *I&EC Proc. Des. Dev.*, **17**, 1, 1978.
3. Moore, C. F., Selected Problems in the Design and Implementation of Direct Digital Control, Ph.D. thesis, Department of Chemical Engineering, Louisiana State University, 1969.
4. Gautam, R. and Mutharasan, R., A General Direct Digital Control Algorithm for a Class of Linear Systems, *A. I. Ch. E. J.*, **24**, 2, 1978, 360–64.

CHAPTER 15

Feedforward Control

15.1. Introduction and Design Fundamentals

Feedforward control is probably one of the more widely used advanced control techniques in the process industries. It is implementable with analog hardware, although somewhat more specialized equipment is needed for its implementation. Its purpose is to protect the control system against the detrimental effects of changing process loads. When properly designed and tuned, feedforward control can produce amazing results. In this chapter we develop the design equations for feedforward control and present two experimental applications that show the benefits of this control technique. As an introduction to feedforward control, consider a simple process consisting of heating a continuous stream of water in a tank, as shown in Figure 15.1.

In this case the process loads are the flow rate and temperature of the supply water. In the usual industrial situation the incoming stream to a process comes from an upstream portion of the plant and is not subject to manipulation or control. If the temperature or flow rate of the incoming water changes, it upsets the controlled variable, the temperature of water in the tank. The system remains disturbed until the feedback system brings the controlled variable back to the set point. Feedforward control can be used to improve the response of the system under these circumstances. The basic principle of this technique is to measure the disturbances as they occur and make adjustments in the manipulated variable so as to prevent

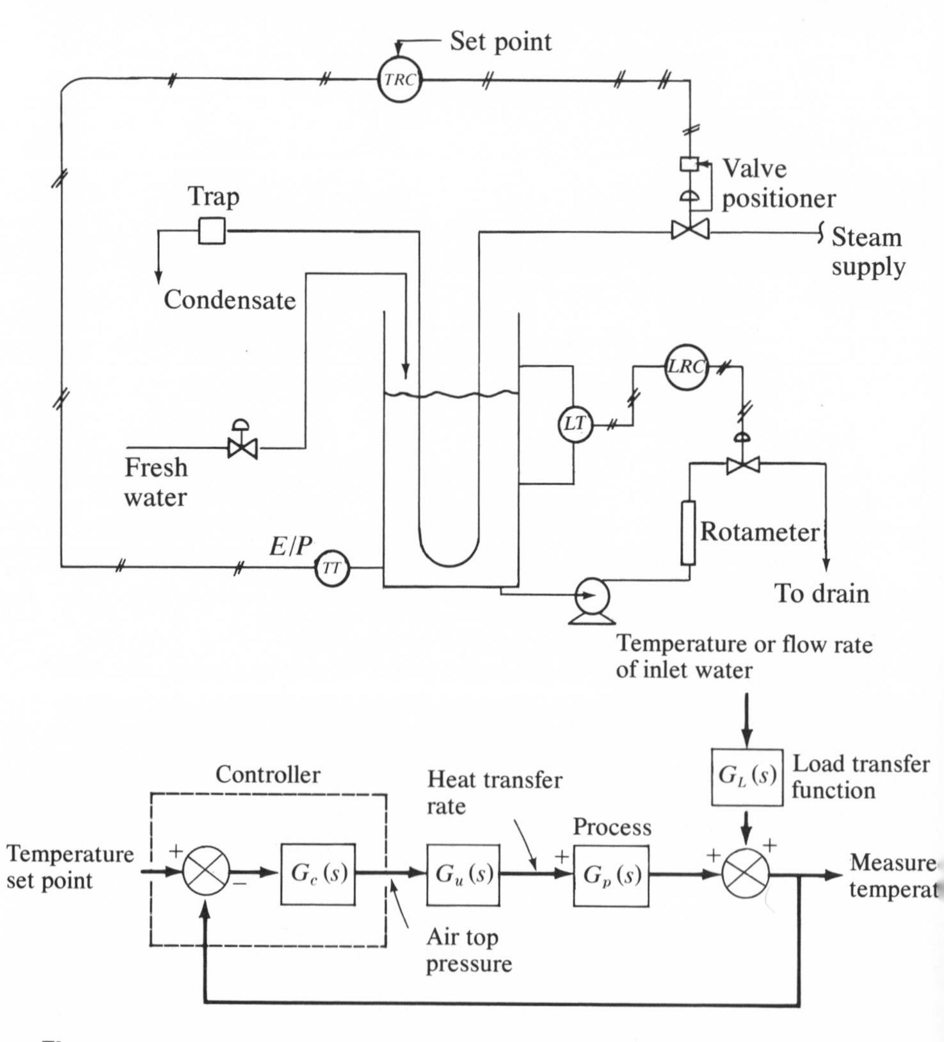

Figure 15.1
Schematic and Block Diagram of Temperature Control System

them from upsetting the controlled variable. The design equations for the feedforward controller are derived as follows:

The response of the feedback control system shown in Figure 15.1 to load changes is

$$C(s) = \frac{G_L(s)}{1 + G_c(s)G_u(s)G_p(s)} L(s) \tag{15.1}$$

Ideally, C should be zero for any load L (note that C is a deviation variable); C is reduced by a large G_c (i.e., high controller gains), but the magnitude of G_c is constrained by physical limitations and system stability.

In feedback control we feed the error signal back to the controller, which in turn updates the manipulated variable so as to reduce the error. But suppose we could measure the signal which has the potential of upsetting the process if no action is taken. Then we would measure and transmit this signal to a controller. This controller would act on this signal and compute the new value of the manipulated variable and forward the output to the final control element. If we did everything correctly, the controlled variable would not be affected if a process load should occur. The block diagram of this control system is shown in Figure 15.2. Note that in this arrangement errors in the controlled variable are not fed back, but the changes in process loads are fed forward. This arrangement is re-

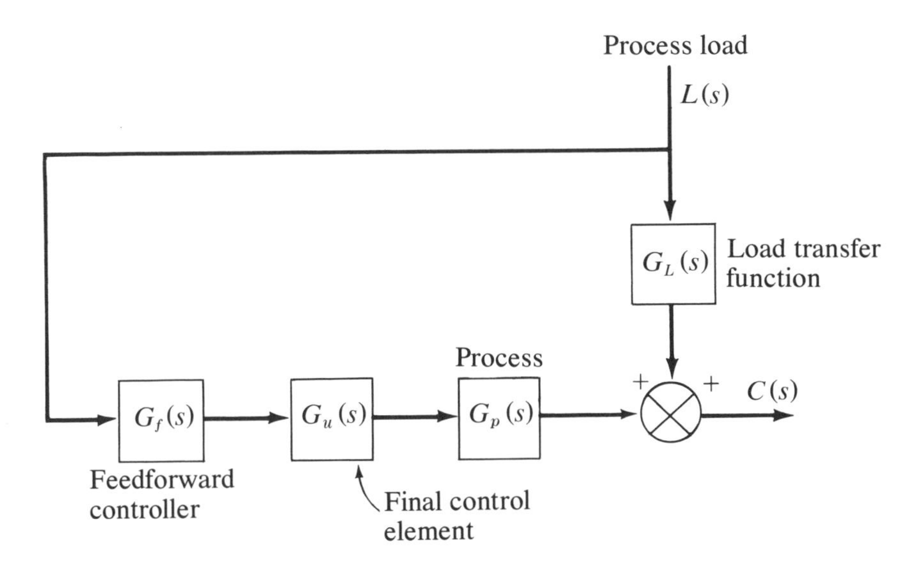

Figure 15.2
Feedforward Control Loop

ferred to as a feedforward control system. From the block diagram of this feedforward control system we can write

$$C(s) = L(s)\{G_L(s) + G_p(s)G_u(s)G_f(s)\} \tag{15.2}$$

where $G_f(s)$ is the transfer function of the feedforward controller. If we set

$$G_f(s) = -\frac{G_L(s)}{G_u(s)G_p(s)} \tag{15.3}$$

Then C will be zero for all L. This equation gives us the basis for feedforward controller design. Equation (15.3) also shows that accurate models of the elements G_L and G_uG_p are required. If the models are not accurate, the terms in the brackets in Equation (15.2) will not vanish, and the controlled variable will show a deviation from set point. Therefore, feedforward control is seldom used alone, but rather in combination with feedback control, as shown in Figure 15.3. Note that this arrangement implies that the sensor dynamics are included in the transfer functions $G_p(s)$, $G_L(s)$, and $G_f(s)$. The transfer function of the feedback/feedforward combination to load changes is

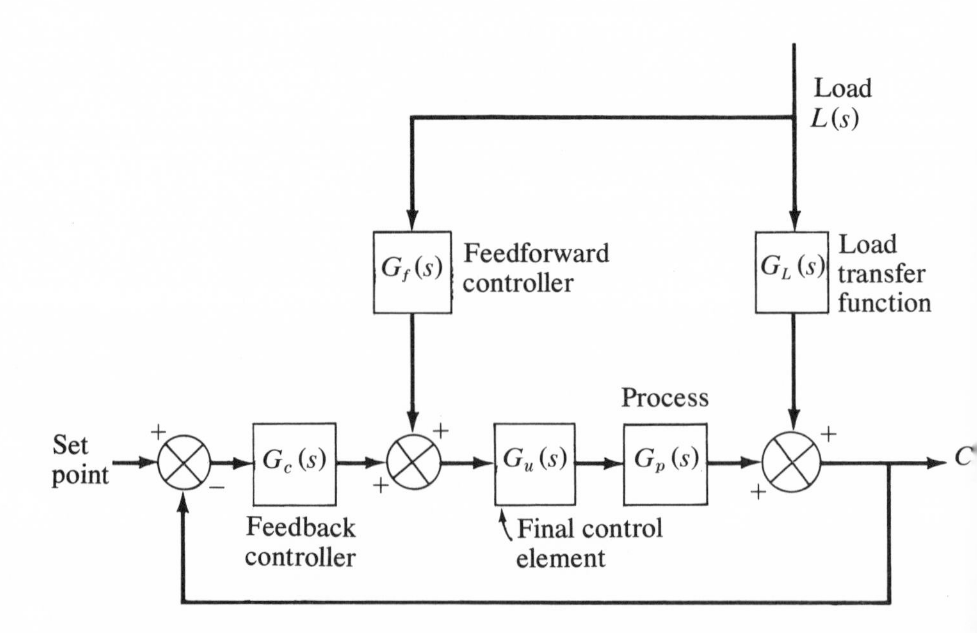

Figure 15.3
Combined Feedback/Feedforward Control System

$$C(s) = \frac{\{G_L(s) + G_p(s)G_u(s)G_f(s)\}}{1 + G_c(s)G_u(s)G_p(s)} L(s) \tag{15.4}$$

The feedforward controller transfer function is still given by Equation (15.3). The feedforward controller eliminates the effects of process loads, whereas the feedback mechanism eliminates the effects of inaccuracies in the feedforward controller and other unmeasured disturbances.

Equation 15.3 shows that to determine the transfer function of the feedforward controller we require the transfer functions $G_p(s)$, $G_u(s)$, and $G_L(s)$. These, of course, are the open-loop transfer functions as shown in Figure 15.4. These transfer functions may be determined from a mathematical model or from experimental pulse or step tests. To determine $[G_u(s)\, G_p(s)]$, for example, by pulse testing, we would employ the following step-by-step procedure:

1. Place the feedback controller in manual and disconnect the feedforward controller as shown in Figure 15.4.
2. Start up the process and adjust the feedback controller output until the desired steady state operation is achieved.
3. Introduce a suitable pulse in M and record the transient response of C.

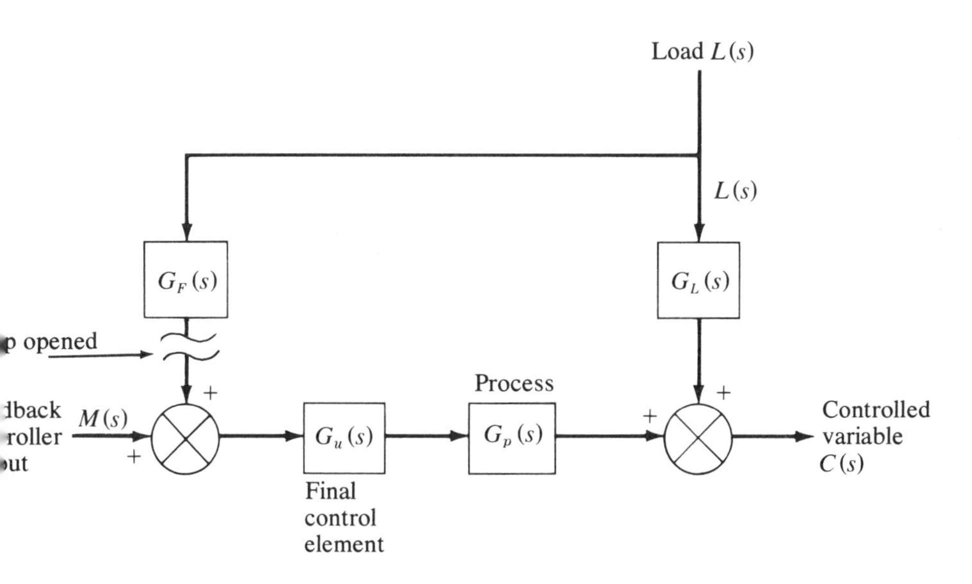

Figure 15.4
Open Loop Transfer Function

4. Analyze the pulse test data by Fourier transforms and generate a frequency-response diagram of the open-loop system.

5. Fit an approximate transfer function to the Bode' plot. This transfer function represents the product of $G_u(s)$ and $G_p(s)$.

6. With the process still operating at steady state and with the controller in manual, introduce a suitable pulse into L and record C.

7. The analysis of the pulse test data as before gives us $G_L(s)$.

Consider the design of a feedforward controller for a system in which the process transfer function as well as the load transfer function can be adequately described as first-order models with dead time, that is, suppose

$$G_u(s)\, G_p(s) = \frac{K_p\, e^{-\theta_d s}}{\tau_p s + 1}$$

and (15.5)

$$G_L(s) = \frac{K_L\, e^{-\theta_L s}}{\tau_L s + 1}$$

Then the feedforward controller will be represented by the equation

$$G_f(s) = -\frac{G_L(s)}{G_u(s)\, G_p(s)} = -\frac{K_L}{K_p}\left[\frac{\tau_p s + 1}{\tau_L s + 1}\right] e^{(\theta_d - \theta_L)s} \tag{15.6}$$

This representation of the feedforward controller shows that this controller has three essential parts:

1. The feedforward controller gain is $-K_L/K_p$.

2. $(\tau_p s + 1)/(\tau_L s + 1)$ represents a lead-lag element. If this element is to be physically realizable, τ_L must be greater than zero. If the load dynamics are very fast, $\tau_L \simeq 0$. Then the lead-lag network reduces to a pure differentiator element, which is physically unrealizable. In this case we would select $\tau_L << \tau_p$ but still greater than zero.

3. If the process dead time θ_d exceeds the load deadtime θ_L, a pure predictor will be needed. This, of course, is also physically unrealizable. In this case we would set $\theta_d = \theta_L$ and eliminate the predictor term.

If we drop the dynamic terms from Equation 15.6, we obtain the design equation of what is referred to as the steady-state feedforward controller, G_{fss}. Thus

$$G_{fss} = -\frac{K_L}{K_p} \tag{15.7}$$

$$= -\left.\frac{\Delta C}{\Delta L}\right|_{M\,=\,\text{constant}} \Big/ \left.\frac{\Delta C}{\Delta M}\right|_{L\,=\,\text{constant}}$$

A steady-state mathematical model or two simple step tests are all that are needed to implement the steady-state feedforward controller. Let us now illustrate the application of feed-forward control to a couple of experimental systems.

15.2. Example 1

Shinskey[1] presented an application of feedforward control to a heat exchanger. A schematic of the system is shown in Figure 15.5.

In this example we are interested in controlling the fluid outlet temperature T_0 by manipulating steam flow F. Let us assume that the fluid inlet temperature T_I is constant and that the primary load disturbance is W, the flow rate of the process fluid.

The design equation for the steady-state feedforward controller can be derived from a steady-state mathematical model of the process. Thus if we apply the basic steady-state energy balance, we will get

$$W\,C_p\,(T_o - T_I) = F\lambda \qquad (15.8)$$

where

W = process fluid flow rate, lb/hr

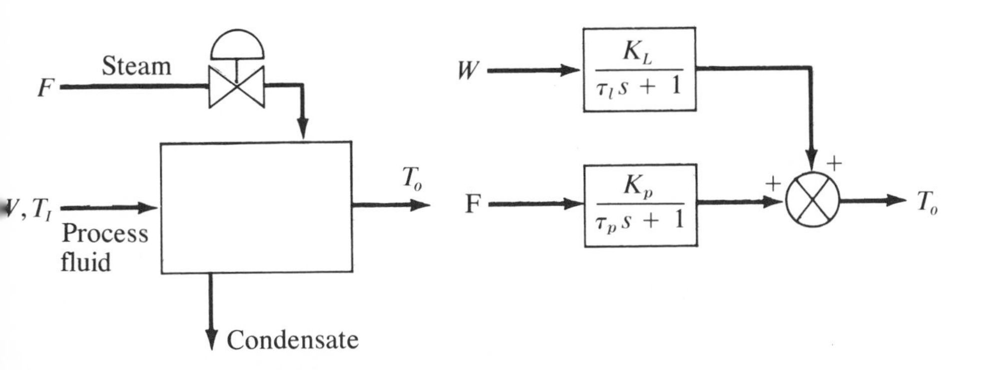

Figure 15.5
Process Schematic of the Heat Exchanger

C_p = specific heat of fluid Btu/1b°F

T_I = fluid inlet temperature

T_o = fluid outlet temperature

F = steam flow, 1b/hr

λ = latent heat of vaporization (i.e., heat released by steam), Btu/lb.

The solution of Equation (15.8) for T_o gives

$$T_o = \frac{F}{W}\frac{\lambda}{c_p} + T_I \tag{15.9}$$

The steady-state gains can be computed as the following partial derivatives at the steady-state operating point F_o and W_o. Thus,

$$\begin{aligned} K_p &= \left.\frac{\partial T_o}{\partial F}\right|_{W_o, F_o} = \frac{\lambda}{W_o c_p} \\ K_L &= \left.\frac{\partial T_o}{\partial W}\right|_{W_o, F_o} = -\frac{F_o}{W_o^2}\frac{\lambda}{c_p} \end{aligned} \tag{15.10}$$

Thus the steady-state feedforward controller will be given by

$$G_{fss} = -\frac{K_L}{K_p} = \frac{F_o}{W_o} \tag{15.11}$$

Note that at steady state, the energy balance, Equation (15.8) becomes

$$W_o c_p (T_o - T_I) = F_o \lambda \tag{15.12}$$

or

$$\frac{F_o}{W_o} = \frac{c_p}{\lambda}(T_o - T_I) \tag{15.13}$$

Thus

$$G_{fss} = \frac{c_p}{\lambda}(T_{sp} - T_I) \tag{15.14}$$

where T_{sp} is the desired value of T_o or set point.

In this investigation Shinskey compared the conventional three-mode feedback control of T_o to feedforward control alone [Equation (15.14)] as well to the combined feedback/feedforward system. Both the steady-state feedforward and the dynamic feedforward controllers were implemented. The dynamic feedforward controller had the transfer function

$$G_f(s) = \frac{F(s)}{W(s)} = \left[\frac{c_p}{\lambda}(T_{sp} - T_I)\right]\frac{\tau_p s + 1}{\tau_L s + 1} \tag{15.15}$$

The time constants τ_p and τ_L were selected by trial and error.

Figure 15.6 shows the results of the investigation for a 40% step change in W. In Figure 15.6*a* the three-mode controller was tuned for optimal recovery at 80% load. The nonlinearity of the process is evident in the overdamped recovery at 40% load. Figure 15.6*b* shows the results of steady-state feedforward control alone, and Figure 15.6*c* shows the results when the dynamics are added. The benefits of feedforward control are quite evident from these plots. Figure 15.6*d*

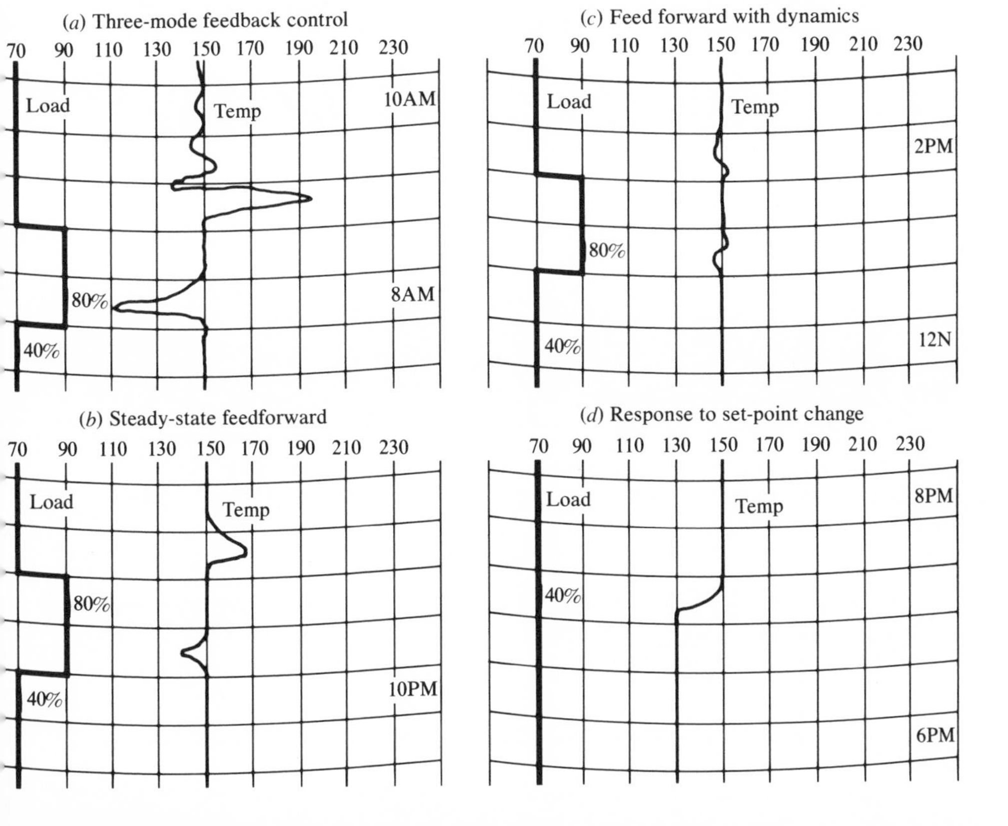

Figure 15.6
Comparison of Feedback and Feedforward Control System for the Heat Exchanger (Reproduced with Permission of Ref. 1)

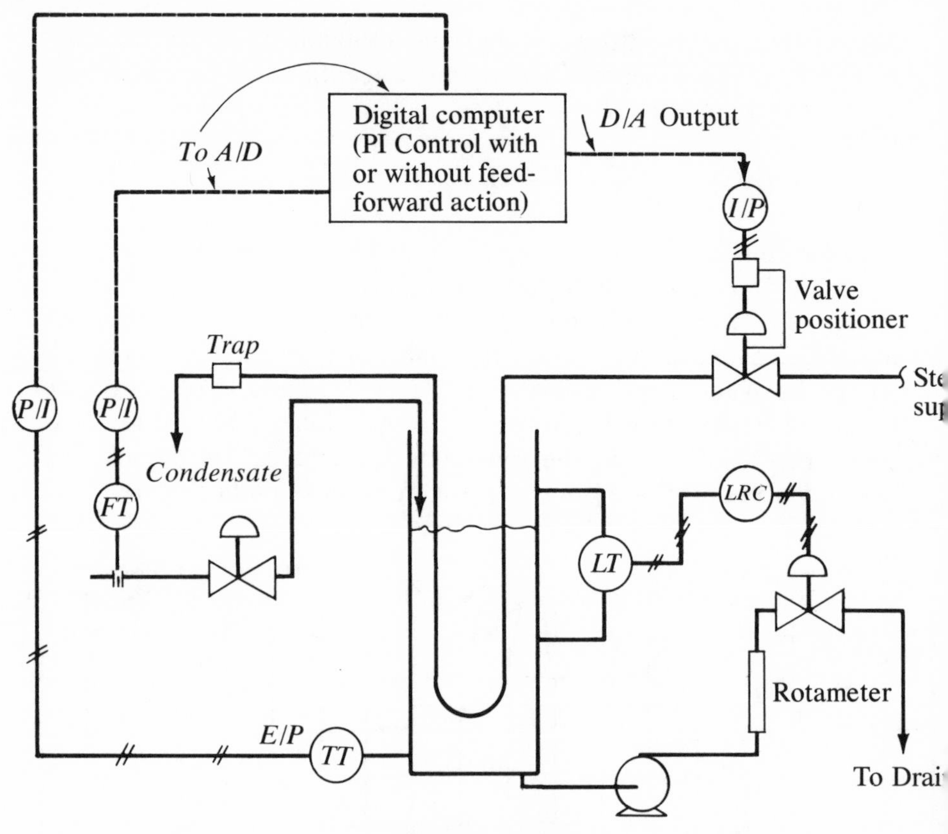

(a) Process Schematic

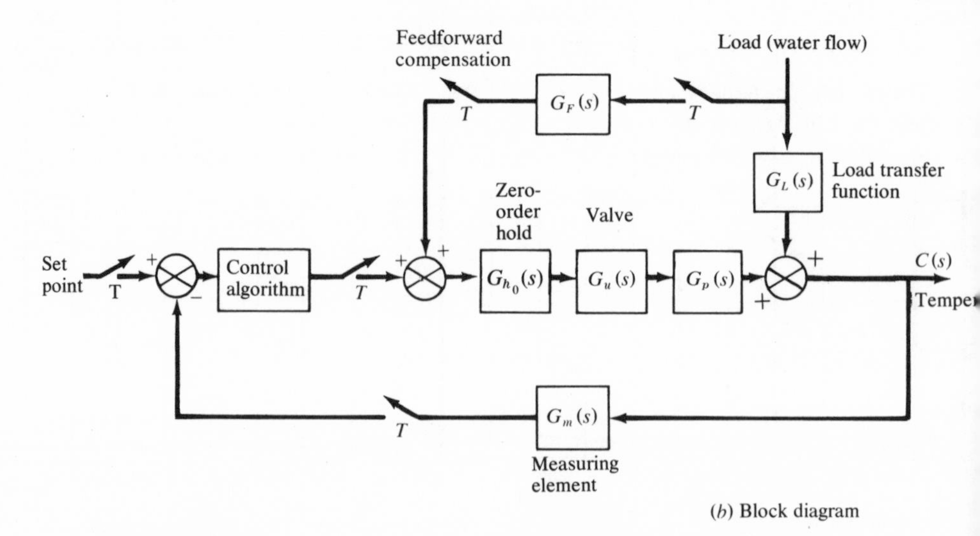

(b) Block diagram

Figure 15.7
The Combined Feedback Feedforward Control System

shows the response of the system having a feedforward-only controller when the set point is changed from 130°F to 150°F. However, for the reasons we mentioned earlier, we would normally use a combined feedback/feedforward system for good control of processes.

15.3. Example 2

This example shows a laboratory application of computer feedback/feedforward control of temperature. The schematic of the process for this computer control application as well as the resulting block diagram are shown in Figure 15.7. The simpler steady-state feedforward control algorithm has been selected to illustrate the technique. In this application the steady-state feedforward control algorithm is given by

$$G_{fss} = -\frac{K_L}{K_p} = -\left.\frac{\Delta T}{\Delta W}\right|_{F\,=\,\text{constant}} \Bigg/ \left.\frac{\Delta T}{\Delta F}\right|_{W\,=\,\text{constant}} \tag{15.16}$$

where

T = temperature transmitter output, psig

W = flow transmitter output, psig

F = air-top pressure to steam control valve, psig

The open-loop step tests suggest that G_{fss} be set equal to 1 psig/psig. That is, for every psig change in the flow transmitter output, the feedforward controller must increment the air-top pressure on the steam control valve by 1 psig. The steady-state operating conditions are shown in Table 15.1.

Table 15.1
Operating Conditions

Tank level set point	18.4 in.
Steady-state water flow	2.54 gal/min
Step changed to	3.3 gal/min
Feedback controller settings:	
Proportional band	2%
Reset, τ_I	0.83 min
Sampling period	5 sec

The records of closed-loop response of the control system, both in the feedback control mode as well as in the combined feedback/feed-forward control mode are shown in Figure 15.8. These results, too, show the excellent performance of the feedforward control system.

Reference

1. Shinskey, F. G., Feedforward Control Applied, *ISA* J., November 1963.

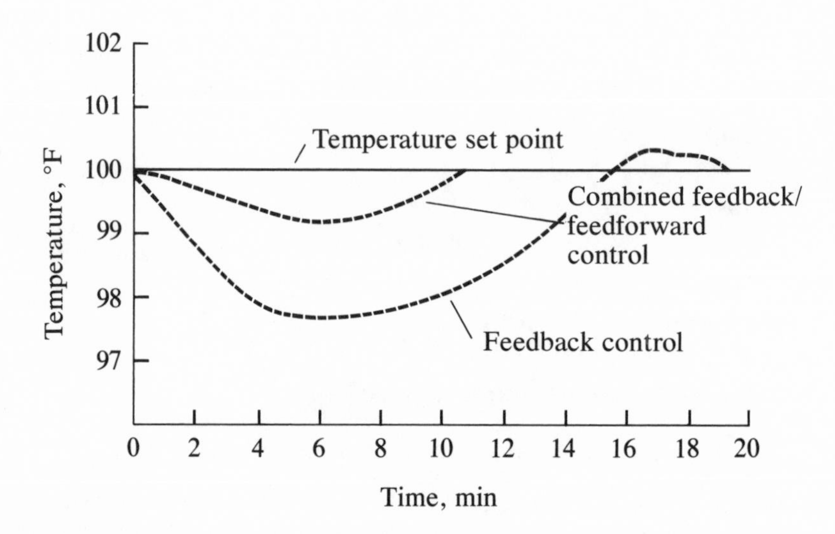

Figure 15.8
Response of Combined Feedback/Feedforward System to a Positive Step Change in Water Flow Rate

CHAPTER 16

Cascade Control

In process-control applications variations in the manipulated variable frequently cause deterioration of the performance of feedback control loops. An example of such a control loop is shown in Figure 16.1. In this application the temperature of the water in the tank is maintained constant by adjusting the flow rate of steam. If a disturbance in the steam-supply pressure occurs, the flow rate of steam changes, which in turn upsets the controlled variable, temperature. Of course, once the temperature measuring device senses the upset, it feeds the information back to the controller, which takes corrective action so as to bring the temperature back to the set point. Meanwhile, the disturbance has entered the process and has upset the controlled variable. One can visualize that if the variations in the steam-supply pressure are frequent, the controlled variable may not remain at the set point for very long.

To correct this problem, a second control loop can be added, as shown in Figure 16.2. In the presence of steam-supply pressure

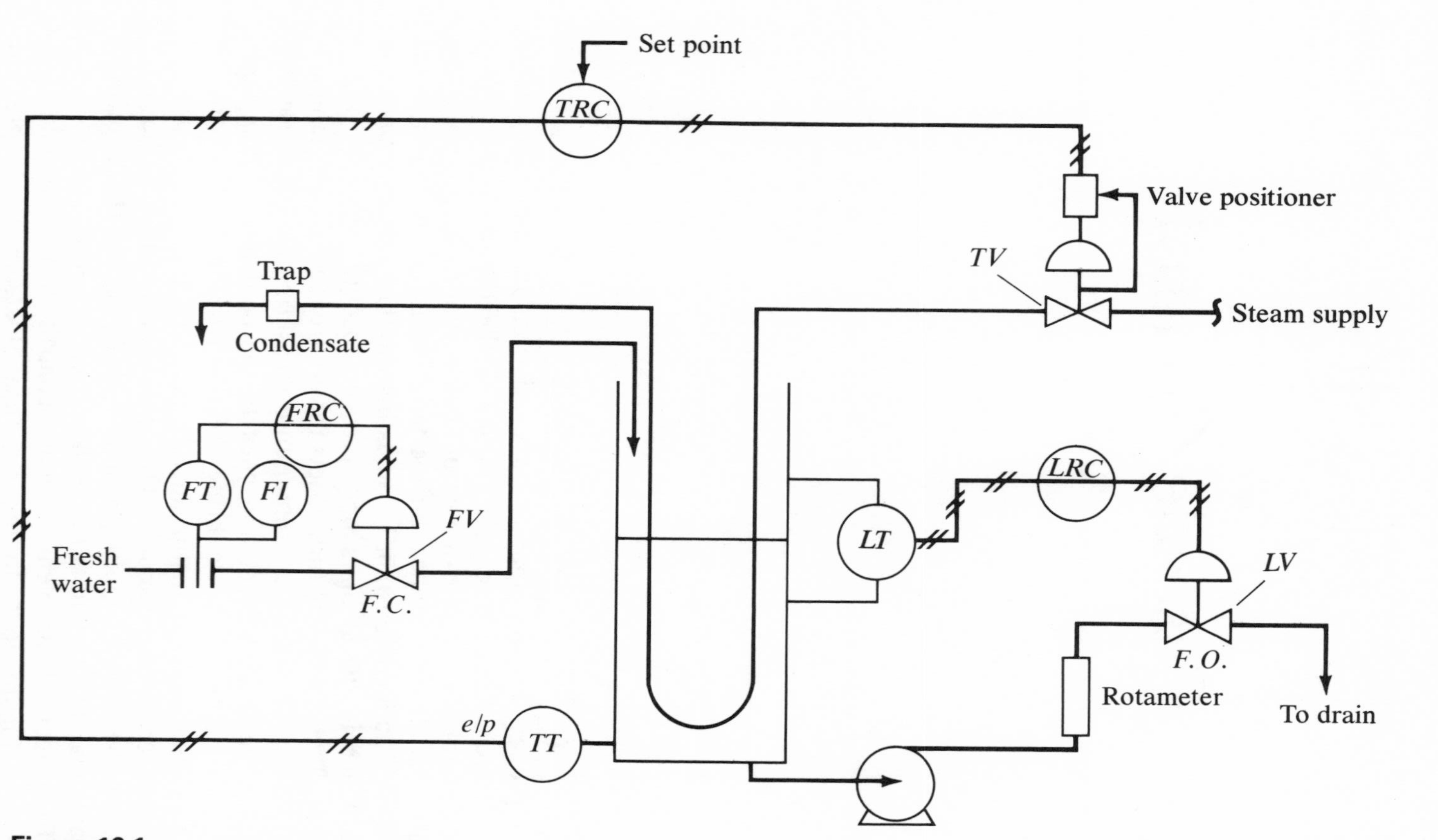

Figure 16.1
Typical Feedback Control System

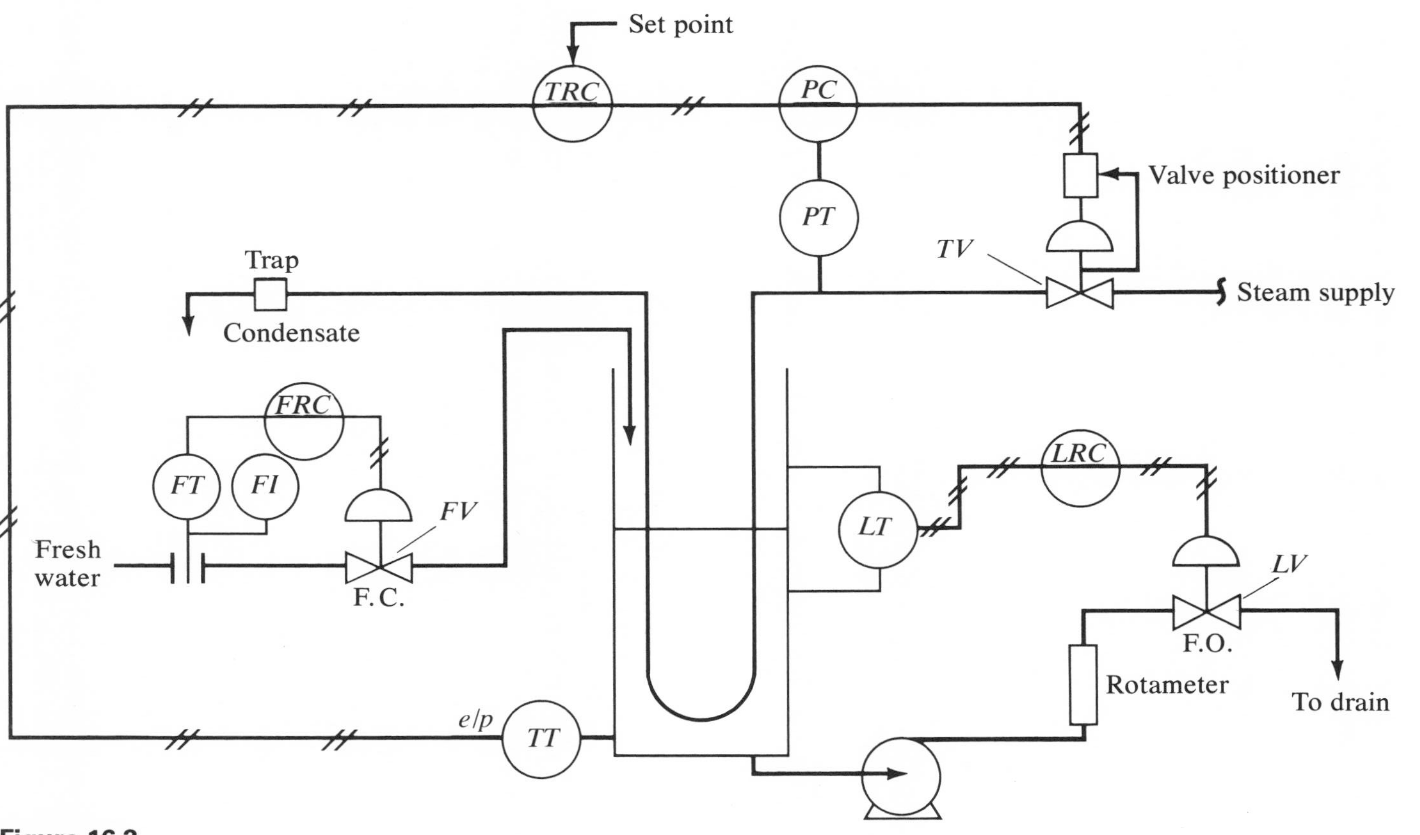

Figure 16.2
Cascade Control of Temperature

fluctuations, the pressure sensor senses the changes, and the pressure controller manipulates the steam valve so as to hold the downstream pressure constant. This way, the effect of supply pressure fluctuations on the steam flow rate can be eliminated. The performance of the feedback control loop is thus greatly improved.

The control situations, such as the one discussed here, in which the output of one controller manipulates the set point of another controller are called *cascade control systems*. The terminology commonly used in describing the cascade control systems is shown on the block diagram of the system in Figure 16.3. It may be noted from this figure that each controller is served by its own measurement device, but only one controller—the primary or master controller—has an independent set point, and only one controller—the secondary or slave controller—has an output to the process. The secondary controller, the manipulated variable, and its measurement device constitute the elements of the *inner* or *secondary loop*. The outer loop consists of all the elements of the cascade control system, including those of the inner loop.

To ensure that the cascade control system functions properly it is necessary that the dynamics of the inner loop be at least as fast as those of the outer loop, and preferably faster. This should be intuitively clear. If the dynamics of the inner loop are much faster than those of the outer loop, the inner controller will correct the effect of disturbances in that loop before they have a chance to upset the controlled variable. If this condition is not met, it is generally impossible to tune the master controller satisfactorily. The commonly encountered process control loops, in order of decreasing speed, are flow, liquid-level and pressure, temperature, and composition. However, these are general observations, and specific process situations must be analyzed thoroughly to assess whether cascade control is needed and if so, which should be the outer loop and which should be the inner loop. Several industrial examples of cascade control are shown in Figure 16.4.

16.1. Controller Design of Cascade Systems

There are two approaches available for determining the tuning constants of cascaded controllers. One is an analytical approach which can be used if the open-loop process transfer functions are

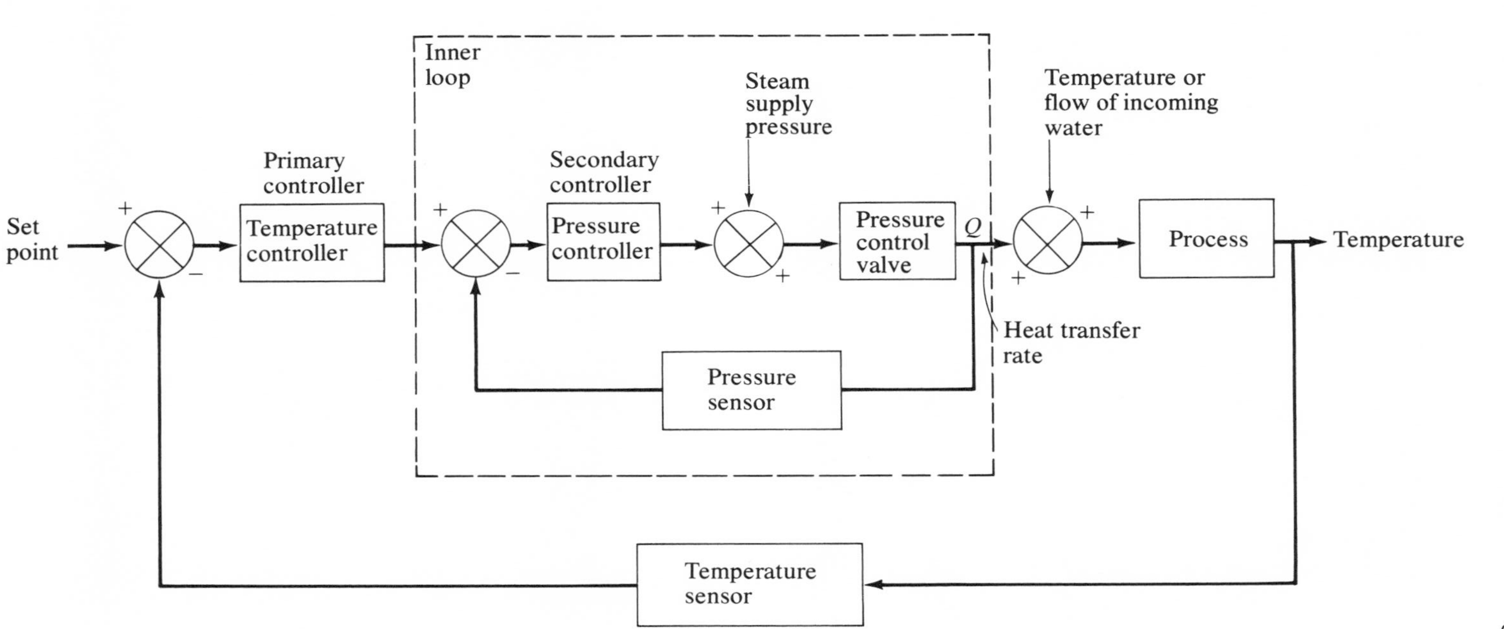

Figure 16.3
Block Diagram of Cascade Control System for Process of Figure 16.2

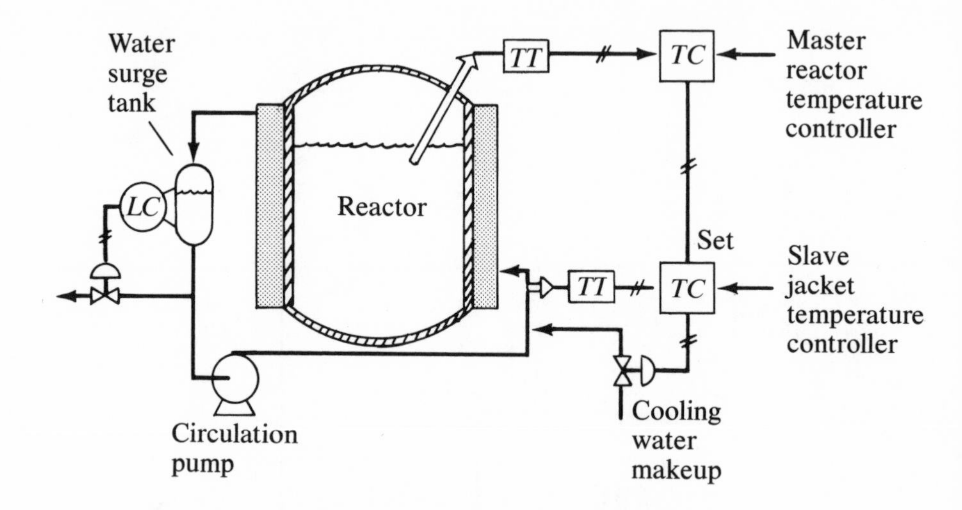

(a) Exothermic batch reactor

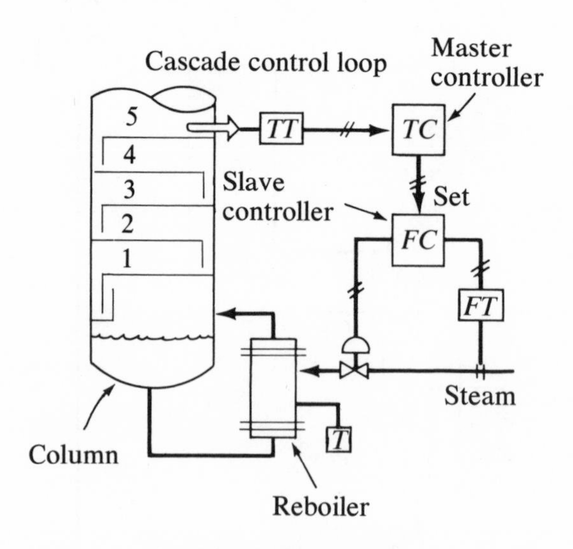

(b) Distillation column

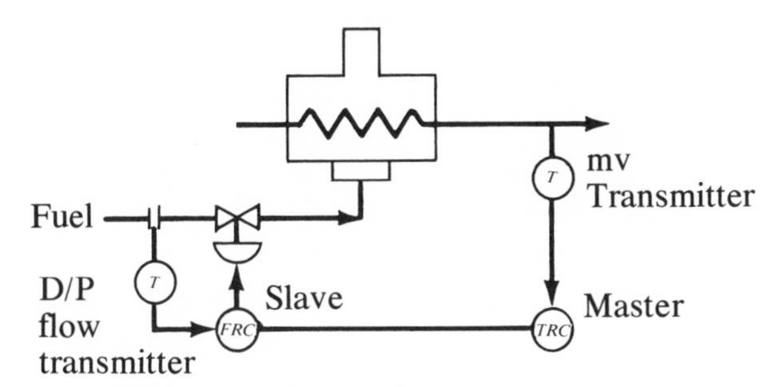

The *TRC-FRC* cascade control loop.

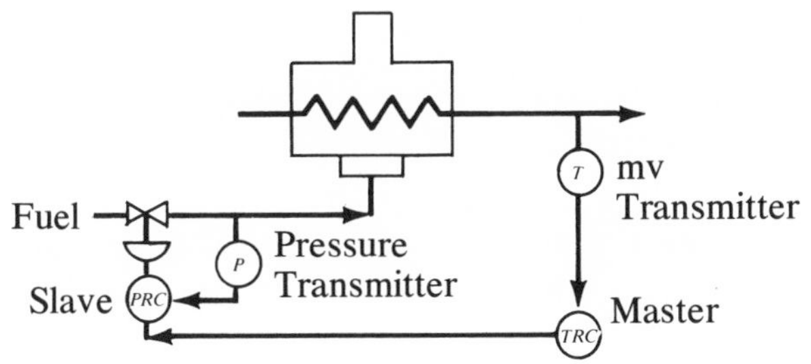

The *TRC-PRC* cascade control loop.

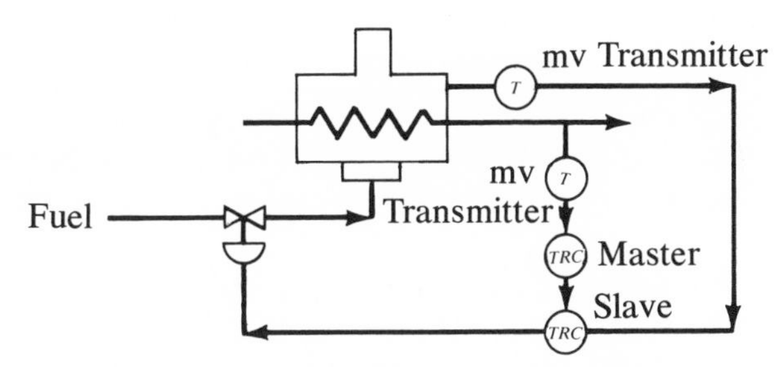

The *TRC-TRC* cascade control loop.

(*c*) Fuel-fired heater

Figure 16.4
Some Industrial Examples of Cascade Control Systems

available. These transfer functions can be developed from a dynamic mathematical model of the process. The second approach is an experimental one in which we would conduct open-loop step or pulse tests on the process, analyze the input-output data, and plot the frequency-response diagram, from which we would determine the tuning constants.

The analytical determination of controller settings for a cascade control system is straightforward once it is recognized that the elements of the inner loop can be reduced to a single block by the block-diagram-reduction techniques. The procedure is as follows:

1. Prepare the frequency-response diagram of all the elements of the inner loop, excluding the secondary controller.

2. Add the curves generated in step 1, graphically, to determine the composite Bode plot of these elements in series.

3. Design the secondary controller according to Cohen and Coon or Ziegler–Nichols criteria or by using Fertik's controller parameter charts (see Chapter 10). Integral action on the secondary controller is often unnecessary, since the inner-loop gain is often large and the effect is eventually corrected by the integral action of the primary controller. If inner-loop gain is low (a frequent occurrence in flow control), integral action is incorporated.

4. Using block-diagram-reduction techniques 1, reduce the closed inner loop to a single block.

5. Prepare the Bode plot of all the elements of the outer loop, excluding the primary controller, but including the block found in step 4.

6. Add the curves found in step 5 to develop the Bode plot of all the elements (except the primary controller) of the outer loop in series.

7. Design the primary controller using the Ziegler–Nichols or some other method.

Let us now illustrate this design method by an example.

Example 1 (with Permission of Ref. 2). Determine the controller settings of the primary controller with and without the inner loop. Measurement lags are negligible.

Solution Part A Cascade Control System.

Step 1. Prepare Bode plot of each element of $G_3(s)$ and $H_2(s)$. In this case the elements are $1/(s + 1)^2$ and $1/(10s + 1)$. The results are shown in Figure 16.6.

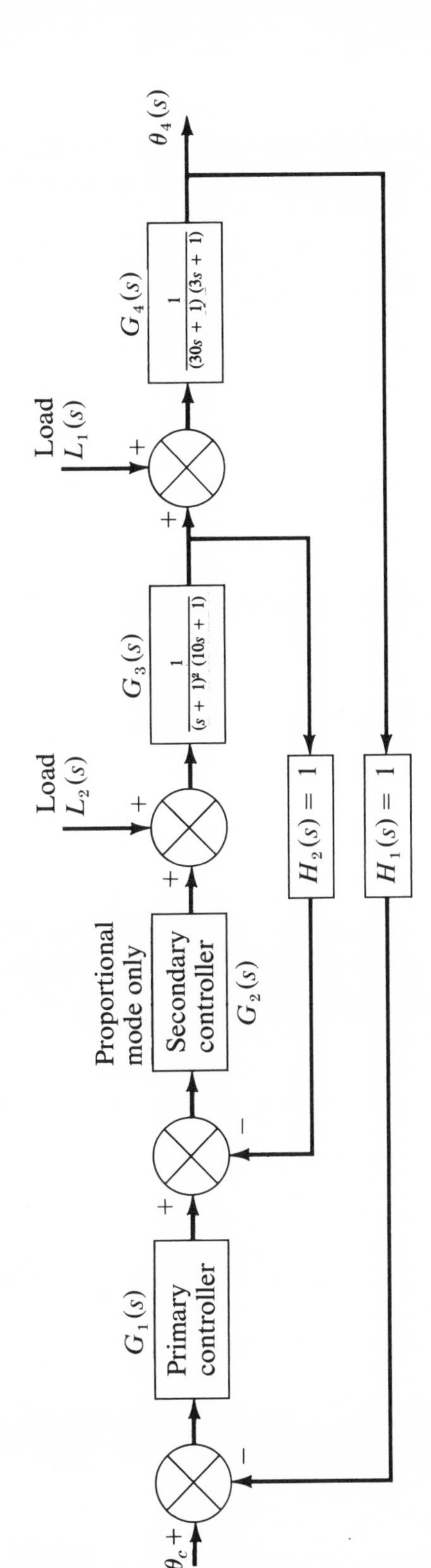

Figure 16.5
Cascade Control System

Step 2. Add the curves generated in step 1 to obtain the Bode plot of $1/((s+1)^2(10s+1))$ as shown in Figure 16.6.

Step 3. From the figure, at $\phi = -180°$, $AR = 0.0416$. Therefore Ultimate gain, Ku = 1/0.0416 = 24. Therefore the Ziegler–Nichols gain for the secondary controller is $K_c = (0.5)(24) = 12$.

Step 4. Consider the inner loop

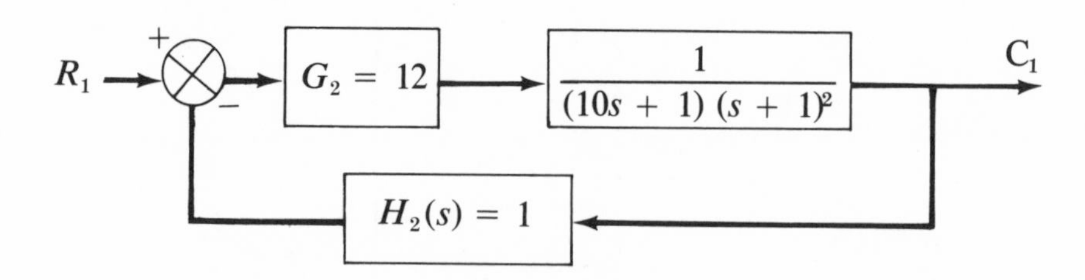

This block diagram can be replaced by its equivalent down below

$$R_1 \longrightarrow \boxed{\frac{G}{1+G}} \longrightarrow C_1$$

where

$$G(s) = G_2G_3(s) = \frac{12}{(10s+1)(s+1)^2}$$

From step 2 we already have a Bode plot of $1/(10s+1)(s+1)^2$. Incorporation of $K_c = 12$ into this Bode plot shifts the entire amplitude ratio portion of the plot upward. The phase angle portion of the plot remains unaffected. The Bode plot of $12/(s+1)^2(10s+1)$ is also shown in Figure 16.6.

Now recall that given a $G(s)$, $G(s)/1 + G(s)$ can be determined from Nichols chart, Figure 16.7.[3] Therefore, read off numerous values of amplitude ratio and phase angles of $G(s) = 12/(s+1)^2(10s+1)$ and go to Nichols chart to determine the amplitude ratios and phase angles of $G(s)/(1+G(s))$. These results are shown in Table 16.1.

Step 5. Determine the amplitude ratios and phase angles of $G_4(s)$ from the relationships

$$AR = 1/\left(\sqrt{1+\omega^2(30)^2}\right)\left(\sqrt{1+\omega^2 3^2}\right)$$

and

$$\phi = \arctan(-3\omega) + \arctan(-30\omega)$$

and enter in Table 16.1.

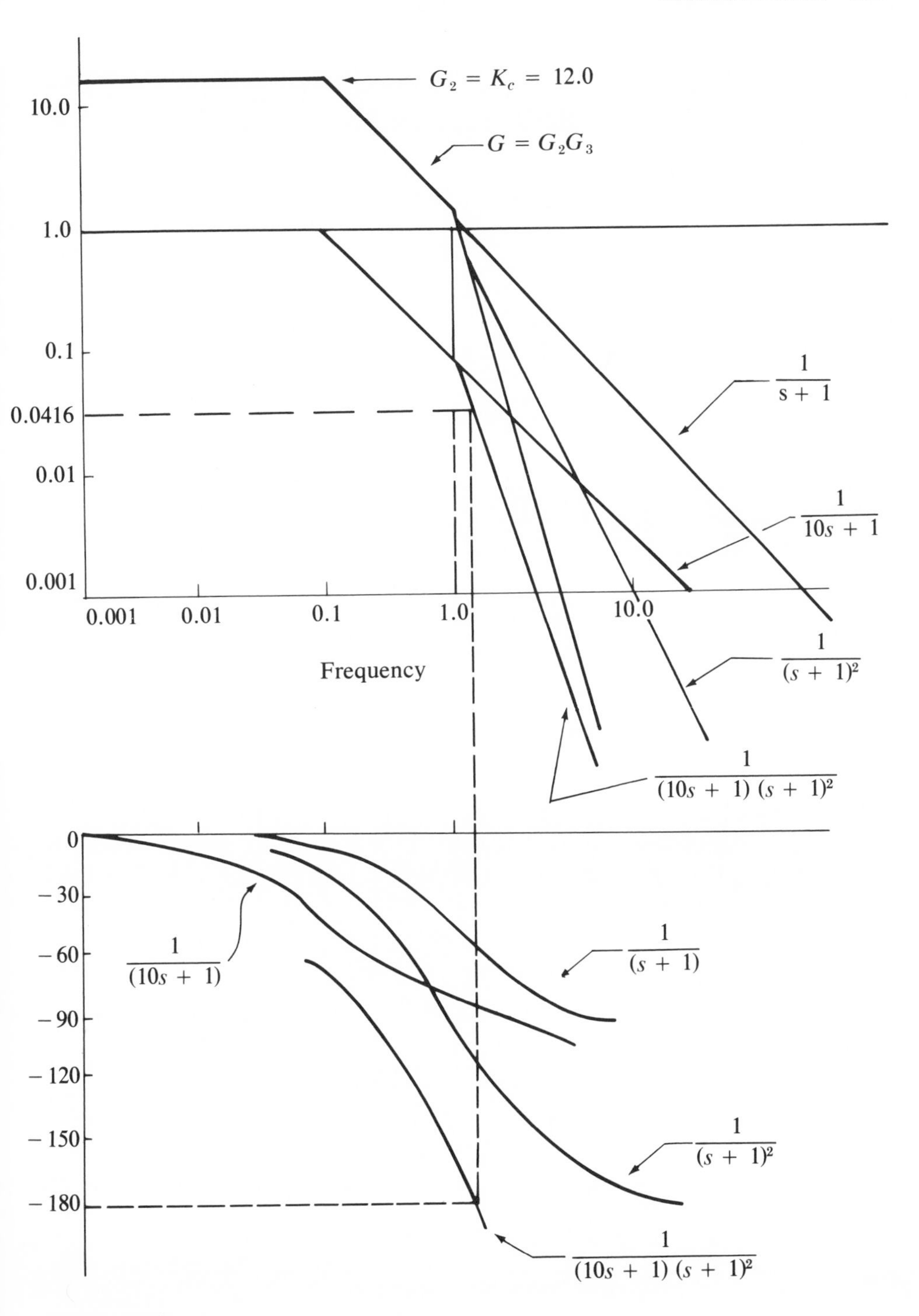

Figure 16.6
Bode Plot of Inner Loop Elements

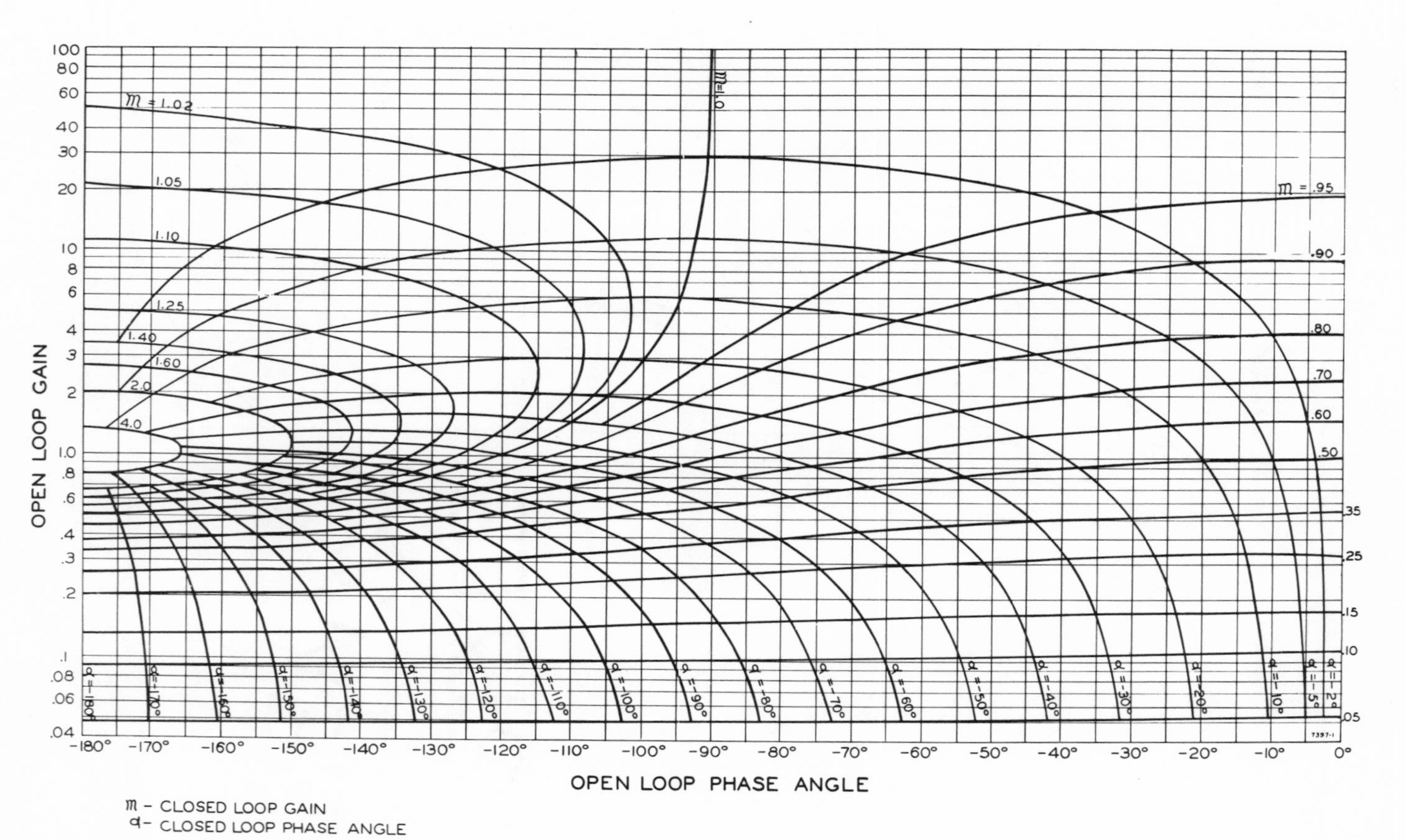

Figure 16.7
Nichols Chart

Table 16.1
Frequency-Response Data For Cascade System

Frequency, ω Radians/min	$G(s) = G_2(s)G_3(s)$		$G(s)/1 + G(s)$		$G_4(s)$		$G_4(s)\left[\frac{G(s)}{1+G(s)}\right]$	
	AR	$\phi°$	*AR*	$\phi°$	*AR*	$\phi°$	*AR*	$\phi°$
0.1	8.4	−55	0.93	−5	0.29	−88		
0.15	6.4	−73	0.945	−8	0.21	−101		
0.2	4.9	−85	0.97	−12	0.14	−112		
0.4	2.4	−120	1.15	−25	0.053	−135	0.061	−160
0.5	1.92	−133	1.33	−32	0.037	−143	0.049	−175
0.6	1.44	−143	1.7	−50	0.027	−148	0.046	−198
1.0	1.20	−170	4.5	−60				
1.2	0.55	−180	1.4	−180				
3.0	0.03	−210	0.04	−220				

Step 6. Combine the amplitude ratios and phase angles of $G_4(s)$ with those of $G(s)/[1 + G(s)]$ to obtain the magnitude ratios and phase angles of $G_4\ G(s)/[1 + G(s)]$, as shown in Table 16.1.

Step 7. At $\phi = -180°$ the amplitude ratio of the combined elements is equal to 0.048. Therefore, the ultimate gain is

$$K_u = \frac{1}{0.048} = 21$$

The Ziegler–Nichols settings for the primary controller are, therefore,

$$K_c = (0.45)(21) = 9.5$$

$$\tau_I = \frac{1}{1.2}\ \frac{2\pi}{\omega|_{\phi=-180°}} = \frac{1}{1.2}\ \frac{2\pi}{(0.53)} = 9.9 \text{ min}$$

$$\text{reset} = \frac{1}{\tau_I} = 0.101$$

Solution Part B: Feedback Control System. For simple feedback control, $G_2(s)$ would not exist. The elements of $G_3(s)$ and $G_4(s)$ are combined to obtain the composite Bode plot of $G_3(s)G_4(s)$, as shown in Table 16.2. From the table

$$\omega|_{\phi=-180°} = 0.16 \text{ and } AR = 0.093$$

Therefore, the Ziegler–Nichols controller settings are

$$K_c = 0.45\ Ku = \frac{0.45}{0.093} = 5$$

Table 16.2
Frequency-Response Data For Feedback System

Frequency, ω Radians/min	$G_3(s)$		$G_4(s)$		$G_3(s)\,G_4(s)$	
	AR	$\phi°$	AR	$\phi°$	AR	$\phi°$
0.1	0.70	−55	0.29	−88	0.2	−143
0.15	0.53	−73	0.21	−101	0.11	−174
0.2	0.41	−85	0.14	−112	0.057	−197
0.4	0.20	−120	0.053	−135		
0.5	0.16	−133	0.037	−143		
0.6	0.12	−143	0.027	−148		
1.0	0.10	−170				
1.2	0.045	−180				
3.0	0.0025	−210				

$$\tau_I = \frac{1}{1.2}\;\frac{2\pi}{\omega|_{\phi=-180°}} = \frac{2\pi}{(1.2)\,(0.16)} = 32.6$$

$$\text{reset} = \frac{1}{32.6} = 0.0306$$

A comparison of the results of the two systems shows that the cascade control system allows much higher gain and reset as compared to the feedback system.

To assess whether integral action is desirable on either controller of the cascade system, let us evaluate the response of the cascade control loop to load changes. Refer to block diagram, Figure 16.5.

a. For a unit step change in L_2 find the offset in θ_4.

$$\frac{\theta_4(s)}{L_2(s)} = \frac{G_3G_4}{1 + H_2G_2G_3 + G_1G_2G_3G_4H_1}$$

Let

$$G_2 = 12 \text{ and } G_1 = 10$$

$$\lim_{t\to\infty}\theta_4(t) = \lim_{s\to 0} s\theta_4(s) = \frac{(1)\,(1)}{1 + (1)\,(12)\,(1) + (10)\,(12)\,(1)\,(1)\,(1)}$$

$$= \frac{1}{1 + 12 + 120} = 0.008$$

$$\text{offset} = 0.008$$

Therefore, integral actional on secondary controller may not be necessary.

b. For a unit change in L_1 find the offset in θ_4

$$\frac{\theta_4(s)}{L_1(s)} = \frac{G_4(1 + H_2G_2G_3)}{1 + H_2G_2G_3 + G_4G_3G_2G_1H_1}$$

$$\text{offset} = \lim_{t \to \infty} \theta_4(t) = \lim_{s \to 0} s\theta_4(s) = \frac{1(1 + 12)}{1 + 12 + 120} = .0978$$

Therefore, integral action on primary controller may be desirable.

Now we describe the experimental approach to determining the tuning constants of cascaded controllers.

Example 2 (by Permission from Ref. 4). The industrial application described here involves a double-cascade control system. The innermost controller is analog, and the two outer control algorithms are executed on a digital computer. The purpose of the double-cascade system is to maintain the controlled variable C at set point despite changes in loads L_1 and L_2. A block diagram of this system is shown in Figure 16.8.

The purpose is to identify the dynamics of the open-loop process and that of each successive level of the closed-loop system. The resulting dynamic data are used to design the innermost analog controller and the two outer digital control algorithms. Then the controller settings are applied to the process operating under closed-loop control, in the presence of load disturbances, to evaluate the adequacy of their design.

The pulse-testing technique has been used in this application. To prepare for the plant tests, the input X_1 and the output Y_1 (see Figure 16.8) were connected to separate pens of a two-pen strip-chart recorder. When the desired steady state was achieved and the process was free from load disturbances, all control loops were switched to manual. A pulse of desired magnitude and duration was introduced into the innermost loop by rotating the manual knob on the innermost controller. By observing the time record of the input pulse, it was possible to return the input to the initial steady-state operation. The test was repeated for different pulse widths and heights (above and below the initial steady-state values). For each run numerous values of the input and output were read from their respective time records into an off-line computer program (Appendix D),

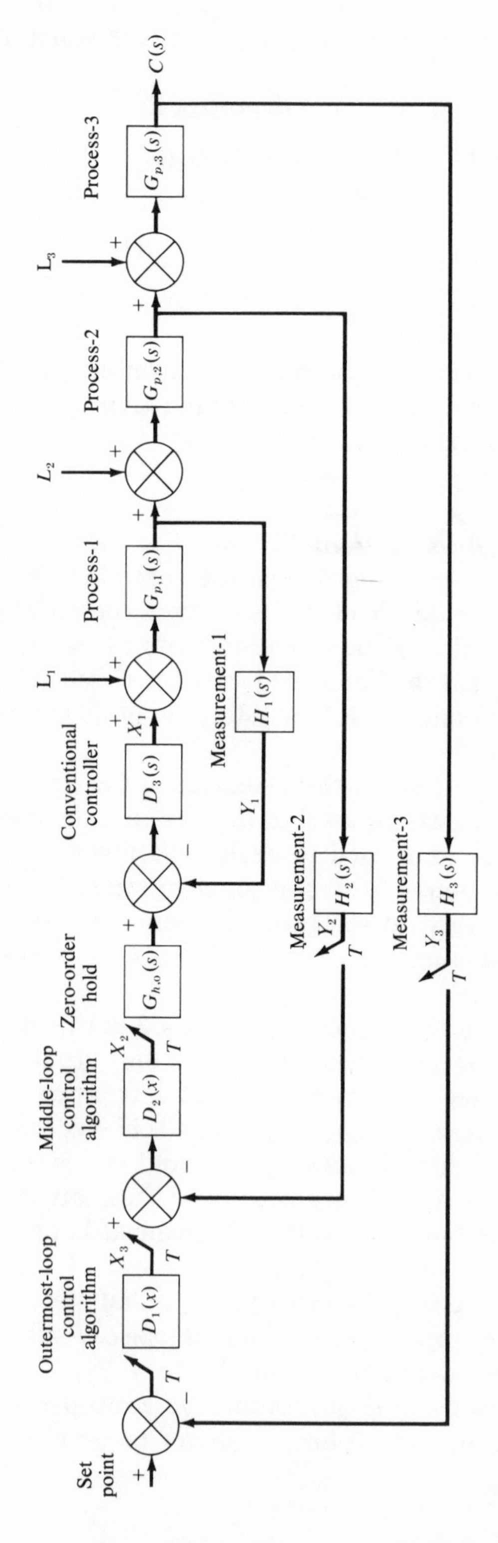

Figure 16.8
The Test System Used in Double-Cascade Control Application

which solved the appropriate equations so as to determine frequency response. From these data the tuning constants for the innermost PI controller were determined by the Ziegler–Nichols method. These settings were implemented on the innermost controller and that loop was switched to automatic. The transients were allowed to disappear, and when the steady state was reached, the second pulse test was conducted.

The innermost controller is a computer-set control station. The controller is equipped with a stepping motor which accepts a pulse train from the process-control computer at a rate of up to 30 pulses per second. The stepping motor moves the setpoint needle of the controller. Each pulse into the stepping motor moves the setpoint needles one thousandth of full scale. The stepping motor responds to discontinuities in voltage between "closure" and "common" terminals. Thus, by disconnecting the computer from the control station and by connecting two leads to the "closure" and "common" terminals, the set point can be pulsed any desired amount by repeatedly contacting the two leads.

In this manner, several pulses of desired magnitude and duration were introduced into the middle loop at X_2, and the transient response Y_2 was recorded. As before, time records for X_2 and Y_2 were processed on the off-line computer as to determine the Ziegler–Nichols settings for the middle-loop PI algorithm. These settings were divided by two (the reason for this is explained in a subsequent paragraph) and then read into the on-line control computer as tuning parameters for the middle-loop algorithm. The middle loop was then switched to automatic by reconnecting the pulse contacts from the computer to the control station.

With the two inner loops in automatic, the outermost loop was switched to manual. A pulse of desired magnitude and duration was generated on the on-line control computer and introduced into the input to the outermost loop at X_3, and the transient response Y_3 was recorded. The time records for Y_3 and X_3 assisted in the determination of the PI control algorithm constants of the outermost loop. The test was repeated using a pulse of different magnitude.

These settings were halved and then read into the control-computer program as the tuning parameters of the outermost algorithm. All three loops were switched to automatic, and the transient response of the process, in the presence of load disturbances, was obtained and analyzed.

Sample time records of input pulse and the associated transient response for each loop are shown in Figures 16.9, 16.10, and 16.11 for one of the tests. The computer-generated Bode plots from the analy-

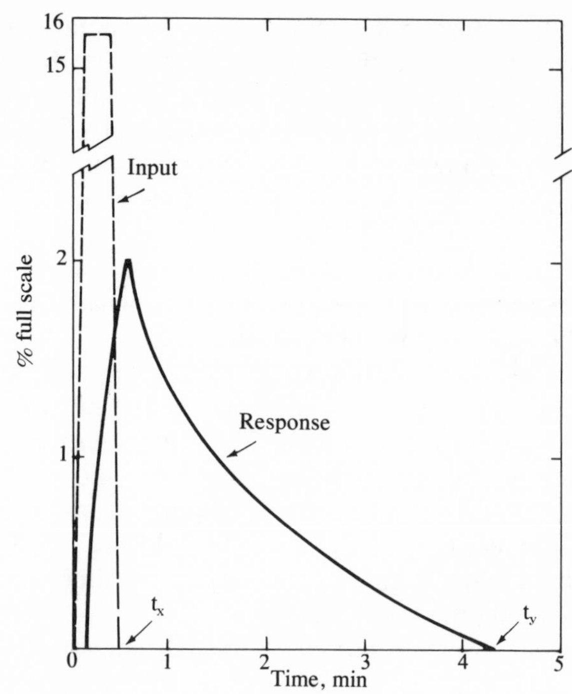

Figure 16.9
The Input and Response for the Inner Loop

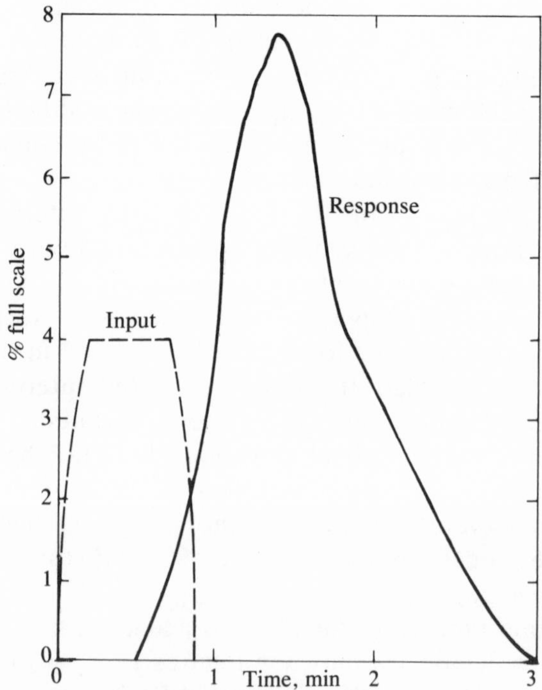

Figure 16.10
The Input and Response for the Middle Loop

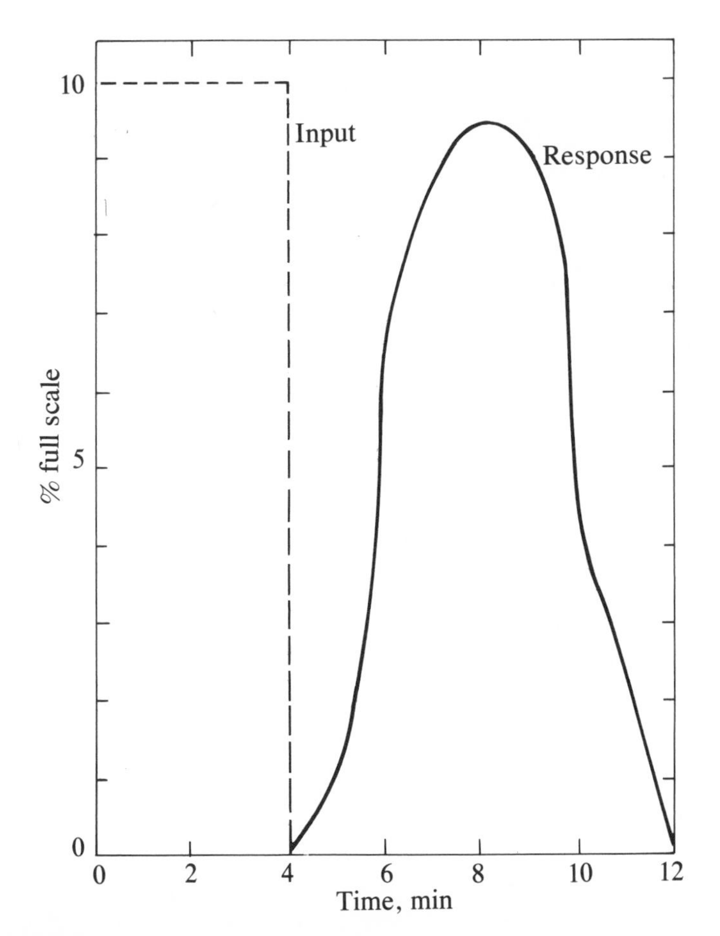

Figure 16.11
The Input and Response for the Outer Loop

sis of these figures are shown in Figures 16.12, 16.13, and 16.14, respectively. The tuning parameters determined from these Bode plots are shown in Table 16.3.

If all three loops of the double-cascade system were analog, these settings would have been satisfactory. However, since the two outer loops are under computer control, their performance depends not only on the tuning constants but also on the sampling period. Indeed, it can be shown by the Z-transform techniques, that a second-order system, which is stable for all values of the gain for a conventional system, can become unstable for some values of the sampling period in a sampled-data loop. The performance of the two outer loops with the Ziegler–Nichols settings, found from pulse testing, was found to exhibit unstable behavior. Since it was not pos-

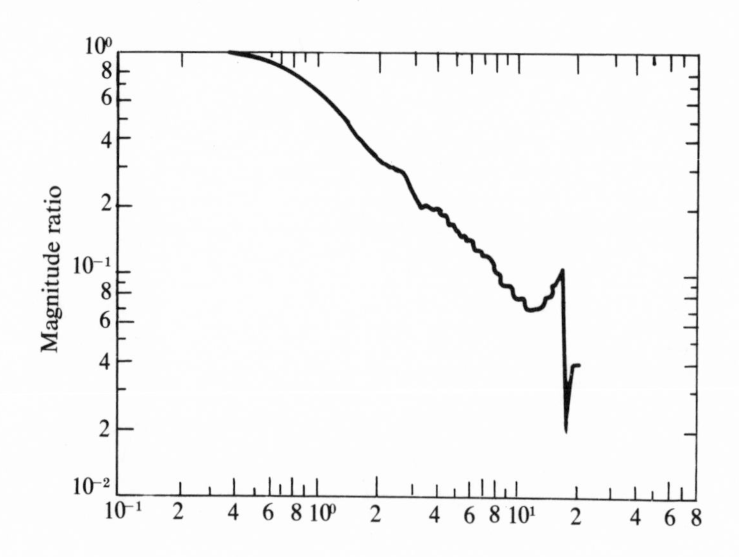

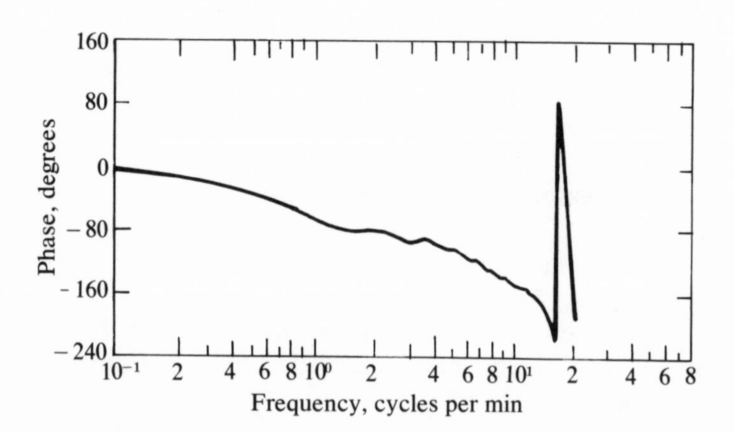

Figure 16.12
Frequency Response for the Inner Loop

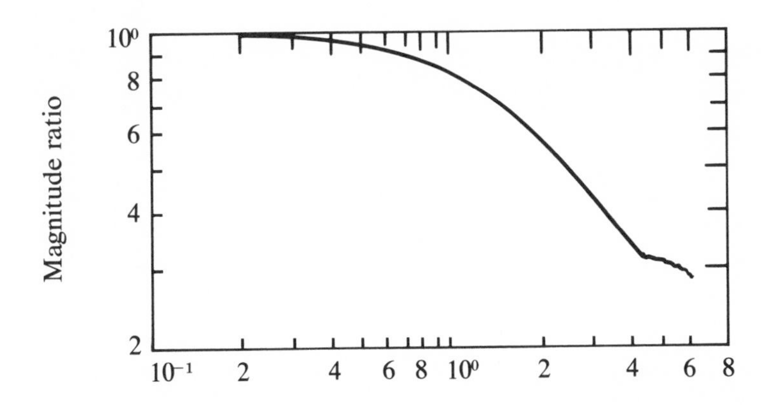

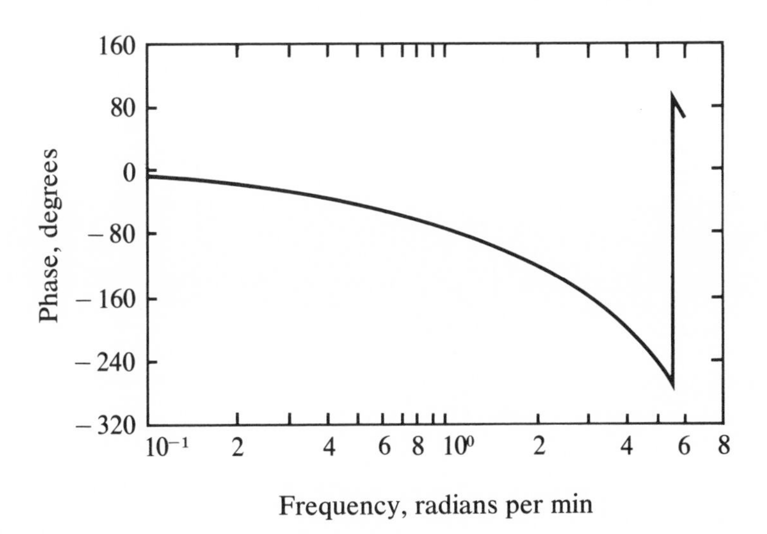

Figure 16.13
Frequency Response for the Middle Loop

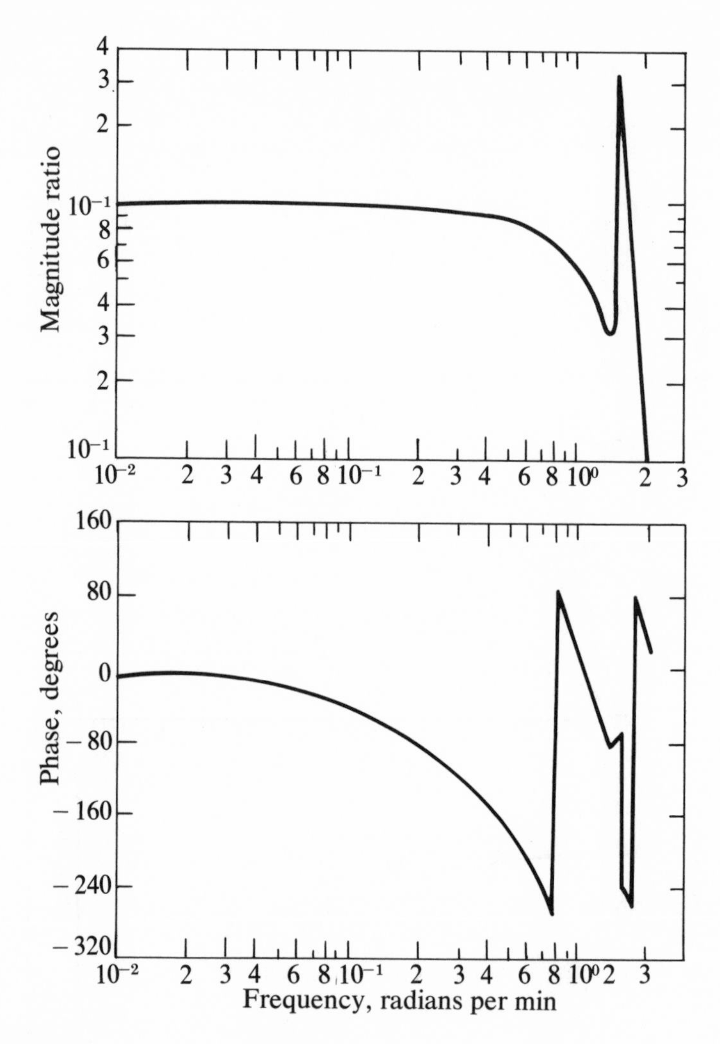

Figure 16.14
Frequency Response for the Outer Loop

Table 16.3
Tuning Parameters for Pulse Data

Loop	*Inner*	*Middle*	*Outer*
Natural period, minutes	0.50	1.88	15.4
Maximum controller, gain	14.00	1.92	8.60
Ziegler–Nichols gain	6.30	0.86	3.87
Actual gain implemented	6.30	0.43	1.94
Reset time, min	0.42	1.57	12.83

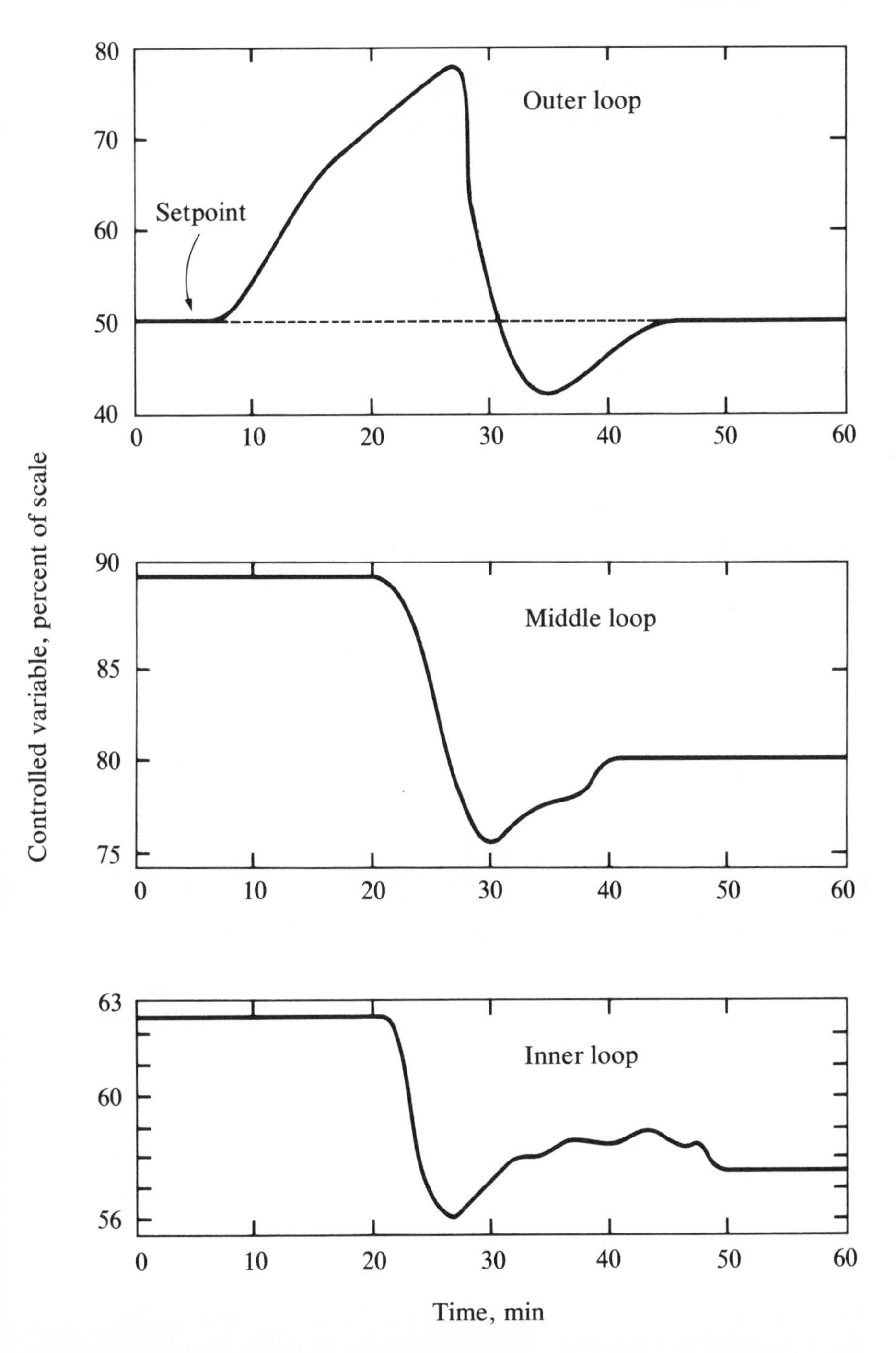

Figure 16.15
The System Response to a Load Change

sible to alter the sampling period due to other constraints, the Ziegler–Nichols settings had to be reduced. Digital computer simulations using Z transforms showed that good transient response could be obtained if Ziegler–Nichols settings were halved.

Finally, the closed-loop response of the system, with all loops in automatic, is shown in Figure 16.15. Subsequent observations have shown these settings to be quite adequate.

16.2. An Industrial Application of Cascade-Control Technique

The block diagram of an industrial polymerization control system is shown in Figure 16.16. In this application the temperature in the jacketed reactor is controlled by manipulating the flow of coolant in the jacket. The temperature response shown in Figure 16.17 has oscillations caused primarily by the variations in supply temperature and pressure of the cooling medium. Thus this system is an ideal candidate for implementing cascade control.

The block diagram of a temperature-on-temperature cascade-control system is shown in Figure 16.18. When this cascade system

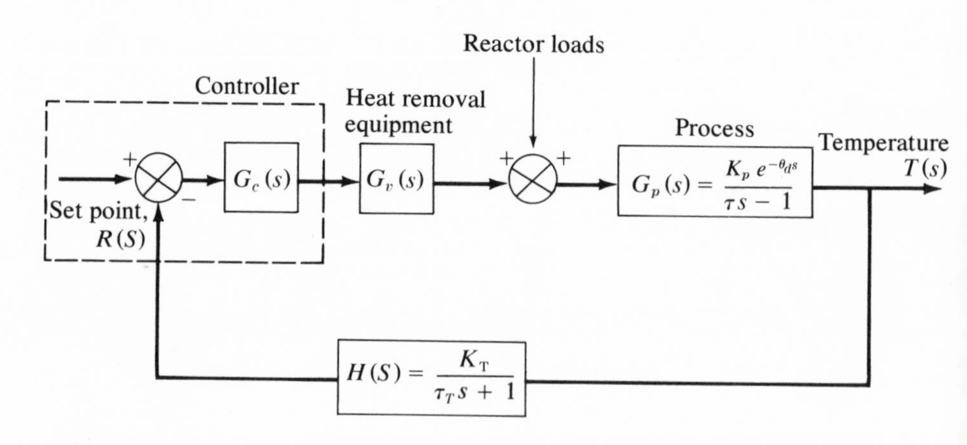

Figure 16.16
Block Diagram of Polymerization Control System (Published by Permission of E. I. Du Pont de Nemours and Company, Louisville, Kentucky)

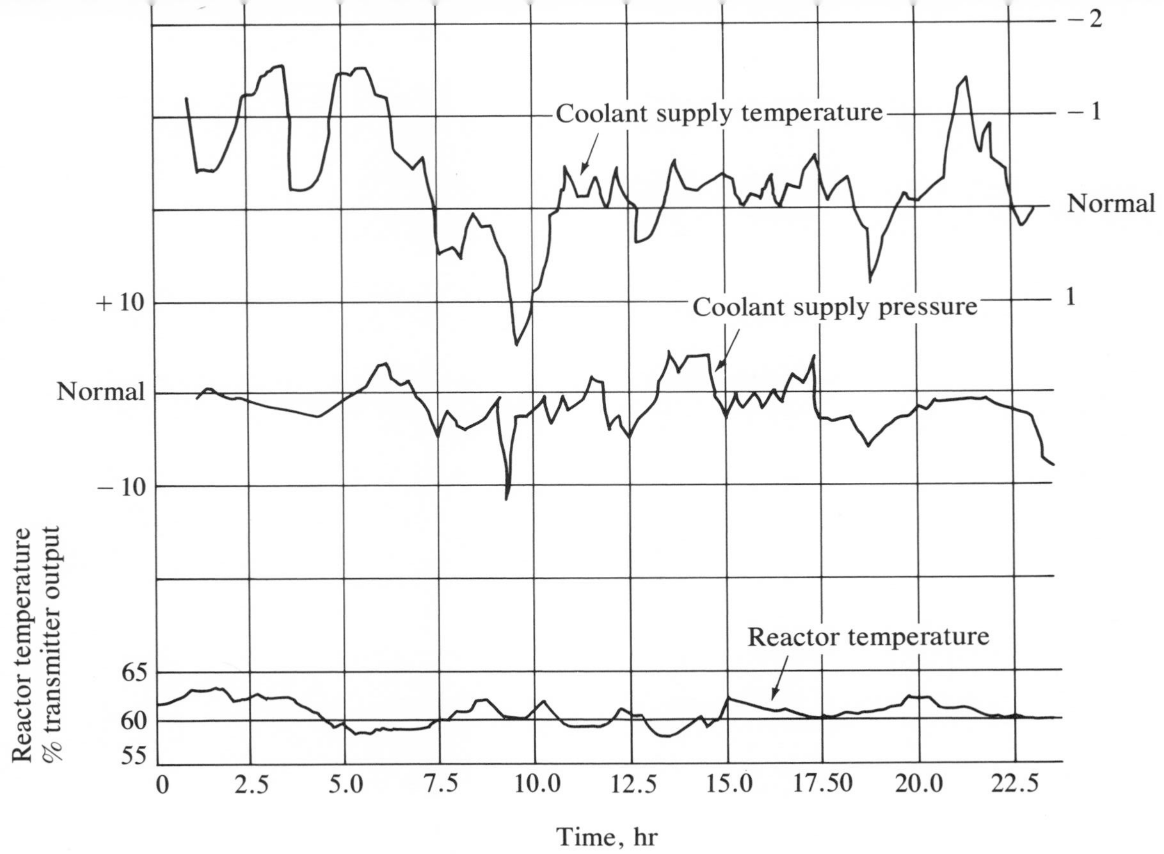

Figure 16.17
Response of Feedback Control System in the Presence of Disturbances

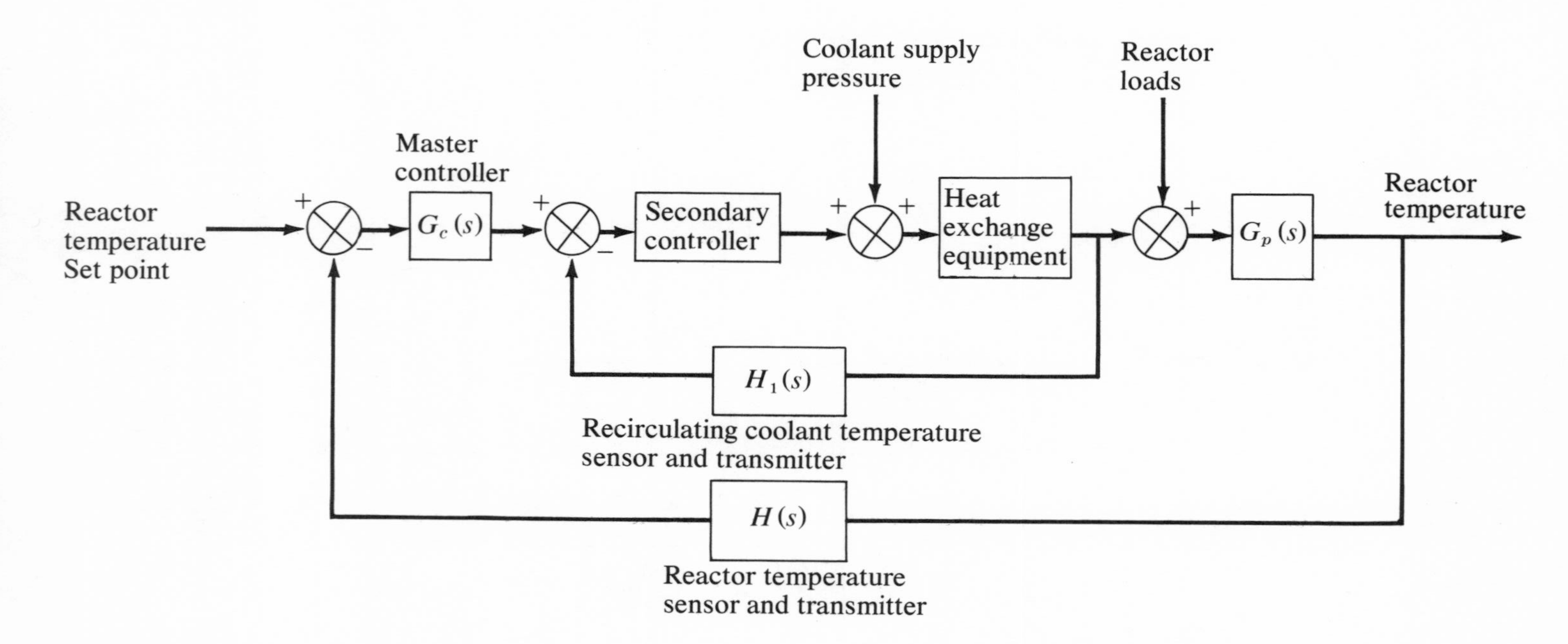

Figure 16.18
Block Diagram of Cascade Control System

was implemented, the temperature response of the system, shown in Figure 16.19, was considerably improved.

From Figure 16.17 note that the supply pressure and temperature disturbances of the coolant system go right through the system and upset the reactor temperature. The feedback system is not able to cope with these upsets. In the cascade-control results shown in Figure 16.19 the secondary loop detects the changes in coolant temperature resulting from these upsets and constantly manipulates the flow of coolant so that the temperature of the reactor is not affected.

16.3. When to Use Cascade Control

Franks and Worley[5] conducted an analog computer simulation study to assess the benefits of cascade control. Figure 16.20 shows the block diagrams and the system time constants. The results of

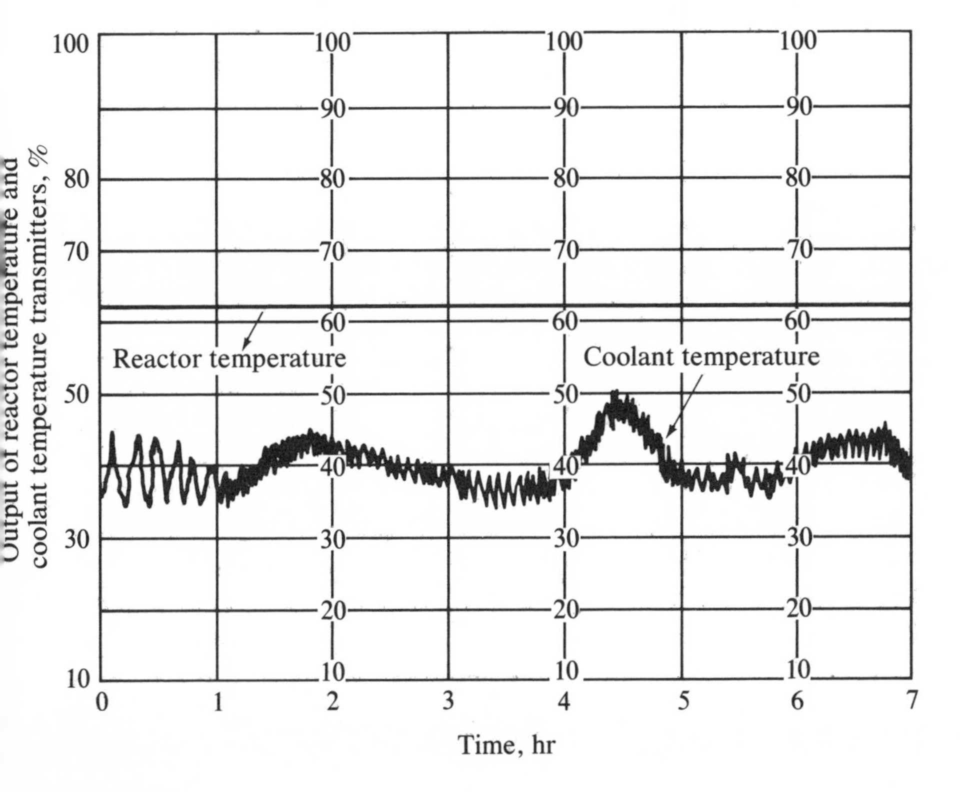

Figure 16.19
Response of Cascade Control System

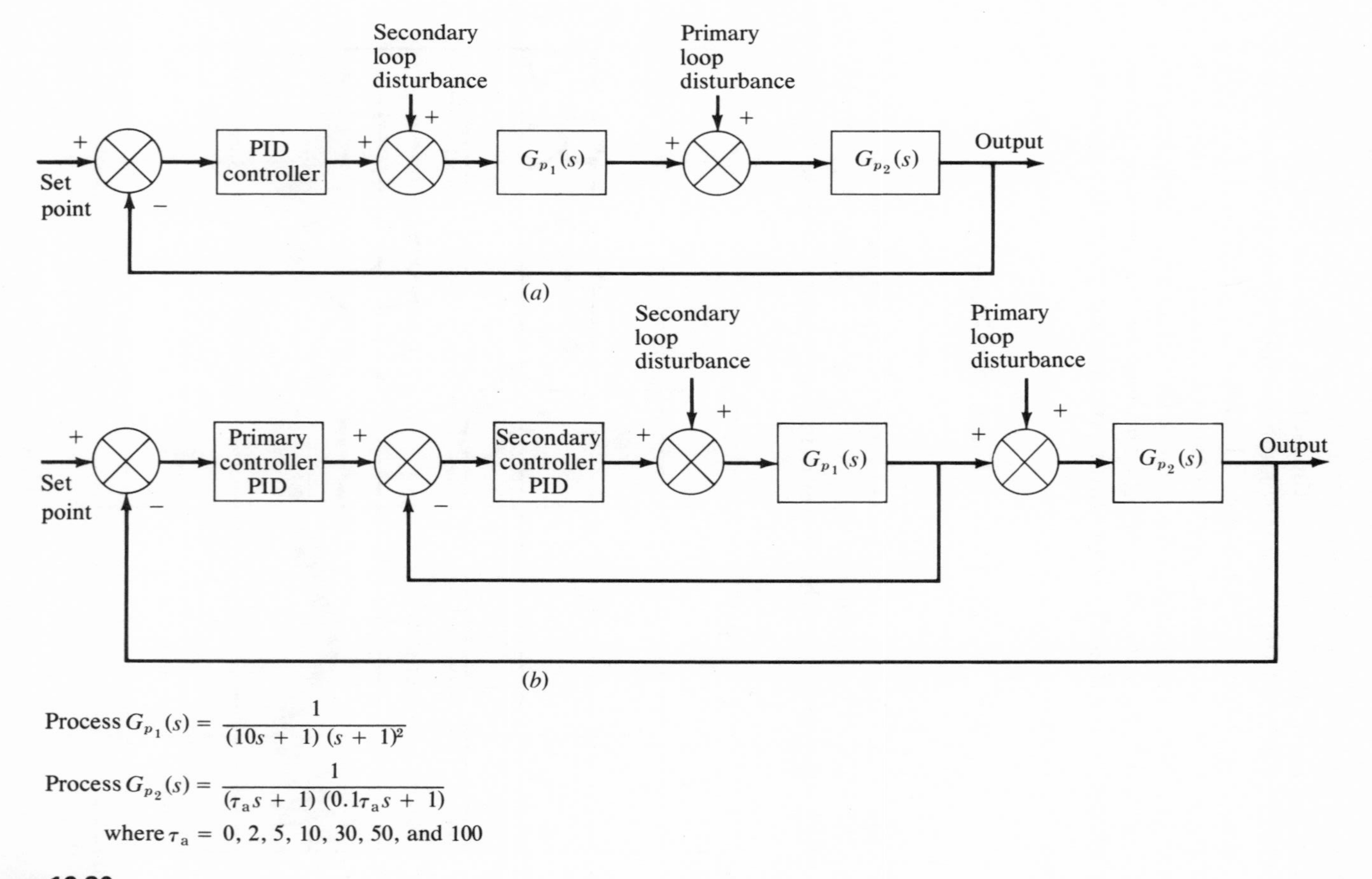

$$\text{Process } G_{p_1}(s) = \frac{1}{(10s + 1)(s + 1)^2}$$

$$\text{Process } G_{p_2}(s) = \frac{1}{(\tau_a s + 1)(0.1\tau_a s + 1)}$$

where $\tau_a = 0, 2, 5, 10, 30, 50, \text{ and } 100$

Figure 16.20
Block Diagrams and System Time Constants Used in the Simulation Study (a) Feedback System (b) Cascade System

their study are summarized in Figure 16.21. This figure shows that the improvement for set-point changes and for master-loop disturbances is much less than for disturbances introduced into the slave loop. Maximum benefit from cascade control for set-point changes and for master-loop disturbances occurs when the dominant time constants of the two loops are roughly equal. The improvement in the response of the control system to slave-loop disturbances increases rapidly as the ratio of the dominant time constants increases.

The applicability of Figure 16.21 is somewhat limited, because it

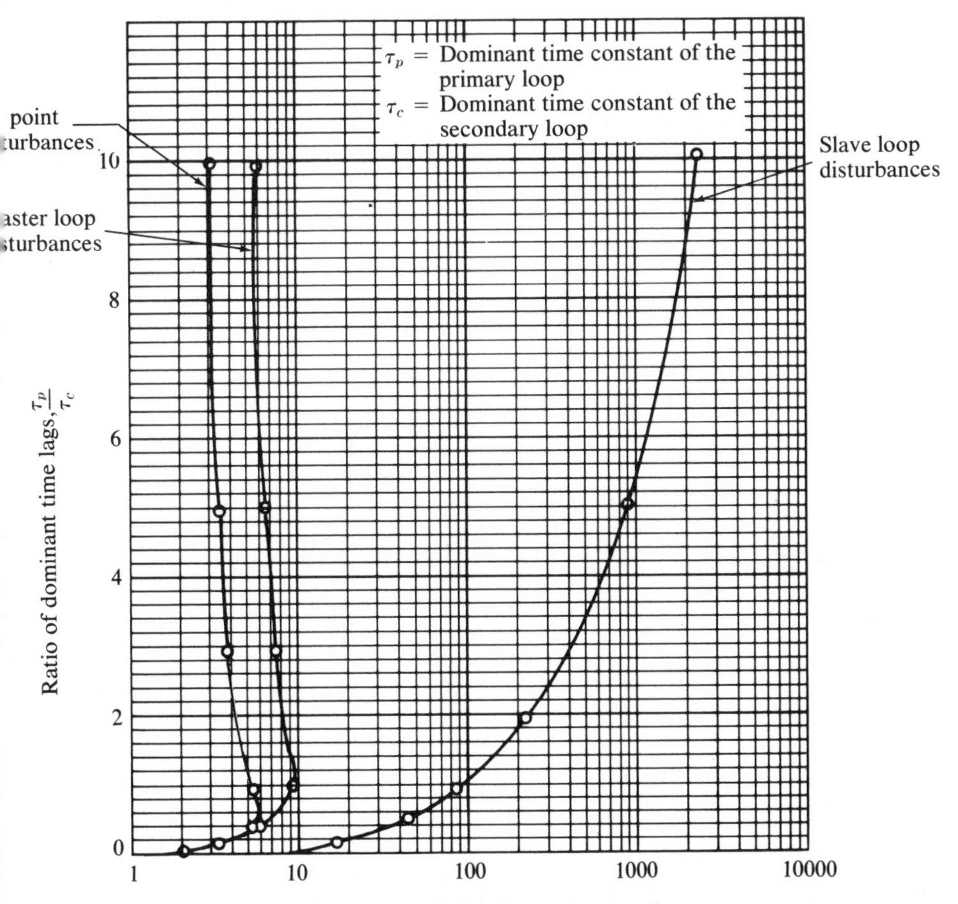

Figure 16.21
Relative Improvement of Cascade Control over Single Loop Control. (Franks, R.G., and C.W. Worley; "Quantitative Analysis of Cascade Control", *Ind. Eng. Chem.*, 48:1074, 1956 (Reproduced with Permission)

does not show the effect of dead time on the control performance of cascade systems. However, it does show quantitatively, the benefits of cascade control for the types of systems that were studied.

References

1. Coughanowr, D. R., Koppel, L. B., *Process Systems Analysis and Control*, McGraw-Hill, New York, 1965.
2. Harriott, P., *Process Control*, McGraw-Hill, New York, 1965.
3. James, et al., *Theory of Servomechanisms*, McGraw-Hill, New York, 1947, pp. 180–182.
4. Schork, F. J., Deshpande, P. B., Double Cascade Controller Tested, *Hydrocarbon Processing*, June 1978, 113–117.
5. Franks, R. G., Worley, C. W., Quantitative Analysis of Cascade Control, *Ind. Eng. Chem.*, **48**, 1956, 1074.

CHAPTER 17

Multivariable Control Systems

Most large processes have many controlled variables and many manipulated variables. Ideally we would like a given manipulated variable to affect only its own controlled variable. Unfortunately, in many cases, a change in one manipulated variable upsets other controlled variables in the process in addition to its own controlled variable. In such multivariable control loops, *coupling* is said to exist. If coupling is severe, a large disturbance will result in the second loop (in a 2×2 system, for example) whenever the manipulated variable of the first loop changes. If, in addition to coupling from the first loop to the second there is coupling from the second loop to the first, *interaction* is said to exist. This interaction can cause oscillations and even instability.

Since coupling or interaction may exist in multivariable control systems, it is first of all important to know which manipulated variables should be connected to which controlled variables. It is possible that one possible combination of controlled and manipulated variables may be better than others. On the other hand, no combination may be satisfactory. The purpose of this chapter, then, is:

1. To develop methods to determine the extent of interaction in a multivariable control system.

2. To determine, based on the information developed in step 1, the proper pairings of manipulated and controlled variables.

3. To design decoupling networks, if no pairing is satisfactory and interaction is severe, so as to achieve noninteracting feedback control of the multivariable system.

17.1. The Interaction Measure

A measure of the extent of interaction in multivariable control systems is obtained by Bristol's method.[1] It is based on the steady-state input-output relationships for the process. The method seeks to determine the best single-input single-output connections (i.e., best pairing of manipulated and controlled variables). It yields a measure of steady-state gain between a given input-output pairing and thus, by using the most sensitive input-output connections, control interaction can be minimized. Consider a system, shown in Figure 17.1, with two controlled variables C_1 and C_2 and two manipulated variables M_1 and M_2 where each C is affected by both Ms. Around some steady-state operating point we can express the steady-state relationships between the Cs and Ms as follows:

$$\Delta C_1 = \left.\frac{\partial C_1}{\partial M_1}\right|_{M_2} \Delta M_1 + \left.\frac{\partial C_1}{\partial M_2}\right|_{M_1} \Delta M_2 = K_{11}\Delta M_1 + K_{12}\Delta M_2 \tag{17.1}$$

$$\Delta C_2 = \left.\frac{\partial C_2}{\partial M_1}\right|_{M_2} \Delta M_1 + \left.\frac{\partial C_2}{\partial M_2}\right|_{M_1} \Delta M_2 = K_{21}\Delta M_1 + K_{22}\Delta M_2 \tag{17.2}$$

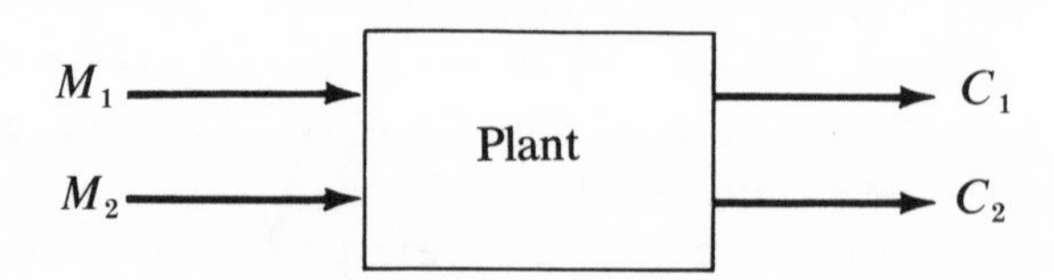

Figure 17.1
A 2 × 2 Multivariable System

The Ks (partial derivatives evaluated at the steady-state operating point) are called the *open-loop steady-state gains* of the process. They quantitatively describe how the Ms affect the Cs. They can be determined from a mathematical model or by experimental tests on the process. To evaluate K_{11}, for example, we would make a small change in M_1 while the process was operating at steady state, while holding M_2 constant. After C_1 and C_2 reached their new steady-state values, we would evaluate

$$K_{11} = \left.\frac{\Delta C_1}{\Delta M_1}\right|_{\substack{M_2 \text{ constant} \\ (\Delta M_2 = 0)}} ; \quad K_{21} = \left.\frac{\Delta C_2}{\Delta M_1}\right|_{\substack{M_2 \text{ constant} \\ (\Delta M_2 = 0)}} \tag{17.3}$$

The gain K_{11}, then, determines the change in C_1 due to a change in M_1 while M_2 is held constant. Suppose, instead of holding M_2 constant while we make a small change in M_1, we simultaneously manipulate M_2 to bring C_2 back to the original value it had before the change in M_1 was made. We can then define another "gain" between C_1 and M_1:

$$a_{11} = \left.\frac{\Delta C_1}{\Delta M_1}\right|_{C_2 \text{ constant}} \tag{17.4}$$

The a_{11} is a measure of how much M_1 would affect C_1 if all other controlled variables were under closed loop-control (i.e., held constant).

The ratio of K_{11} to a_{11} is called the *relative gain*, λ_{11}. Thus

$$\lambda_{11} = \frac{K_{11}}{a_{11}} = \frac{\left.\frac{\Delta C_1}{\Delta M_1}\right|_{M_2 \text{ constant}}}{\left.\frac{\Delta C_1}{\Delta M_1}\right|_{C_2 \text{ constant}}} \tag{17.5}$$

By comparing the relative gains for each manipulated variable, we can get a quick quantitative assessment of which M has the most effect on a given controlled variable C, and thus how we should pair the Ms and Cs.

The a_{ij} gains can be computed from the K_{ij} gains as follows. Rewriting the open-loop relationship between M s and C s from Equations (17.1) and (17.2) we have

$$\Delta C_1 = K_{11}\,\Delta M_1 + K_{12}\,\Delta M_2 \tag{17.6}$$

$$\Delta C_2 = K_{21}\,\Delta M_1 + K_{22}\,\Delta M_2 \tag{17.7}$$

The a_{11}, by definition, is

$$\frac{\Delta C_1}{\Delta M_1} \text{ with } \Delta C_2 = 0.$$

Thus, from Equation (17.7)

$$\Delta C_2 = 0 = K_{21}\,\Delta M_1 + K_{22}\,\Delta M_2$$

solving for ΔM_2:

$$\Delta M_2 = -\frac{K_{21}}{K_{22}}\,\Delta M_1 \qquad (17.8)$$

Substituting Equation (17.8) into (17.6) we get

$$\Delta C_1 = K_{11}\,\Delta M_1 - \frac{K_{12}\,K_{21}}{K_{22}}\,\Delta M_1 = \left(K_{11} - \frac{K_{12}\,K_{21}}{K_{22}}\right)\Delta M_1$$

Thus (17.9)

$$a_{11} = \left.\frac{\Delta C_1}{\Delta M_1}\right|_{\Delta C_2 = 0} = K_{11} - \frac{K_{12}\,K_{21}}{K_{22}} = \frac{K_{11}K_{22} - K_{12}K_{21}}{K_{22}}$$

The relative gain λ_{11} is then

$$\lambda_{11} = \frac{K_{11}}{a_{11}} = \frac{K_{11}\,K_{22}}{K_{11}K_{22} - K_{12}K_{21}} \qquad (17.10)$$

The other three gains a_{ij} can be computed similarly; they are:

$$a_{12} = \left.\frac{\Delta C_1}{\Delta M_2}\right|_{\Delta C_2 = 0} = \frac{K_{12}K_{21} - K_{11}K_{22}}{K_{21}}$$

$$\Rightarrow \lambda_{12} = \frac{K_{12}\,K_{21}}{K_{12}K_{21} - K_{11}K_{22}}$$

$$a_{21} = \left.\frac{\Delta C_2}{\Delta M_1}\right|_{\Delta C_1 = 0} = \frac{K_{12}K_{21} - K_{11}K_{22}}{K_{12}} \qquad (17.11)$$

$$\Rightarrow \lambda_{21} = \frac{K_{12}\,K_{21}}{K_{12}K_{21} - K_{11}K_{22}}$$

$$a_{22} = \left.\frac{\Delta C_2}{\Delta M_2}\right|_{\Delta C_1 = 0} = \frac{K_{11}K_{22} - K_{12}K_{21}}{K_{11}}$$

$$\Rightarrow \lambda_{22} = \frac{K_{11}\,K_{22}}{K_{11}K_{22} - K_{12}K_{21}}$$

To facilitate the pairing of manipulated and controlled variables, it is convenient to arrange the relative gains as follows:

$$
\begin{array}{c|cccc}
 & M_1 & M_2 & \cdots & M_n \\
\hline
C_1 & \lambda_{11} & \lambda_{12} & \cdots & \lambda_{1n} \\
C_2 & \lambda_{21} & \lambda_{22} & \cdots & \lambda_{2n} \\
\vdots & \vdots & \vdots & \vdots & \vdots \\
C_n & \lambda_{n1} & \lambda_{n2} & \cdots & \lambda_{nn}
\end{array}
\qquad (17.12)
$$

For each controlled variable C_i the manipulated variable selected is the one having the largest positive relative gain. For example, the manipulated variable for C_1 would be the one corresponding to the largest relative gain λ_{1j} in the first row of the relative gain matrix above. The relative gain matrix for the two-variable system is as follows:

$$
\begin{array}{c|cc}
 & M_1 & M_2 \\
\hline
C_1 & \dfrac{K_{11}K_{22}}{K_{11}K_{22} - K_{12}K_{21}} & \dfrac{K_{12}K_{21}}{K_{12}K_{21} - K_{11}K_{22}} \\
C_2 & \dfrac{K_{12}K_{21}}{K_{12}K_{21} - K_{11}K_{22}} & \dfrac{K_{11}K_{22}}{K_{11}K_{22} - K_{12}K_{21}}
\end{array}
\qquad (17.13)
$$

A useful property of the relative gain matrix is that each row and each column sums to 1. Thus, in a 2×2 system, only one of the four relative gains need to be explicitly computed.

It is possible for relative gains to be negative. If an m and a C with a negative relative gain are paired, the system will be uncontrollable and unstable—each variable will be driven to its limit. Poor control may also result if $\lambda_{ij} >> 1$.

For systems with more than two Cs, the a_{ij}s can be conveniently computed from the K_{ij}s as follows. First, arrange the Ks in a matrix as

$$
\mathbf{K} = \begin{bmatrix}
K_{11} & K_{12} & \cdots & K_{1n} \\
K_{21} & K_{22} & \cdots & K_{2n} \\
\vdots & & & \\
K_{n1} & K_{n2} & \cdots & K_{nn}
\end{bmatrix}
\qquad (17.14)
$$

Then, compute a *complimentary matrix* **C**, by first inverting, then transposing the matrix **K**:

$$\mathbf{C} = (\mathbf{K}^{-1})^T = \begin{bmatrix} C_{11} & C_{12} & \cdots & C_{1n} \\ C_{21} & C_{22} & \cdots & C_{2n} \\ C_{n1} & C_{n2} & \cdots & C_{nn} \end{bmatrix} \tag{17.15}$$

The element on the ith row and jth column of **C** is the *reciprocal of* a_{ij}; that is,

$$C_{ij} = \frac{1}{a_{ij}} \tag{17.16}$$

Thus each relative gain term λ_{ij} is found by multiplying each element in matrix **K** by its corresponding term in matrix **C**:

$$\lambda_{ij} = \frac{K_{ij}}{a_{ij}} = K_{ij}\, C_{ij} \tag{17.17}$$

Example 1 (By Permission from Ref. 2). As shown in Figure 17.2, two liquids are mixed in line to produce a mixture of desired composition X^{set}. The total flow rate is also to be controlled. The problem is to find the proper pairing of the controlled and manipu-

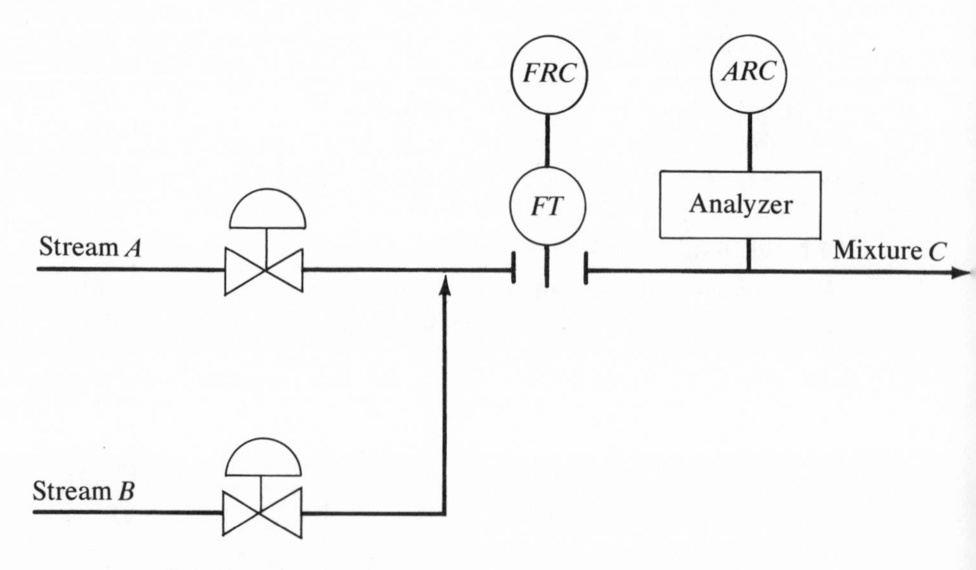

Figure 17.2
Schematic of Blending System

lated variables if the composition X is to be controlled at 0.3 mass fraction A.

The equations relating controlled variables to the manipulated variables are:

$$C = A + B$$

and

$$X = \frac{A}{A + B}$$

The steady-state gains are

$$K_{11} = \frac{\partial C}{\partial A} = \frac{\partial (A + B)}{\partial A} = 1; K_{12} = \frac{\partial C}{\partial B} = \frac{\partial (A + B)}{\partial B} = 1$$

$$K_{21} = \frac{\partial X}{\partial A} = \frac{\partial [A/(A + B)]}{\partial A} = \frac{(A + B) - A}{(A + B)^2} = \frac{B}{C^2} = \frac{1 - X}{C}$$

$$K_{22} = \frac{\partial X}{\partial B} = \frac{\partial [A/(A + B)]}{\partial B} = \frac{O - A}{(A + B)^2} = -\frac{A}{C^2} = -\frac{X}{C}$$

Therefore, with reference to Equation (17.13), λ_{11} is computed as

$$\lambda_{11} = \frac{K_{11} K_{22}}{K_{11}K_{22} - K_{12}K_{21}} = \frac{(1)\left(-\frac{X}{C}\right)}{(1)\left(-\frac{X}{C}\right) - (1)\left(\frac{1 - X}{C}\right)} = X$$

As has been pointed out earlier, only one element of the relative gain matrix need be explicitly computed, since the rows and columns of the matrix sum to 1. Thus

$\lambda =$

	A	B
C	X	$1 - X$
X	$1 - X$	X

If $X^{set} = 0.3$. then

$\lambda =$

	A	B
C	0.3	0.7
X	0.7	0.3

Since the elements with the largest positive relative gain are to be selected, this matrix suggests that we control the total flow by manipulating B and composition by manipulating A.

Example 2.[3] We now present a laboratory application of Bristol's approach to multivariable pairing. We operate the closed-loop process with correct pairing and with incorrect pairing of the controlled and manipulated variables to assess the benefits of proper pairing. The hardware for this experiment is shown in Figure 17.3.

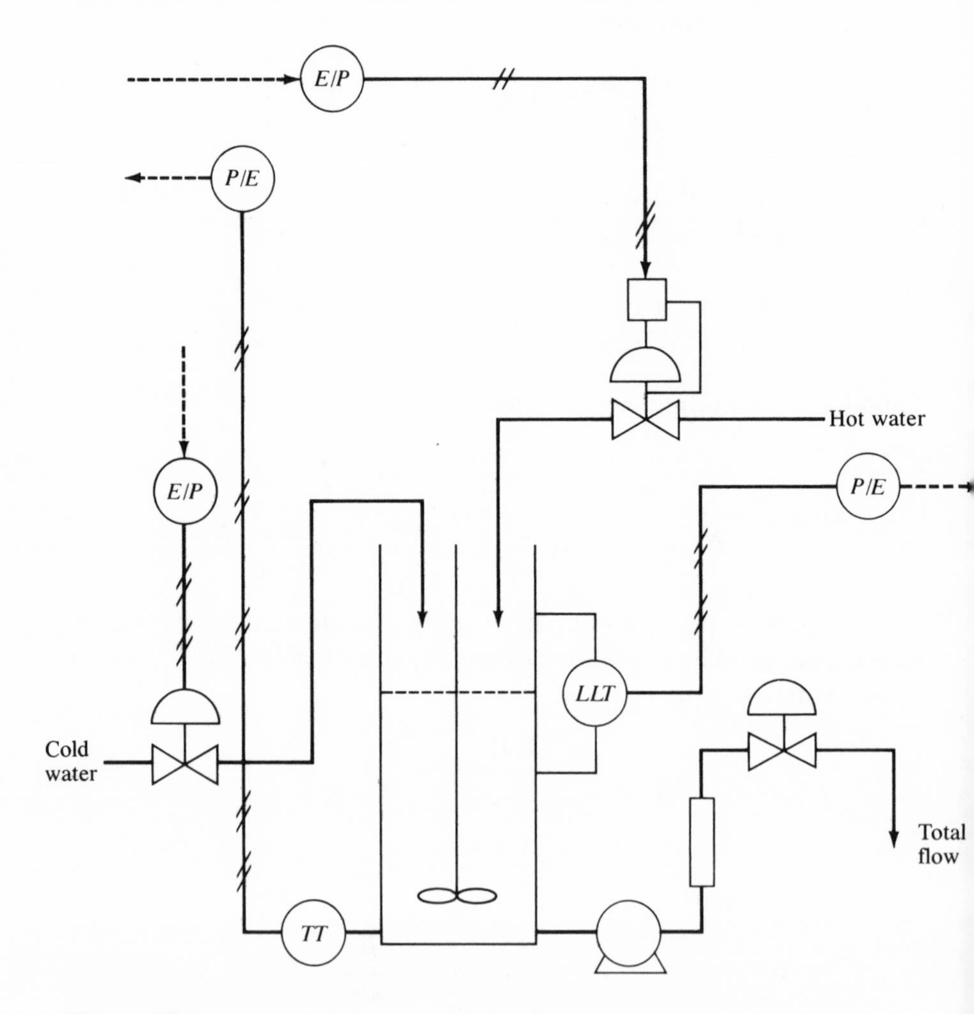

Figure 17.3
Schematic of Multivariable System (Dashed Lines Indicate Signal Transmission Between Process and Computer

The process objective is to control the level (in effect, total flow) and temperature of water in the tank. There are two inputs to the process, namely, the flow of cold water and the flow of hot water into the tank. So, the controlled variables are temperature and total flow and the manipulated variables are the cold-water and hot-water flow rates. The question is, should the temperature be controlled by manipulating hot water flow and level (i.e., total flow) by cold-water flow or vice versa? Bristol's method provides the answer.

Bristol's Relative Gains Analysis. The functional steady-state relationship between temperature, total flow and the flow streams is

$$T = f(m_c, m_h) = (m_c T_c + m_h T_h)/m_t \tag{17.18}$$

$$m_t = f(m_c, m_h) = m_c + m_h$$

Around some steady-state operating point these relationships can be expressed as

$$\Delta T = \left.\frac{\partial T}{\partial m_c}\right|_{m_h = \text{constant}} \Delta m_c + \left.\frac{\partial T}{\partial m_h}\right|_{m_c = \text{constant}} \Delta m_h$$

$$= K_{11}\Delta m_c + K_{12}\Delta m_h$$

and (17.19)

$$\Delta m_t = \left.\frac{\partial m_t}{\partial m_c}\right|_{m_h = \text{constant}} \Delta m_c + \left.\frac{\partial m_t}{\partial m_h}\right|_{m_c = \text{constant}} \Delta m_h$$

$$= K_{21}\Delta m_c + K_{22}\Delta m_h$$

Recall that the relative gains matrix for a 2 × 2 system is

$$\lambda = \begin{array}{c|cc} & m_c & m_h \\ \hline T & \lambda_{11} & \lambda_{12} \\ m_T & \lambda_{21} & \lambda_{22} \end{array} \tag{17.20}$$

where

$$\lambda_{11} = \frac{K_{11} K_{22}}{K_{11}K_{22} - K_{12}K_{21}}$$

Since each row and column sums to one, only one λ need be computed in this application. Thus, taking the partial derivatives of the terms in Equation (17.18) as indicated in Equation (17.19), we obtain the following relative gains matrix.

$$\lambda = \begin{array}{c|cc} & m_c & m_h \\ \hline T & \dfrac{m_h}{m_t} & \dfrac{m_c}{m_t} \\ & & \\ m_T & \dfrac{m_c}{m_t} & \dfrac{m_h}{m_t} \end{array} = \begin{array}{c|cc} & m_c & m_h \\ \hline T & 0.172 & 0.828 \\ & & \\ m_t & 0.828 & 0.172 \end{array} \tag{17.21}$$

This equation shows that T should be controlled by manipulating m_h and m_t by manipulating m_c. In this application both loops use a PI control algorithm on the digital computer as the control element. The algorithm was tuned by trial and error. The steady-state operating conditions were: level set point, 50% (which corresponded to total outlet flow of 11.6 lit/min); temperature set point, 24.4°C; cold-water flow, 9.61 lit/min; hot water flow, 1.99 lit/min. The process was operated with correct pairing as well as with incorrect pairing. The benefits of proper pairing are clearly evident in the set-point responses shown in Figures 17.4 and 17.5. These results show that Bristol's approach is a simple and powerful tool in the control systems design of multi-variable processes.

17.2. Interaction and Decoupling[4,5]

If the relative gains are numerically close to each other, interaction ("fighting loops") in a multivariable control system is likely to be a problem, particularly if the response times of the loops are comparable. In cases in which cross coupling between loops is severe, the system can become unstable, and decoupling will be required.

To successfully implement a decoupler, the model or frequency response of the process should be known to a high degree of accuracy.

A decoupler is a device that eliminates interaction between manipulated and controlled variables by, in effect, changing all the manipulated variables in such a manner that only the desired controlled variable changes.

For example, in a two-variable interacting system, shown in Figure 17.6, we can write:

$$\begin{aligned} \Delta C_1(s) &= G_{11}(s)\Delta M_1(s) + G_{12}(s)\Delta M_2(s) \\ \Delta C_2(s) &= G_{21}(s)\Delta M_1(s) + G_{22}(s)\Delta M_2(s) \end{aligned} \tag{17.22}$$

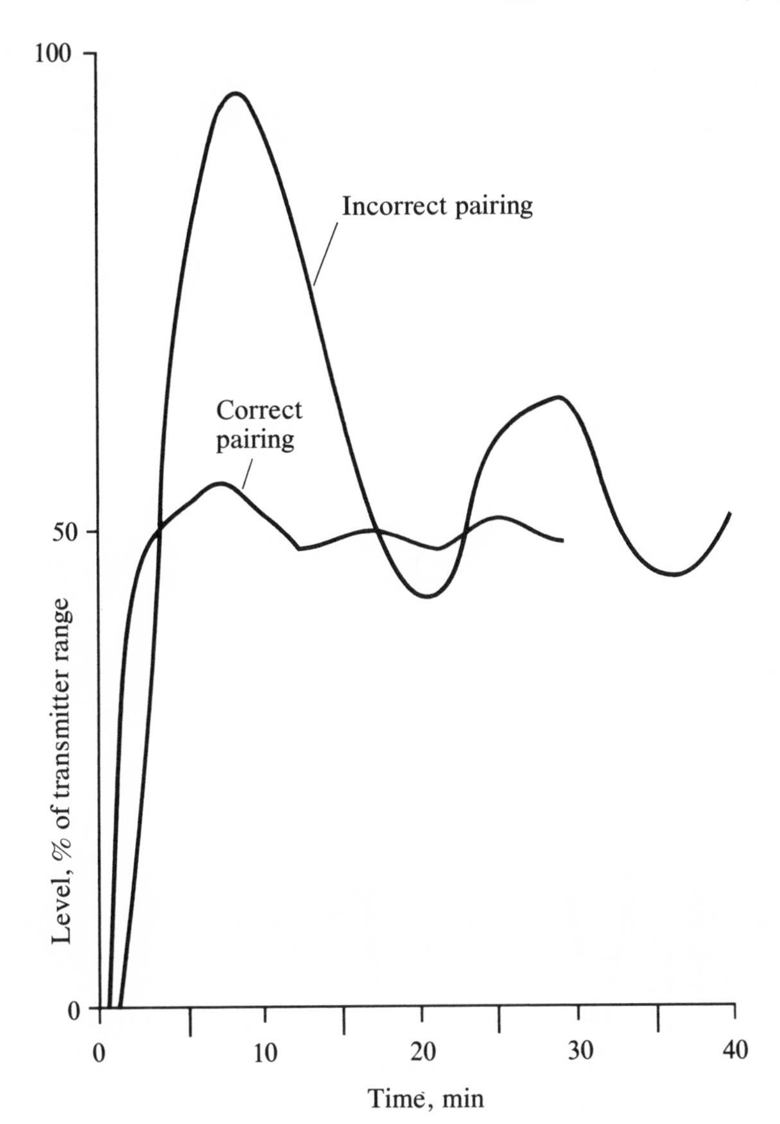

Figure 17.4
Transient Response of Level

The decoupler inputs are two new manipulated variables u_1 and u_2, and its outputs are the original manipulated variables M_1 and M_2, as shown in Figure 17.7. We want to design the decoupler so that u_1 affects only C_1 and u_2 affects only C_2.

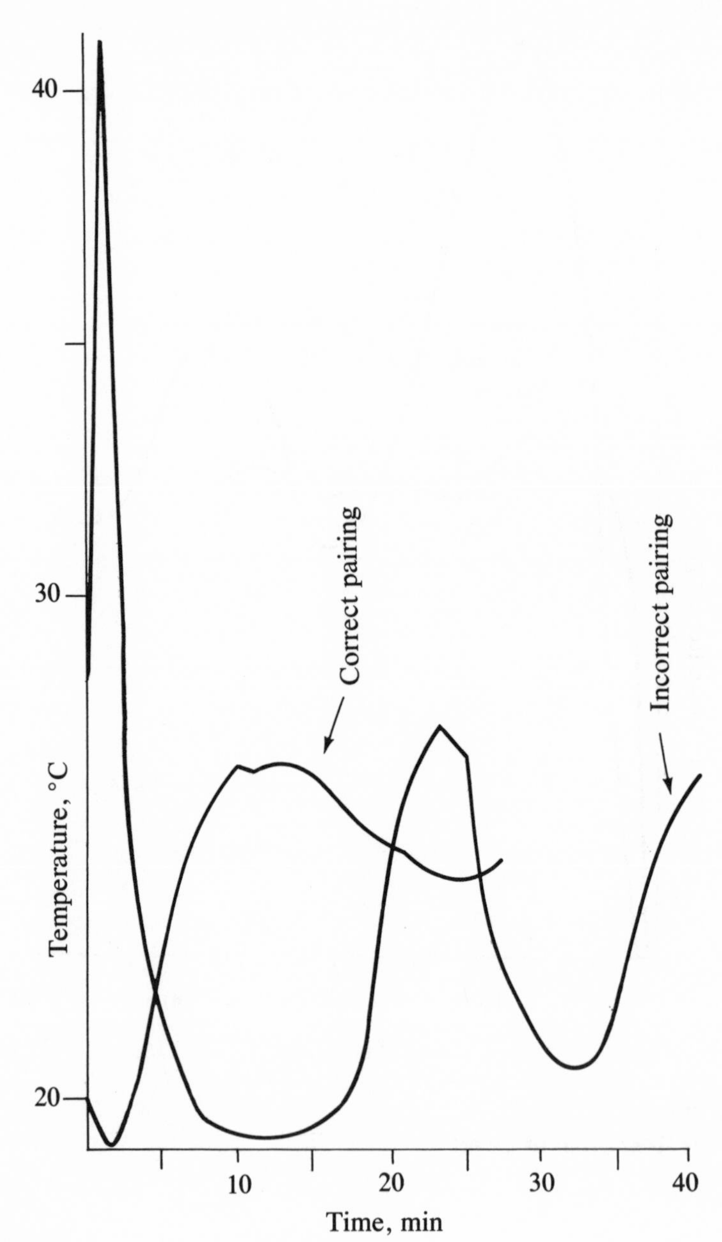

Figure 17.5
Transient Response of Temperature

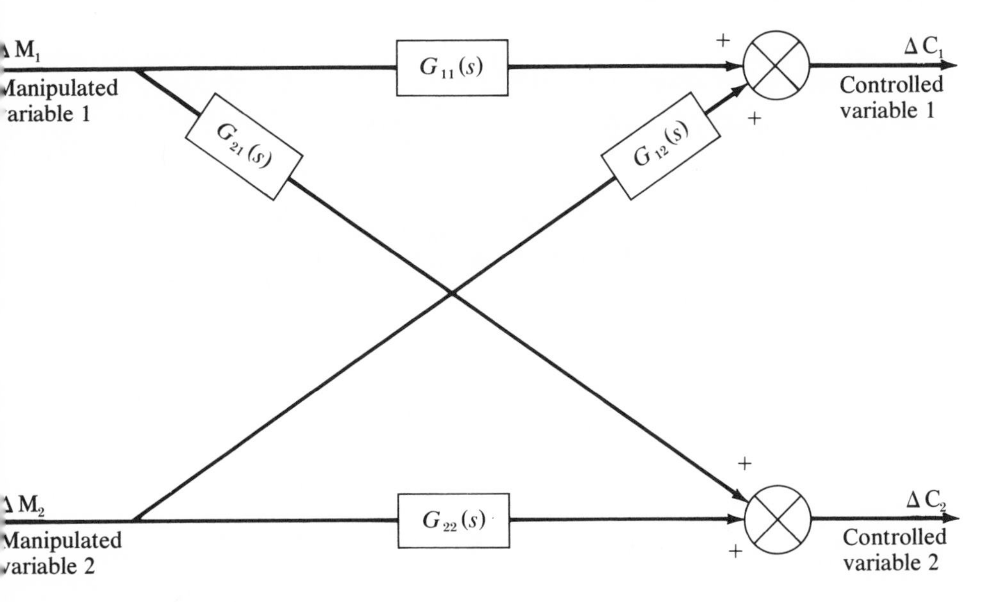

Figure 17.6
2 × 2 Multivariable System

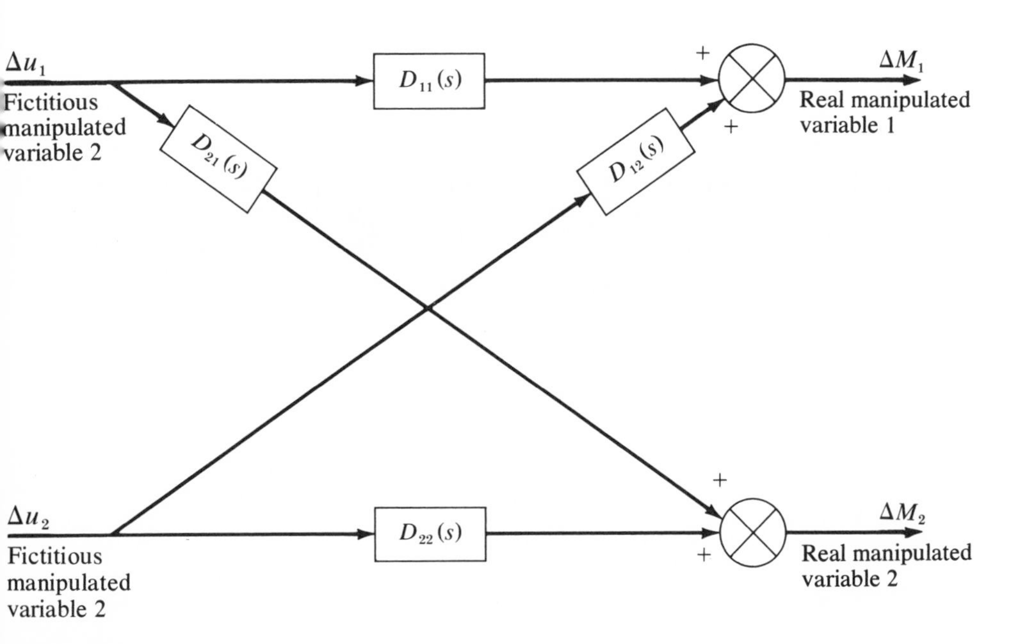

Figure 17.7
Decoupler for 2 × 2 System

From Figure 17.7 it can be seen that the decoupler equations can be written as

$$\begin{aligned} \Delta M_1(s) &= D_{11}(s)\Delta u_1(s) + D_{12}(s)\Delta u_2(s) \\ \Delta M_2(s) &= D_{21}(s)\Delta u_1(s) + D_{22}(s)\Delta u_2(s) \end{aligned} \tag{17.23}$$

For ease in building the decoupler, let us specify that $D_{11}(s) = D_{22}(s) = 1$. In this case, the decoupler equations become:

$$\begin{aligned} \Delta M_1(s) &= \Delta u_1(s) + D_{12}(s)\Delta u_2(s) \\ \Delta M_2(s) &= D_{21}(s)\Delta u_1(s) + \Delta u_2(s) \end{aligned} \tag{17.24}$$

Substituting Equation (17.24) into Equation (17.22) gives the equations for the process-decoupler combination:

$$\begin{aligned} \Delta C_1 &= G_{11}(\Delta u_1 + D_{12}\Delta u_2) + G_{12}(D_{21}\Delta u_1 + \Delta u_2) \\ &= (G_{11} + G_{12}D_{21})\,\Delta u_1 + (G_{11}D_{12} + G_{12})\,\Delta u_2 \\ \Delta C_2 &= G_{21}(\Delta u_1 + D_{12}\Delta u_2) + G_{22}(D_{21}\Delta u_1 + \Delta u_2) \\ &= (G_{21} + G_{22}D_{21})\,\Delta u_1 + (G_{21}D_{12} + G_{22})\,\Delta u_2 \end{aligned} \tag{17.25}$$

For complete decoupling, we want ΔC_1 to be affected *only* by Δu_1 and ΔC_2 *only* by Δu_2; that is,

$$\begin{aligned} \Delta C_1 &= H_1\,\Delta u_1 \\ \Delta C_2 &= H_2\,\Delta u_2 \end{aligned} \tag{17.26}$$

Comparing Equations (17.25) and (17.26) gives four equations, which can be solved for D_{12}, D_{21}, H_1, and H_2;

$$\begin{aligned} G_{11} + G_{12}D_{21} &= H_1 \\ G_{11}D_{12} + G_{12} &= 0 \\ G_{21} + G_{22}D_{21} &= 0 \\ G_{21}D_{12} + G_{22} &= H_2 \end{aligned}$$

From which,

$$\begin{aligned} D_{12}(s) &= -\frac{G_{12}(s)}{G_{11}(s)} \\ D_{21}(s) &= -\frac{G_{21}(s)}{G_{22}(s)} \\ H_1(s) &= G_{11}(s) - \frac{G_{12}(s)G_{21}(s)}{G_{22}(s)} \\ H_2(s) &= G_{22}(s) - \frac{G_{12}(s)G_{21}(s)}{G_{11}(s)} \end{aligned} \tag{17.27}$$

From Equation (17.27) it can be seen why good process models are necessary to effectively decouple a system. If the models are inaccurate, then the cross term of the process-decoupler combination will not be zero.

In theory, at least, a decoupler can be built with analog hardware, particularly when simple (first-order) models are used for the process. However, success is not likely with such simple models—success is much more likely if a digital computer is used where some form of algorithm is available to make use of actual on-line performance to update the process models.

The design equations for a general decoupler for a $n \times n$ system are conveniently summarized using matrix notation. The block diagram of a general multivariable system is shown in Figure 17.8.

The terms are defined as follows:

$$\mathbf{G} = \begin{bmatrix} G_{11}(s) \cdot \cdot G_{1n}(s) \\ \cdot \\ \cdot \\ \cdot \\ G_{n1}(s) \cdot \cdot G_{nn}(s) \end{bmatrix} \qquad \mathbf{D} = \begin{bmatrix} D_{11}(s) \cdot \cdot \cdot D_{1n}(s) \\ \cdot \\ \cdot \\ \cdot \\ D_{n1}(s) \cdot \cdot \cdot D_{nn}(s) \end{bmatrix} \tag{17.28}$$

Transfer function matrix — Decoupler matrix

$$\mathbf{H} = \begin{bmatrix} H_{11}(s) & \cdot & \cdot & \cdot & 0 \\ \cdot & & & & \\ \cdot & & H_{22}(s) & & \\ 0 & & & & \\ \cdot & & & & \\ \cdot & & & & H_{nn}(s) \end{bmatrix}$$

Diagonal matrix of process decoupler combination; i.e. C = Hu

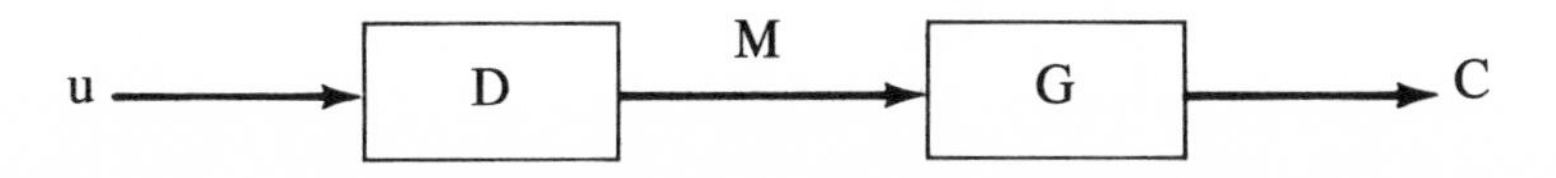

Figure 17.8
General $n \times n$ Multivariable System

$$\mathbf{u} = \begin{bmatrix} u_1 \\ u_2 \\ \cdot \\ \cdot \\ \cdot \\ \cdot \\ \cdot \\ u_n \end{bmatrix} \qquad \mathbf{M} = \begin{bmatrix} M_1 \\ M_2 \\ \cdot \\ \cdot \\ \cdot \\ \cdot \\ \cdot \\ M_n \end{bmatrix} \qquad \mathbf{C} = \begin{bmatrix} C_1 \\ C_2 \\ \cdot \\ \cdot \\ \cdot \\ \cdot \\ \cdot \\ C_n \end{bmatrix}$$

Vector of new manipulated variables — Vector of original manipulated variables — Output vector

From Figure 17.8 we can write:

$$\mathbf{C} = \mathbf{G}\,\mathbf{M}$$

$$\mathbf{M} = \mathbf{D}\,\mathbf{u}$$

Thus

$$\mathbf{C} = \mathbf{G}\,\mathbf{D}\,\mathbf{u}$$

But we want

$$\mathbf{C} = \mathbf{H}\,\mathbf{u} \tag{17.29}$$

Thus

$$\mathbf{G}\,\mathbf{D} = \mathbf{H}$$

or

$$\mathbf{D} = \mathbf{G}^{-1}\mathbf{H}$$

which defines the decoupler.

Example 2. Luyben[6] has presented an application of decoupling to a 2×2 distillation control system. In this application the objectives are to control the bottoms and overhead composition of the binary column by manipulating the reboiler heat duty and the reflux flow, respectively. Since a change in either reboiler heat duty or reflux flow upsets both compositions, we have an interacting system. The purpose is to design decouplers so as to achieve noninteracting feedback control of the multivariable system.

The block diagram of the control system without the decoupling elements is shown in Figure 17.9. The block diagram of the control system with the decoupling elements is shown in Figure 17.10. The open-loop transfer functions relating the controlled variables to the manipulated variables are:

$$X_D(s) = G_{11}(s)R(s) + G_{12}(s)V_B(s) \tag{17.30}$$

and

$$X_B(s) = G_{21}(s)R(s) + G_{22}(s)V_B(s) \tag{17.31}$$

where

X_D = distillate composition, mole fraction

X_B = bottoms composition, mole fraction

V_B = vapor boil up rate moles/min (indicative of reboiler heat duty)

R = reflux flow rate, moles/min

$G_{11}(s)$ = open-loop transfer function relating X_D to R

$G_{12}(s)$ = open-loop transfer function relating X_D to V_B

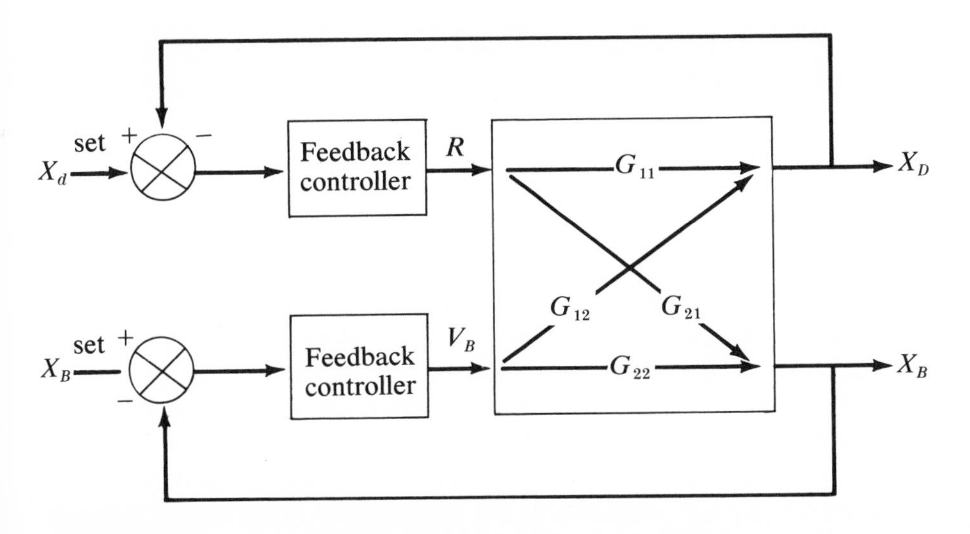

Figure 17.9
Distillation Column Control System without Decoupler

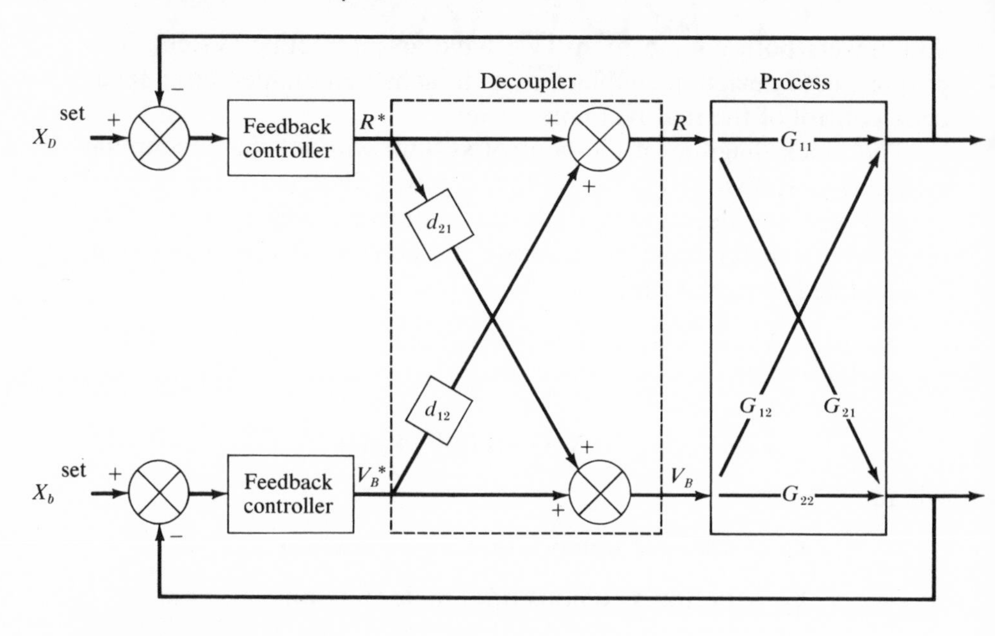

Figure 17.10
Distillation Control Scheme with Decouplers

$G_{21}(s) =$ open-loop transfer function relating X_B to R

$G_{22}(s) =$ open-loop transfer function relating X_B to V_B

In accordance with Equation (17.27), the decoupler design equations are

$$D_{12}(s) = -\frac{G_{12}(s)}{G_{11}(s)} \tag{17.32}$$

$$D_{21}(s) = -\frac{G_{12}(s)}{G_{22}(s)} \tag{17.33}$$

In this application, the open-loop transfer functions were available in the frequency domain in the form of Bode plots. Thus graphical manipulations via Equations (17.32) and (17.33) resulted in the Bode plot of $D_{12}(s)$ and $D_{21}(s)$ to which transfer functions were fitted. Since the decouplers turned out to be lead-lag networks, conventional hardware could be used to implement them. The computer simulations of the control system resulted in the closed-loop responses which are shown in Figures 17.11*a* and 17.11*b*. These figures show the benefits of decoupling the multivariable system.

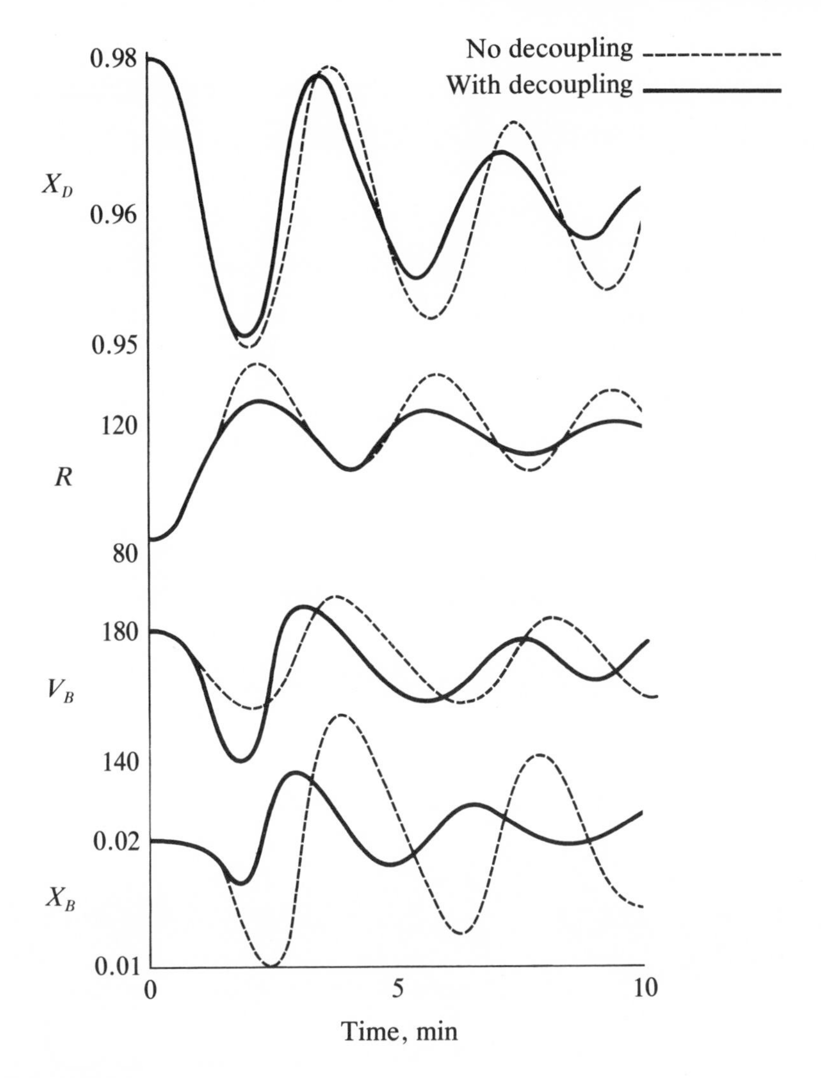

Figure 17.11*a*
Transient Response (X_B = 0.02, X_D = 0.98) with X_D^{set} Disturbance (Reproduced with Permission from Ref. 6)

The approach to multivariable controller design described in this chapter consisted of selecting the proper pairing of manipulated and controlled variables and if necessary designing decouplers so as to achieve noninteracting feedback control. This is only one of the several approaches available for the design of multivariable control systems. Edgar[7] has surveyed the various other methods available. The interested reader is referred to his article for details.

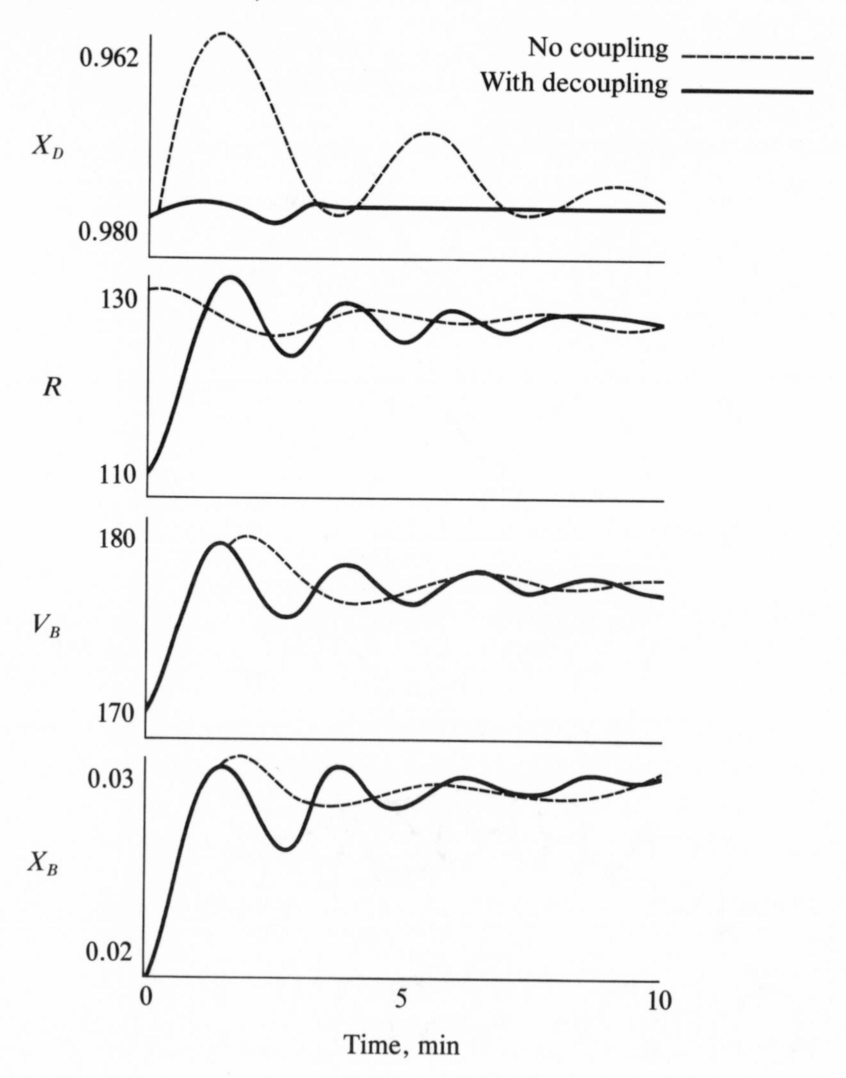

Figure 17.11*b*
Transient Response (X_B = 0.02, X_D = 0.98) with X_B^{set} Disturbance (Reproduced with Permission from Ref. 6)

References

1. Bristol, E. H., On a New Measure of Interaction for Multivariable Process Control, *IEEE Transactions on Automatic Control*, AC-11, 1966, p. 133.

2. Nisenfeld, A. E., Schultz, H. M., Interaction Analysis Applied to Control Systems Design, *Instrumentation Technology*, April 1971.
3. Desphande, P. B., Laukhuf, W. L. S., Patke, N. G., Advanced Process Control Experiments, *Chemical Engineering Education*, **XIV**, 2, 1980.
4. Boksenbom, A. S., Hood, R., *General Albebraic Method Applied to Control Analysis of Complex Engine Types,* Report NCA-TR-980, Washington, DC, 1949.
5. Greenfield, G. G., Ward, T. J., Structural Analysis for Multivariable Process Control, *I & EC Fundamentals*, **6**, 1967, p. 564.
6. Luyben, W. L., *Distillation Decoupling, AIChE J.*, **2**, 1970, 198–203.
7. Edgar, T. F., *Status of Design Methods for Multivariable Control, Chemical Process Control, AIChE* Symposium Series, Vol. 2, No. 159, 1976, pp. 99–111.

Authors' Update on Interaction

The interaction measure has been extended by T. J. McAvoy[1] to take into account process dynamics. The interested reader may consult the following reference for details.

1. McAvoy, T. J., Dynamic Interaction Analysis of Complex Control Systems, Annual A.I.Ch.E. meeting, New Orleans, LA, November 8-12, 1981. (Paper No. 57d)

Appendix A

Table of Z Transforms and Modified Z Transforms

Reprinted with Permission from B.C. Kuo's *Analysis and Synthesis of Sampled-data Control Systems*, Prentice-Hall, 1963.

Laplace transform $E(s)$	*Time function* $e(t)$	*z-transform* $E(z)$	*Modified z-transform* $E(z, m)$
1	$\delta(t)$	1	0
e^{-nTs}	$\delta(t - nT)$	z^{-n}	z^{-n-1+m}
$\frac{1}{s}$	$u(t)$	$\frac{z}{z-1}$	$\frac{1}{z-1}$
$\frac{1}{s^2}$	t	$\frac{Tz}{(z-1)^2}$	$\frac{mT}{z-1} + \frac{T}{(z-1)^2}$
$\frac{2!}{s^3}$	t^2	$\frac{T^2 z(z+1)}{(z-1)^3}$	$T^2\left[\frac{m^2}{z-1} + \frac{2m+1}{(z-1)^2} + \frac{2}{(z-1)^3}\right]$
$\frac{(n-1)!}{s^n}$	t^{n-1}	$\lim_{a\to 0} (-1)^{n-1} \frac{\partial^{n-1}}{\partial a^{n-1}}\left(\frac{z}{z - e^{-aT}}\right)$	$\lim_{a\to 0} (-1)^{n-1} \frac{\partial^{n-1}}{\partial a^{n-1}}\left(\frac{e^{-amT}}{z - e^{-aT}}\right)$
$\frac{1}{s+a}$	e^{-at}	$\frac{z}{z - e^{-aT}}$	$\frac{e^{-amT}}{z - e^{-aT}}$
$\frac{1}{(s+a)(s+b)}$	$\frac{1}{b-a}(e^{-at} - e^{-bt})$	$\frac{1}{b-a}\left(\frac{z}{z - e^{-aT}} - \frac{z}{z - e^{-bT}}\right)$	$\frac{1}{b-a}\left(\frac{e^{-amT}}{z - e^{-aT}} - \frac{e^{-bmT}}{z - e^{-bT}}\right)$
$\frac{1}{s(s+a)}$	$\frac{1}{a}(u(t) - e^{-at})$	$\frac{1}{a}\frac{(1 - e^{-aT})z}{(z-1)(z - e^{-aT})}$	$\frac{1}{a}\left(\frac{1}{z-1} - \frac{e^{-amT}}{z - e^{-aT}}\right)$
$\frac{1}{s^2(s+a)}$	$\frac{1}{a}\left(t - \frac{1 - e^{-at}}{a}\right)$	$\frac{1}{a}\left[\frac{Tz}{(z-1)^2} - \frac{(1 - e^{-aT})z}{a(z-1)(z - e^{-aT})}\right]$	$\frac{1}{a}\left[\frac{T}{(z-1)^2} + \frac{amT - 1}{a(z-1)} + \frac{e^{-amT}}{a(z - e^{-aT})}\right]$
$\frac{(s+b)}{s^2(s+a)}$	$\frac{a-b}{a^2}u(t) + \frac{b}{a}t + \frac{1}{a}\left(\frac{b}{a} - 1\right)e^{-at}$	$\frac{1}{a}\left[\frac{bTz}{(z-1)^2} + \frac{(a-b)(1 - e^{-aT})z}{a(z-1)(z - e^{-aT})}\right]$	$\frac{1}{a}\left[\frac{bT}{(z-1)^2} + \left(bmT + 1 - \frac{b}{a}\right)\frac{1}{z-1} + \frac{b-a}{a}\frac{e^{-amT}}{(z - e^{-aT})}\right]$
$\frac{1}{s(s+a)(s+b)}$	$\frac{1}{ab}\left(u(t) + \frac{b}{a-b}e^{-at} - \frac{a}{a-b}e^{-bt}\right)$	$\frac{1}{ab}\left[\frac{z}{z-1} + \frac{bz}{(a-b)(z - e^{-aT})} - \frac{az}{(a-b)(z - e^{-aT})}\right]$	$\frac{1}{ab}\left[\frac{1}{z-1} + \frac{be^{-amT}}{(a-b)(z - e^{-aT})} - \frac{ae^{-bmT}}{(a-b)(z - e^{-bT})}\right]$
$\frac{1}{(s+a)^2}$	te^{-at}	$\frac{Tze^{-aT}}{(z - e^{-aT})^2}$	$\frac{Te^{-amT}[e^{-aT} + m(z - e^{-aT})]}{(z - e^{-aT})^2}$

$\frac{1}{s^3(s+a)}$	$\frac{1}{2a}\left(t^2 - \frac{2}{a}t + \frac{2}{a^2}u(t) - \frac{2}{a^2}e^{-at}\right)$	$\frac{1}{a}\left[\frac{T^2z}{(z-1)^3} + \frac{(aT-2)Tz}{2a(z-1)^2} + \frac{z}{a^2(z-1)} - \frac{z}{a^2(z-e^{-aT})}\right]$	$\frac{1}{a}\left[\frac{T^2}{(z-1)^3} + \frac{T^2(m+\frac{1}{2}) - T/a}{(z-1)^2} + \frac{(amT)^2/2 - amT + 1}{a^2(z-1)} - \frac{e^{-amT}}{a^2(z-e^{-aT})}\right]$
$\frac{a}{s^2+a^2}$	$\sin at$	$\frac{z \sin aT}{z^2 - 2z\cos aT + 1}$	$\frac{z \sin amT + \sin(1-m)aT}{z^2 - 2z\cos aT + 1}$
$\frac{s}{s^2+a^2}$	$\cos at$	$\frac{z(z - \cos aT)}{z^2 - 2z\cos aT + 1}$	$\frac{z\cos amT - \cos(1-m)aT}{z^2 - 2z\cos aT + 1}$
$\frac{a}{s^2-a^2}$	$\sin h\,at$	$\frac{z \sin h\,aT}{z^2 - 2z\cos h\,aT + 1}$	$\frac{z\sin h\,amT + \sin h(1-m)aT}{z^2 - 2z\cos h\,aT + 1}$
$\frac{s}{s^2-a^2}$	$\cos h\,at$	$\frac{z(z - \cos h\,aT)}{z^2 - 2z\cos h\,aT + 1}$	$\frac{z\cos h\,amT - \cos h(1-m)aT}{z^2 - 2z\cos h\,aT + 1}$
$\frac{a}{s(s^2+a^2)}$	$\frac{1}{a}(u(t) - \cos at)$	$\frac{1}{a}\left[\frac{z}{z-1} - \frac{z(z-\cos aT)}{z^2 - 2z\cos aT + 1}\right]$	$\frac{1}{a}\left[\frac{1}{z-1} - \frac{z\cos amT - \cos(1-m)aT}{z^2 - 2z\cos aT + 1}\right]$
$\frac{a^2}{s^2(s^2+a^2)}$	$t - \frac{1}{a}\sin at$	$\frac{Tz}{(z-1)^2} - \frac{1}{a}\frac{z\sin aT}{z^2 - 2z\cos aT + 1}$	$\frac{mT}{z-1} + \frac{T}{(z-1)^2} - \frac{z\sin amT + \sin(1-m)aT}{a(z^2 - 2z\cos aT + 1)}$
$\frac{1}{s(s+a)^2}$	$\frac{1}{a^2}[u(t) - (1+at)e^{-at}]$	$\frac{1}{a^2}\left[\frac{z}{z-1} - \frac{z}{z-e^{-aT}} - \frac{aTe^{-aT}z}{(z-e^{-aT})^2}\right]$	$\frac{1}{a^2}\left[\frac{1}{z-1} - \left(\frac{1+amT}{z-e^{-aT}} + \frac{aTe^{-aT}}{(z-e^{-aT})^2}\right)e^{-amT}\right]$
$\frac{1}{s^2(s+a)^2}$	$\frac{t}{a^2} - \frac{2}{a^3}u(t) + \left(\frac{t}{a^2} + \frac{2}{a^3}\right)e^{-at}$	$\frac{1}{a^3}\left[\frac{(aT+2)z - 2z^2}{(z-1)^2} + \frac{2z}{z-e^{-aT}} + \frac{aTe^{-aT}z}{(z-e^{-aT})^2}\right]$	$\frac{1}{a^3}\left[\frac{aT}{(z-1)^2} + \frac{amT-2}{z-1} + \left(\frac{aTe^{-aT}}{(z-e^{-aT})^2} - \frac{amT-2}{z-e^{-aT}}\right)e^{-amT}\right]$
$\frac{1}{(s+a)^2+b^2}$	$\frac{1}{b}e^{-at}\sin bt$	$\frac{1}{b}\left(\frac{ze^{-aT}\sin bT}{z^2 - 2ze^{-aT}\cos bT + e^{-2aT}}\right)$	$\frac{1}{b}\frac{e^{-amT}[z\sin bmT + e^{-aT}\sin(1-m)bT]}{z^2 - 2ze^{-aT}\cos bT + e^{-2aT}}$
$\frac{s+a}{(s+a)^2+b^2}$	$e^{-at}\cos bt$	$\frac{z^2 - ze^{-aT}\cos bT}{z^2 - 2ze^{-aT}\cos bT + e^{-2aT}}$	$\frac{e^{-amT}[z\cos bmT + e^{-aT}\sin(1-m)bT]}{z^2 - 2ze^{-aT}\cos bT + e^{-2aT}}$

Table of Z Transforms and Modified Z Transforms (*continued*)

Laplace transform $E(s)$	*Time function* $e(t)$	*z-transform* $E(z)$	*Modified z-transform* $E(z, m)$
$\frac{1}{s[(s+a)^2+b^2]}$	$\frac{1}{a^2+b^2}[1 - e^{-at}\sec\phi\cos(bt+\phi)]$ $\phi = \tan^{-1}\left(\frac{-a}{b}\right)$	$\frac{1}{a^2+b^2}\left[\frac{z}{z-1} - \frac{z^2 - ze^{-aT}\sec\phi\cos(bT-\phi)}{z^2 - 2ze^{-aT}\cos bT + e^{-2aT}}\right]$	$\frac{1}{a^2+b^2}\left[\frac{1}{z-1} - \frac{e^{-amT}\sec\phi\{z\cos(bmT+\phi) - e^{-aT}\cos[(1-m)bT-\phi]\}}{z^2 - 2ze^{-aT}\cos bT + e^{-2aT}}\right]$

Some Useful Properties and Theorems of Z Transforms

1. $Z\{f(t-kt)\} = Z^{-k}F(Z)$
2. $Z\{f_1(t) + f_2(t)\} = F_1(Z) + F_2(Z)$
3. $Z\{e^{-at}f(t)\} = \sum_{n=0}^{\infty} f(nT)\, Z^{-n}\, e^{-anT}$
4. $\lim_{t\to\infty} f(t) = \lim_{Z\to 1}\left(\frac{Z-1}{Z}F(Z)\right)$
5. $\lim_{t\to 0} f(t) = \lim_{Z\to\infty} F(Z)$

Appendix B

Exercises

Chapter 1 Review of Conventional Process Control

1. For the three transfer functions given show that K_p is indeed the steady-state gain for a unit step disturbance.

a. First-order: $G(s) = \dfrac{K_p}{\tau s + 1}$

b. Second-order overdamped: $G(s) = \dfrac{K_p}{(\tau_1 s + 1)(\tau_2 s + 1)}$

c. Second-order underdamped: $G(s) = \dfrac{K_p}{\tau^2 s^2 + 2\zeta\tau s + 1}$

Hint: Use the final value theorem.

2. (a) Determine the transient closed-loop response of the process shown in the following block diagram to a change in set point $R(t) =$

$5(1 - e^{-t/2})$. Design the PID controller using Bode plot and the Ziegler–Nichols method.

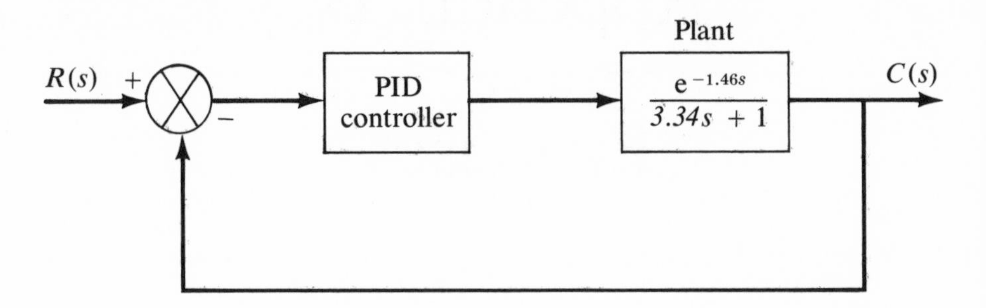

Hint: Represent the dead time by the fourth-order Pade′ approximation.

$$e^{-\tau s} = \frac{e^{-\tau s/2}}{e^{\tau s/2}} \approx \frac{1 - \tau s/2 + \tau^2 s^2/12 - \tau^3 s^3/120 + \tau^4 s^4/1680}{1 + \tau s/2 + \tau^2 s^2/12 + \tau^3 s^3/120 + \tau^4 s^4/1680}$$

(b). Simulate the system shown in part (a) on the analog computer using the fourth-order Pade′ approximation and the Ziegler–Nichols tuning constants and obtain the transient response to the same change in set point. If computation and simulation are done correctly, the results of parts (a) and (b) should agree.

3. Two process-control systems are to be tuned experimentally. The derivative mode on the controller is turned off, and the integral mode is set at the lowest setting. The gain of the controller is gradu-

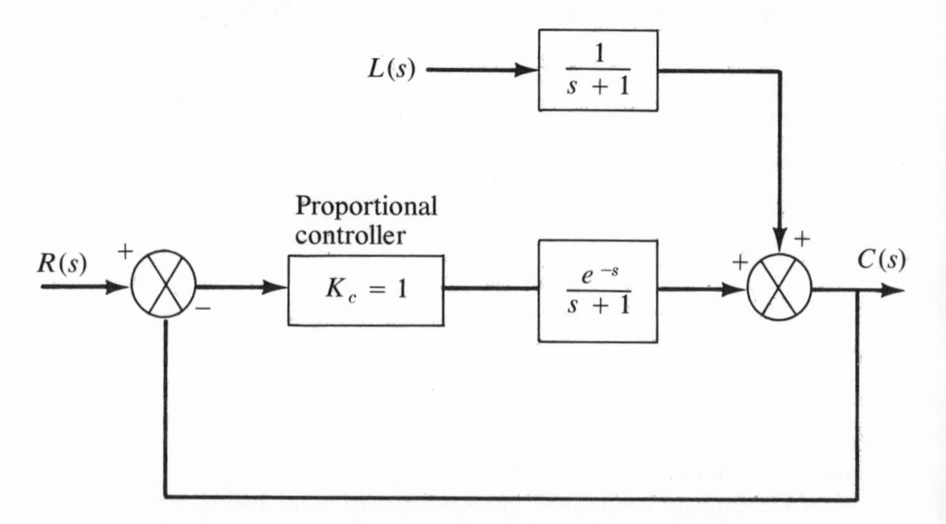

Hint: Use second-order Pade′ approximation for dead time.

Figure B1. Block Diagram for Problem No. 4

ally increased (proportional band decreased) until the controlled variable begins to oscillate with constant amplitude. The final gain setting and the period of oscillation of each system is given below. Determine the two- and three-mode controller settings for each system.

System	Ultimate Gain	Ultimate Period
1	0.4	20
2	2.0	0.5

4. Determine the transient response of the controlled variable shown in the block diagram, Figure B1, to a unit step change in load L
Hint: Use second-order Pade′ approximation for dead time.

5. For the overdamped second-order plus dead-time process shown in the following block diagram, how much better will the control be if the dead time were halved? Answer in terms of allowable controller gains and the associated step responses

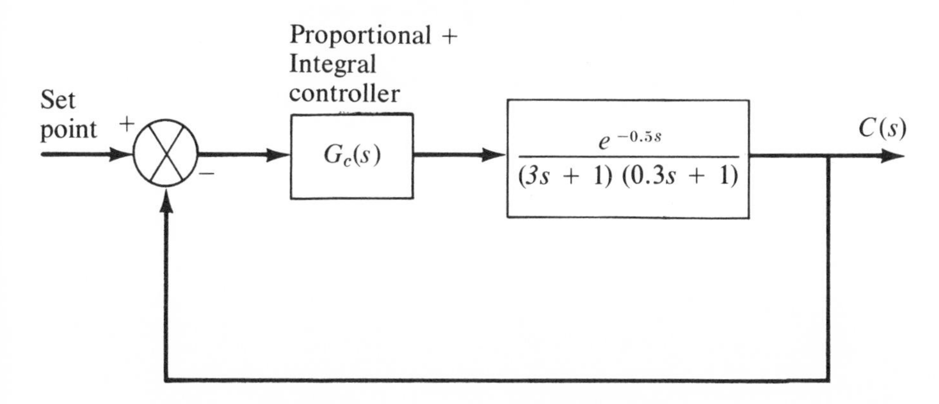

6. For the block diagram shown in the following figure (a) prepare the frequency response diagram and determine the ultimate gain and Ziegler–Nichols setting for the proportional controller; (b) apply Routh Criteria and determine the ultimate gain; and (c) determine if the closed-loop system would be stable if K_c were equal to 6.

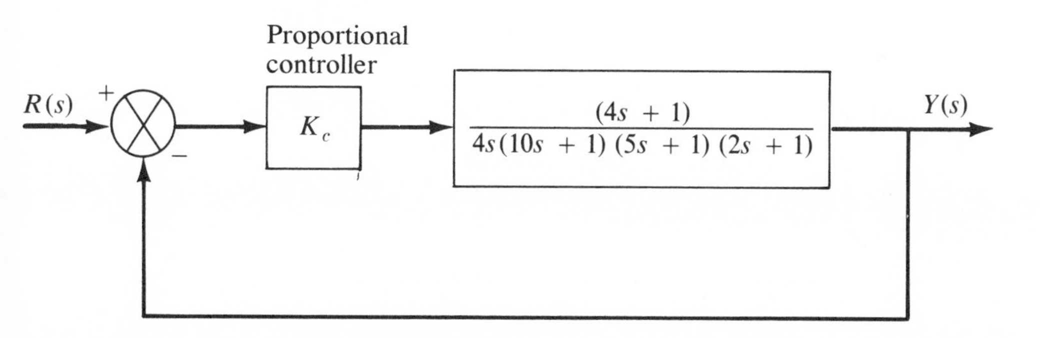

7. The block diagram of an open-loop furnace control system is shown below. Approximate this system by a first-order lag with dead time. Compare the step response of the process with that of the approximating model.

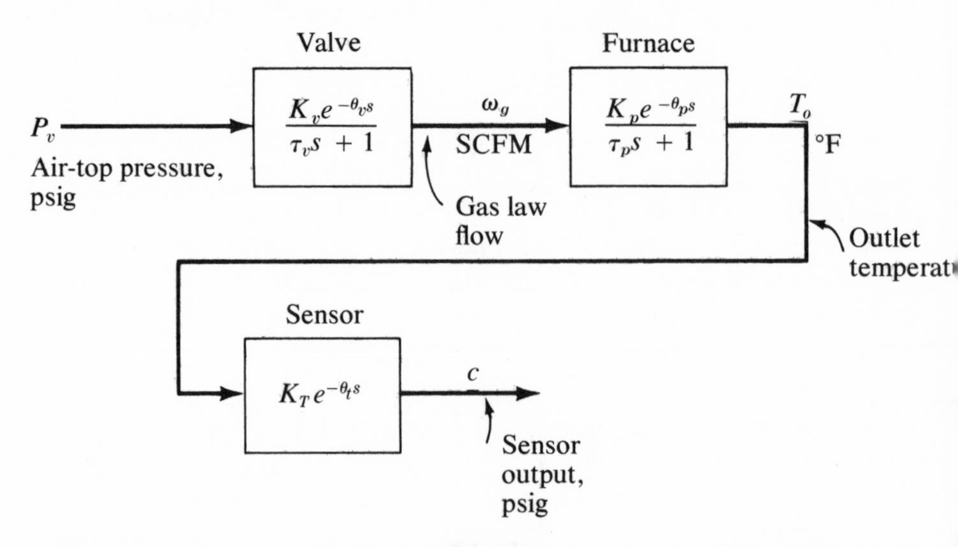

Parameters: $K_v = 100 \dfrac{\text{SCFM}}{\text{psig}}$; $K_p = 2.5 \dfrac{°\text{F}}{\text{SCFM}}$;

$K_T = 0.005 \dfrac{\text{psig}}{°\text{F}}$;

$\theta_v = 0.02$ sec;

$\theta_T = 0.13$ sec;

$\theta_p = 0.15$ sec;

$\tau_p = 0.5$ sec

$\tau_v = 0.088$ sec;

8. The block diagram of a heat exchanger control system is shown in the following block diagram

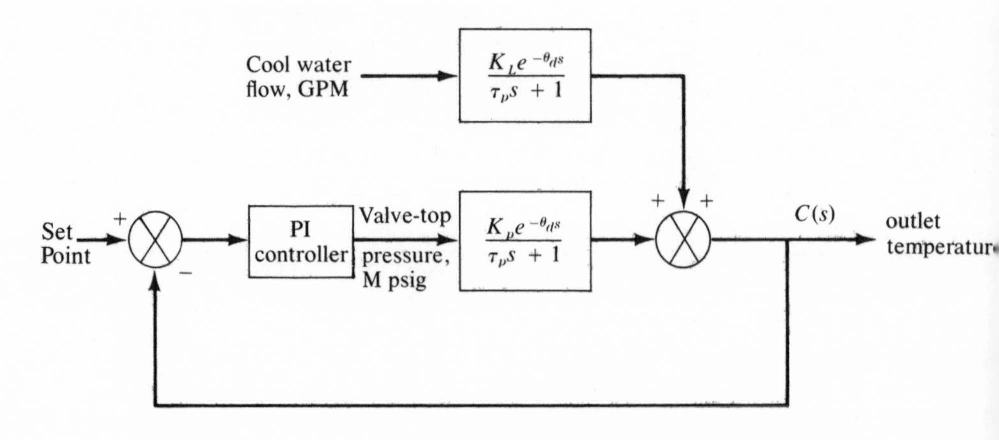

$K_p = 2$ psig/psig; $K_L = 0.5$ psig/gpm

Case No.	τ_p min	θ_d gpm	ΔL gpm
1	2.0	1.0	+10
2	5.0	1.0	− 5
3	3.0	3.0	+ 3
4	10.0	0	−30
5	10.0	7.5	−10

(a) For each of the five cases above find

(1) Ziegler–Nichols controller gain K_c

(2) Cohen–Coon controller gain K_c

(3) ΔC_{peak} in psig if Ziegler–Nichols and Cohen–Coon controller settings are used

(4) Period of oscillation of response *Pu*

(5) settling time *TS*

(6) time to first peak *TP*

(b) If the controller gain is cut in half, the overshoot to a step-load disturbance, when a proportional-only controller is used, is reduced from about 50% to about 15%. When integral mode is added, the peak is about 10% higher than with proportional-only controller. For the first case, PI control, estimate ΔC_{peak} if K_c were set at half the Ziegler–Nichols gain.

9. Show that the block diagram of the control system shown in the first figure will reduce to that shown in the second figure if the load transfer function $G_L(s)$ equals the process transfer function $G_p(s)$.

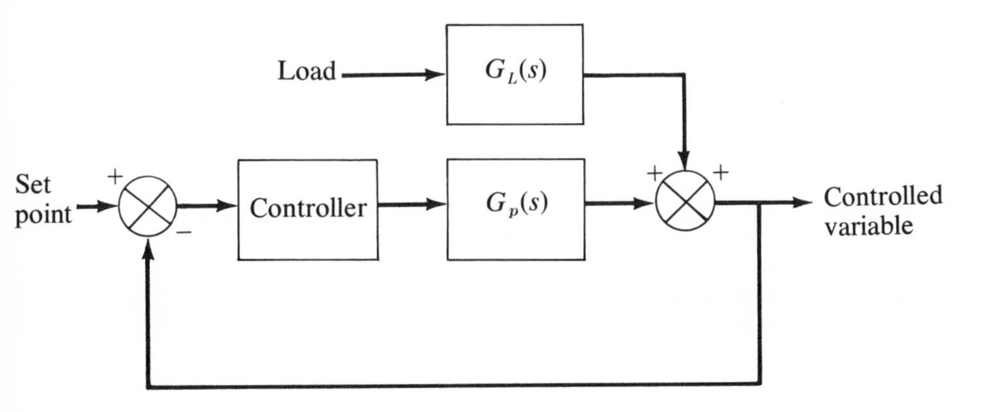

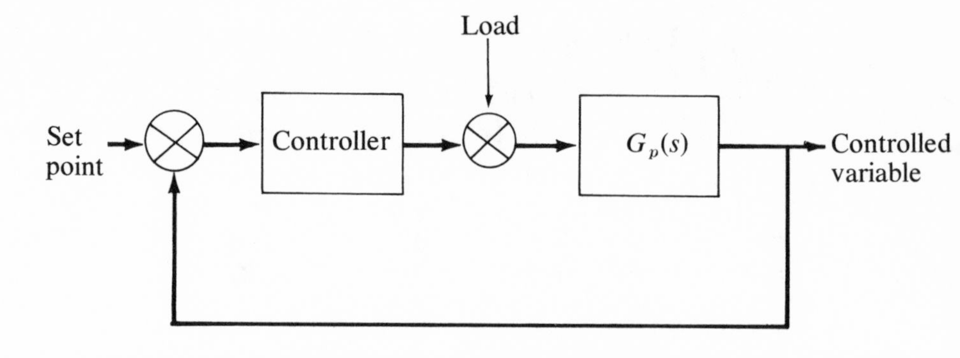

Chapter 5 *Z* Transforms

1. Determine the Z transform of the following functions by both the method of residues and from the definition of Z transforms (the first part will require outside reading).

(a) $\cos \beta t$ (b) $\dfrac{1}{s(s+1)}$

(c) $\dfrac{s}{(s+1)(s+2)}$ (d) $\dfrac{s+2}{s^2+2s+1}$

2. Determine the inverse Z transform of

(1) $\dfrac{Z^3+Z^2+Z}{(Z^2-Z+1)(Z^2-2Z+1)}$ (2) $\dfrac{Z}{Z^2-3Z+2}$

(3) $\dfrac{1}{Z^2-3Z+2}$ (4) $\dfrac{Z}{(Z-0.5)}$

(5) $\dfrac{0.5Z}{(Z-1)(Z-0.5)}$ (6) $\dfrac{1}{(Z-1)^2(Z-2)}$

by the following methods:
(a) partial fraction expansion
(b) power series expansion
(c) method of residues (this would require outside reading)
(d) using the computer program in the appendix

3. Determine the initial and final value of $F(t)$ whose Z transform is given by

$$F(Z) = \frac{2Z}{(Z-1)(1-0.4Z^{-1})}$$

4. Apply the definition of Z transforms and the translation of the function theorem to determine the Z transform of

$$F(s) = \frac{1 - e^{-sT}}{s}$$

where T is a constant.

Chapter 6 Pulse Transfer Functions

1. For the following systems determine the pulse transfer function $G(Z)$. For Parts (c) and (d) let $\theta_d = NT$ where N is an integer.

(a) $G(s) = \dfrac{K_p(1 - e^{-sT})}{s(\tau_p s + 1)}$

(b) $G(s) = \dfrac{K_p(1 - e^{-sT})}{s(\tau_1 s + 1)(\tau_2 s + 1)}$

(c) $G(s) = \dfrac{K_p e^{-\theta_d s}(1 - e^{-sT})}{s(\tau_p s + 1)}$

(d) $G(s) = \dfrac{K_p e^{-\theta_d s}(1 - e^{-sT})}{s(\tau_1 s + 1)(\tau_2 s + 1)}$

2. Determine the pulse transfer function $G(Z)$ of the following system

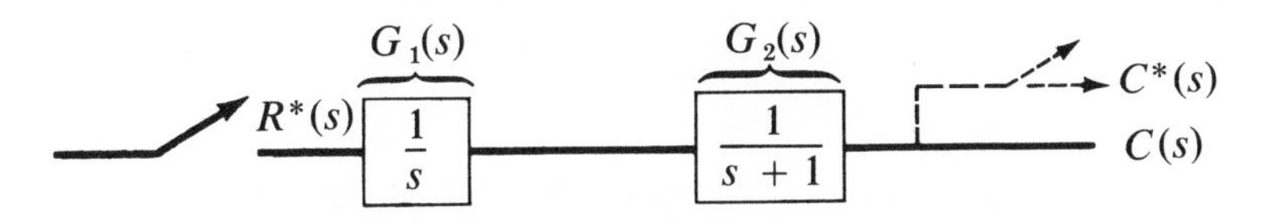

Calculate $Z\{G_1(s)\}Z\{G_2(s)\}$. Is it equal to $Z\{G_1(s)G_2(s)\}$?

Chapter 8 Open-Loop Response

1. For the sampled-data control systems shown below determine

the open-loop response to a unit step change in input, $X(t)$ at several sampling instants.

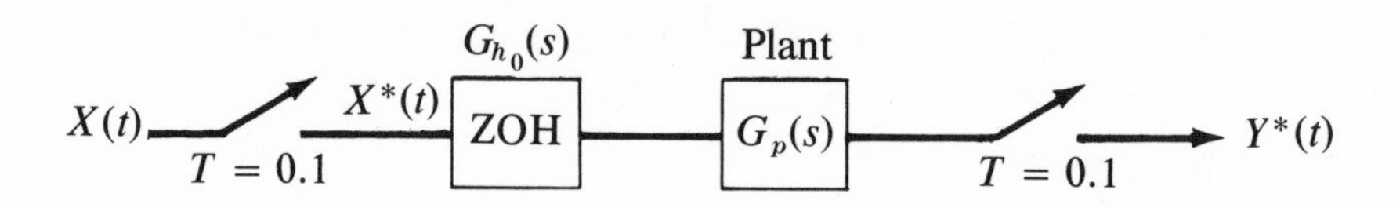

(a) $G_p(s) = \dfrac{1}{(s + 1)}$ (b) $G_p(s) = \dfrac{e^{-0.2s}}{s + 1}$

(c) $G_p(s) = \dfrac{1}{(s + 1)(\frac{1}{2}s + 1)}$ (d) $G_p(s) = \dfrac{e^{-0.2s}}{(s + 1)(0.2s + 1)}$

2. An approximating model for an industrial process is shown below. Determine the open-loop response to a step change in input.

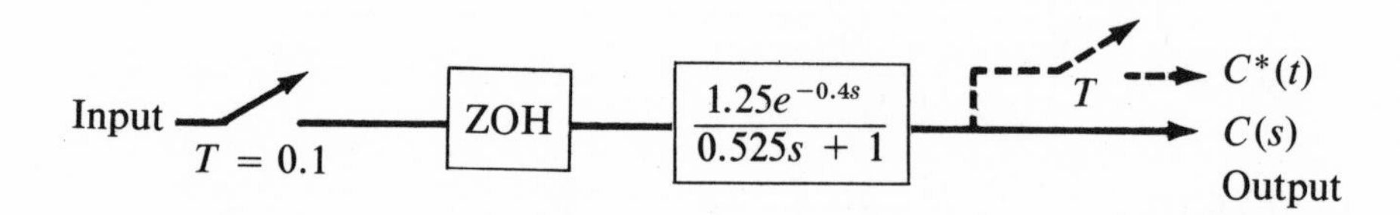

3. Moore, et al. (Instrument Practice, January 1969) have shown that under certain circumstances a sampled-data system can be approximated as a continuous data system as shown below

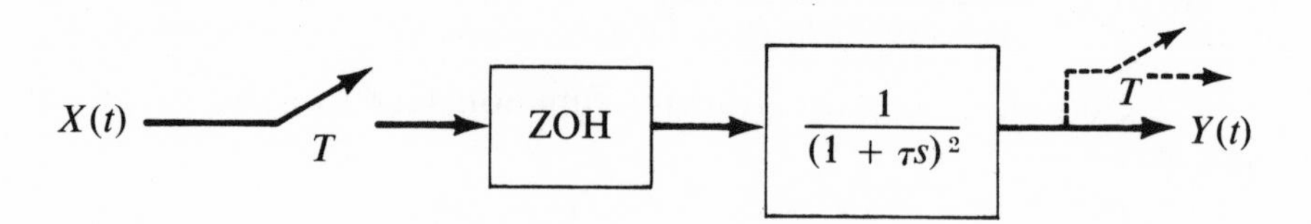

(a) Sampled-Data System

Evaluate the open-loop response of system (a) by Z-transform method and that of (b) by Laplace transform method to a unit step change in input X. Given: $\tau = 1$ and $T = 0.1$.

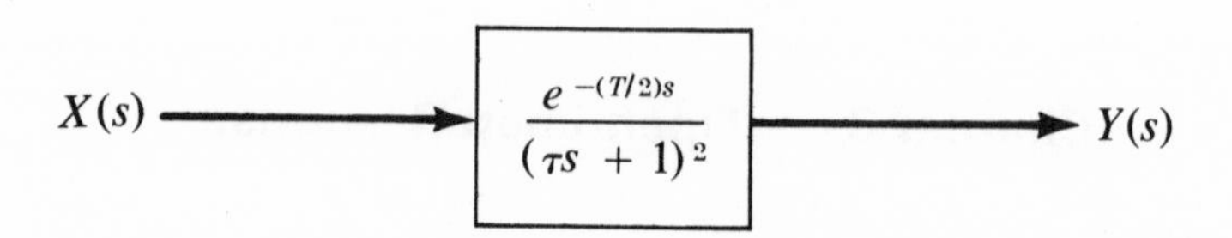

(b) Equivalent Continuous System

Chapter 9 Closed-Loop Response

1. For the digital control system shown below determine $C(Z)/R(Z)$ by (a) the method described in this chapter and (b) by the signal flow graph method (Part b will require outside reading). Determine the transient response of the control system to $R(t) = (1 - e^{-0.5t})$. Assume $T = 0.1$.

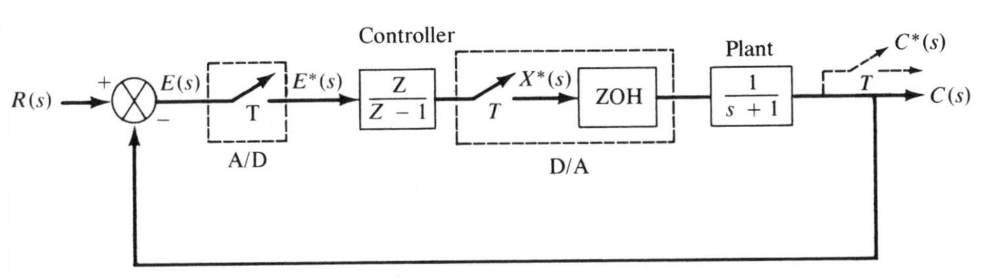

2. Determine $C_M(Z)/R(Z)$ for the following control system and obtain the transient response to a unit step change in set point. Assume $T = 0.1$.

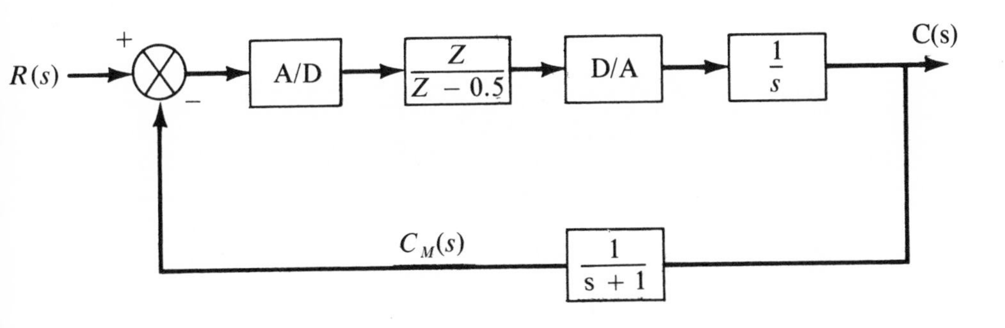

3. Shown in the following figure is the sampled-data version of the control system of Problem 2, Chapter 1.
Use the computer program in the appendix to determine the closed-loop transient response to $R(t) = 5(1 - e^{-t/2})$ for
(a) $T = 0.2$
(b) $T = 0.4$
(c) $T = 0.6$
(d) $T = 0.8$
(e) $T = 1.0$

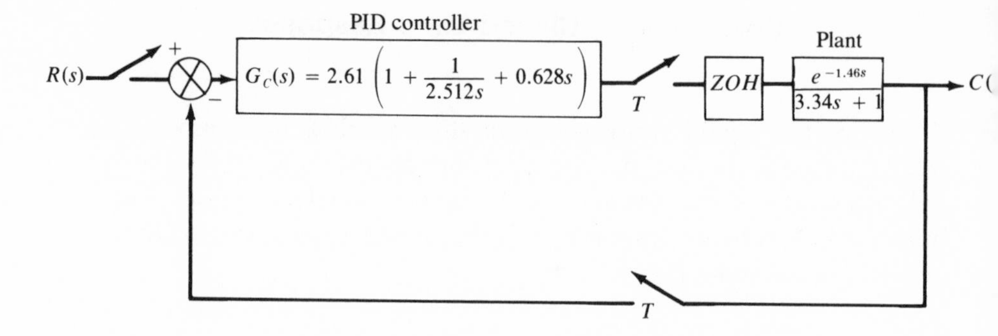

How does the selection of sampling period affect stability?

4. Moore, et al. (Instrument Practice, January 1969), have shown that under certain circumstances, the sampled-data control system can be approximated as a continuous data system as shown in the following block diagram.

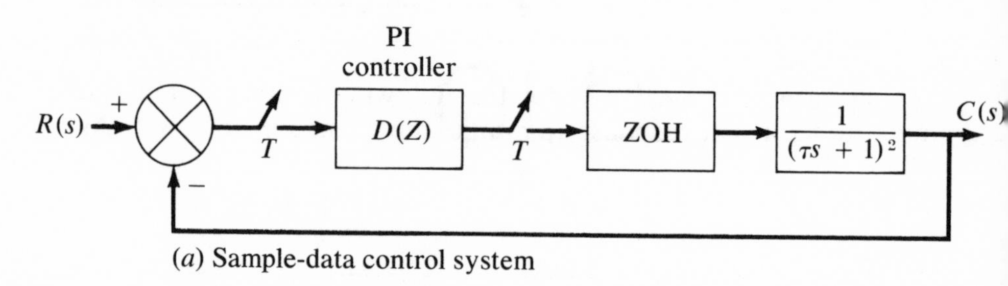

(*a*) Sample-data control system

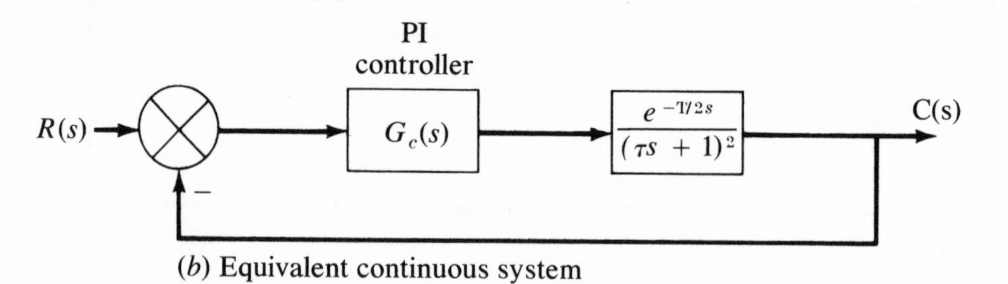

(*b*) Equivalent continuous system

For $\tau = 1$ and $T = 0.2$ find the following

(i.) Ziegler–Nichols tuning constants for system *b*

(ii.) Determine transient response of system *b* to a unit step change in set point (either use analog/hybrid computer for simulation or solve analytically with a fourth order Pade′ approximation)

(iii.) Use the computer program in the appendix and determine the transient response of system (*a*)

(iv.) Plot and compare the transient response of the two systems.

Chapter 10 Control Systems Design

1. For the sampled-data control system shown below design the control algorithm that will respond to a unit step change in set point in minimum settling time having zero steady-state error at the sampling instants. Then evaluate the transient response analytically or through computer simulation.

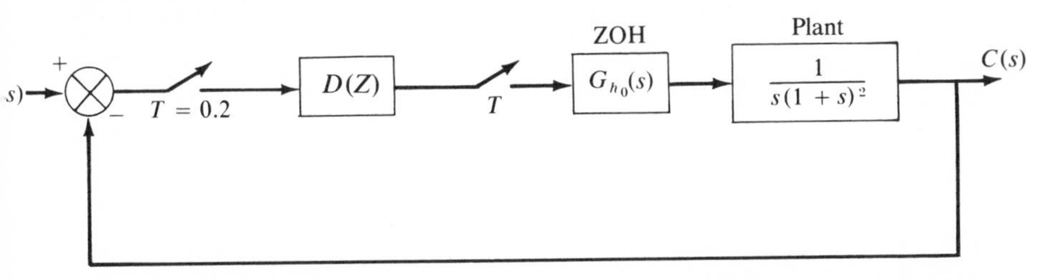

2. The following sampled-data control system shows an overdamped second-order plus dead-time process. Design (a) a deadbeat controller (b) Dahlin control algorithm, and (c) PI control algorithm. Compare the transient response of the three systems to a unit step change in set point.

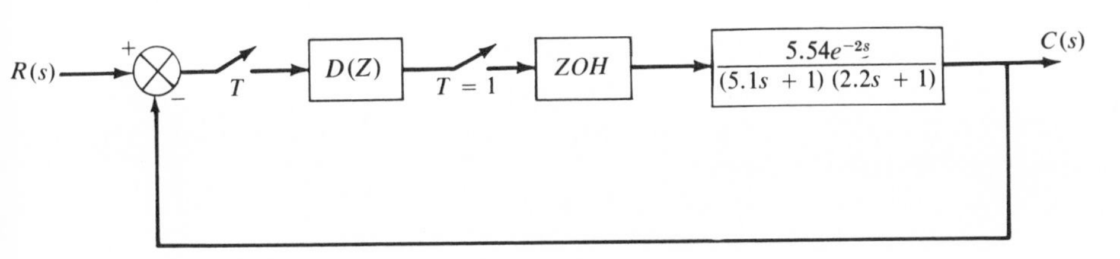

3. Design the Dahlin algorithm for the following control system

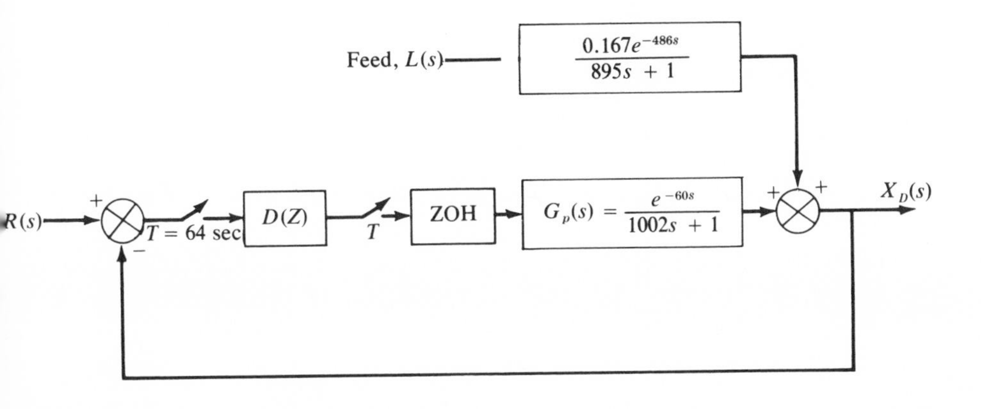

Simulate the control system on the computer and plot the transient response of the distillate composition (X_D) to (a) negative 0.01 step change in set point and (b) step change of 0.22 in load L.

4. Design a PI controller algorithm for the control system of Problem 7, Chapter 1. Use pneumatic/electrical transducers (3 to 15 psig/0-10 volts DC), a zero-order hold, and $T = 0.1$. Use the computer program in the appendix to determine the transient closed-loop response to a unit step change in set point.

5. The block diagram of a sampled-data control system is shown below (from C. L. Smith, *Digital Computer Process Control*, Intext Publishers, 1973 with permission.)

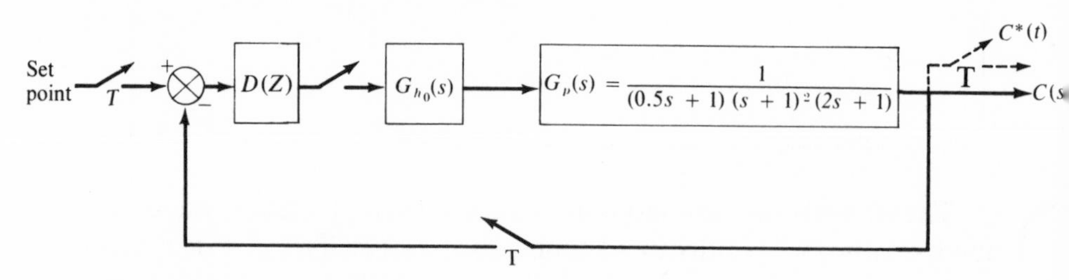

(a) Approximate the process transfer function by a second-order model with dead time

(b) For $T = 5$ determine the deadbeat algorithm, $D(Z)$

(c) For $T = 1$ determine the deadbeat algorithm, $D(Z)$

(d) Determine the closed-loop response of $C(t)$ of the original fourth-order system to a unit step change in set point using $D(Z)$ developed in parts b and c.

Chapter 11 Stability

1. Determine the proportional gain K_c for which the following systems become unstable:

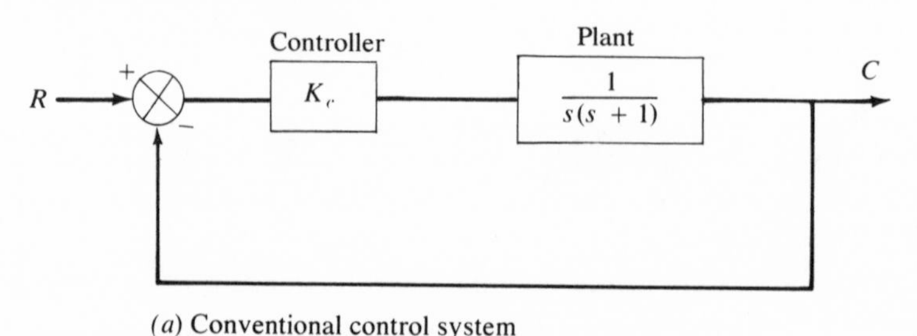

(*a*) Conventional control system

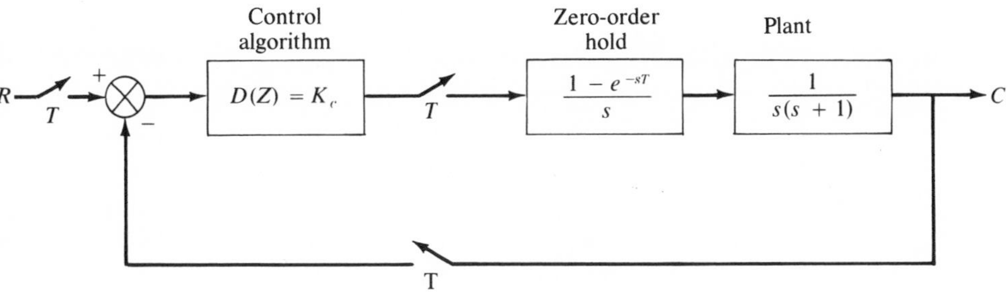

(*b*) Sampled-data control system

2. In the following figure is shown a second-order sampled-data control system. Using the Schur-Cohn determinants, determine the range of K values for which the system is stable.

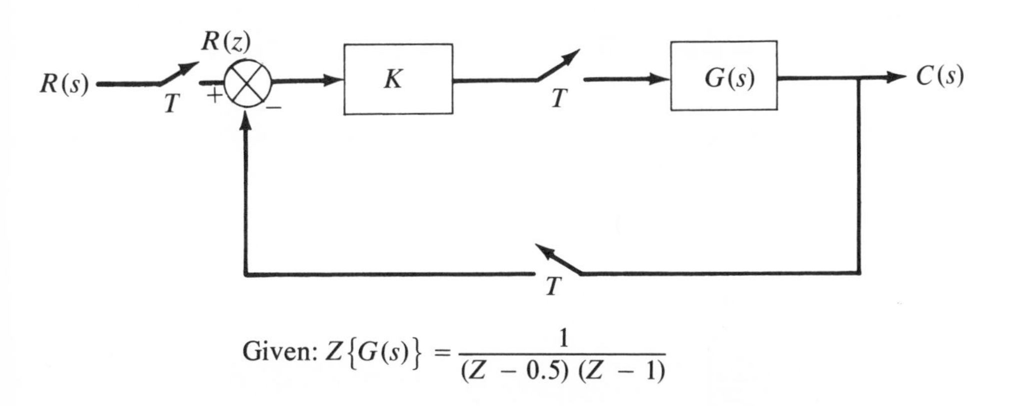

$$\text{Given: } Z\{G(s)\} = \frac{1}{(Z - 0.5)(Z - 1)}$$

Chapter 12 Modified *Z* Transforms

1. Find the modified Z transforms of the following functions:

(a) $\dfrac{1}{s^2}$

(b) $\dfrac{1}{s^2(s + 1)}$

(c) $\dfrac{2}{s(s + 1)(s + 2)}$

2. Determine the complete time response of the following control system to a unit step change in set point

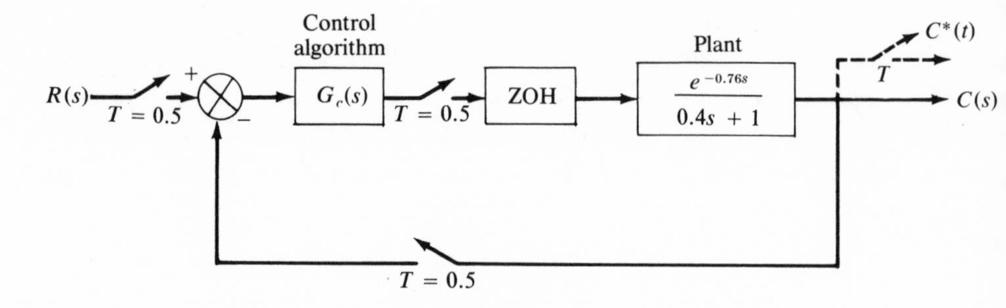

(a) Proportional control algorithm: $G_c(s) = 0.3$

(b) PI algorithm: $G_c(s) = 0.3\left(1 + \dfrac{1}{0.5s}\right)$

Chapter 13 Process Identification

1. The results from a pulse test are shown in the following table. Use the computer program from this chapter to develop the parameters of the process model, assuming that the data can be adequately described by a second-order plus dead-time model.

Time	Input	Time	Output
0.03300	1.74000	0.16700	0.04000
0.06700	1.95000	0.20000	0.07000
0.30000	1.95000	0.23300	0.13000
0.33000	1.09000	0.27000	0.17000
0.36700	0.39000	0.30000	0.23000
0.40000	0.20000	0.33000	0.27000
0.43300	0.10000	0.40000	0.37000
0.46700	0.00000	0.43000	0.43000
		0.47000	0.46000
		0.50000	0.49000
		0.53000	0.50000
		0.57000	0.50000
		0.63000	0.48000
		0.67000	0.45000
		0.73000	0.43000
		0.80000	0.39000
		0.83000	0.37000
		0.87000	0.36000
		0.93000	0.32000
		1.00000	0.30000
		1.07000	0.27000
		1.13000	0.27000
		1.30000	0.25000
		1.50000	0.23000
		1.67000	0.18000
		1.83000	0.18000
		1.97000	0.16000
		2.03000	0.14000
		2.37000	0.14000
		3.70000	0.05000
		5.03000	0.05000
		6.37000	0.00000

2. Fit an approximate transfer function to each of the frequency-response data shown in the following figure. (From W.L., Luyben, *Process Modeling, Simulation, and Control for Chemical Engineers*, McGraw-Hill, 1973 with permission.)

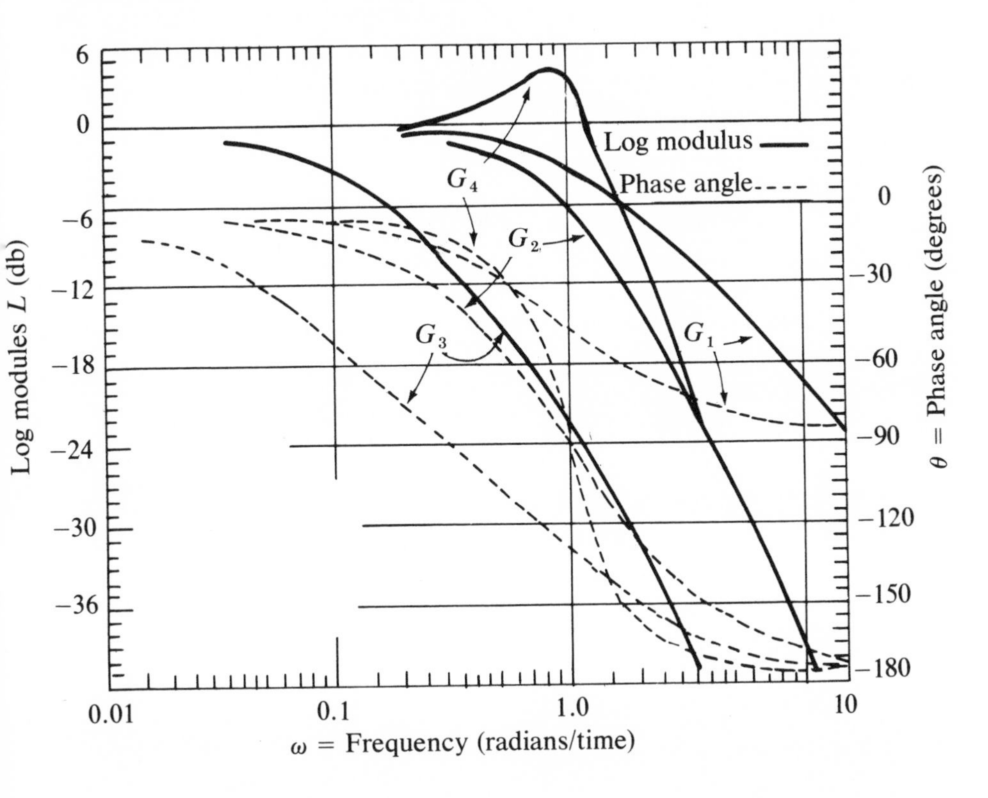

Chapter 14 Algorithms for Processes with Dead Time

1. Design the Smith predictor control system and the analytical predictor control system for the first-order plus dead-time process whose transfer function is given by

$$G_p(s) = \frac{-1e^{-0.76s}}{0.4s + 1}$$

The sampling period $T = 0.5$. Determine the closed-loop response to a unit step change in set point and load. The problem may require trial-and-error tuning of the digital controller.

2. For the furnace control system shown in problem 7, Chapter 1 design the analytical predictor and the Smith predictor control algorithms. Assume that $T = 0.1$ and pneumatic/electrical transducers

(0 to 10 volts DC/3 to 15 psig) are used in the equivalent system. Determine the transient closed-loop responses, through computer simulation, to a unit step change in set point. Use a zero-order hold.

3. Design Gautam and Mutharasan algorithm for case 3, Problem 8, Chapter 1 with $T = 0.5$ and a zero-order hold. Determine the transient response, through simulation, to a unit step change in set point and to a step change of +3 gpm in load.

Chapter 15 Feedforward Control

1. Design a feedforward controller for the following heat exchanger system (from P. Harriott, *Process Control*, McGraw Hill, 1965 with permission).

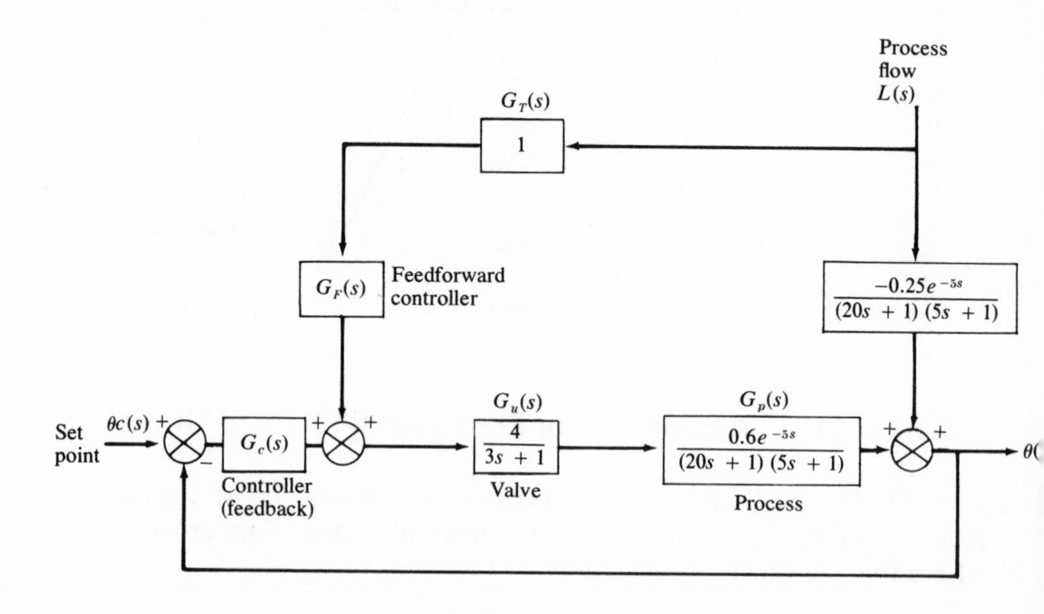

Chapter 16 Cascade Control

1. The schematic and block diagram of a reactor control system is shown below. For this control system determine (a) Ziegler–Nichols tuning of the two controllers (use three-mode master con-

troller and a PI slave controller) of the cascade system, (b) Ziegler–Nichols setting of the equivalent single-loop PID controller, (c) determine the transient response to a unit step change in set point, and to a step change of 0.2 in L_1 and L_2 respectively, and (d) determine ITAE for each case and assess the benefits of cascade control.

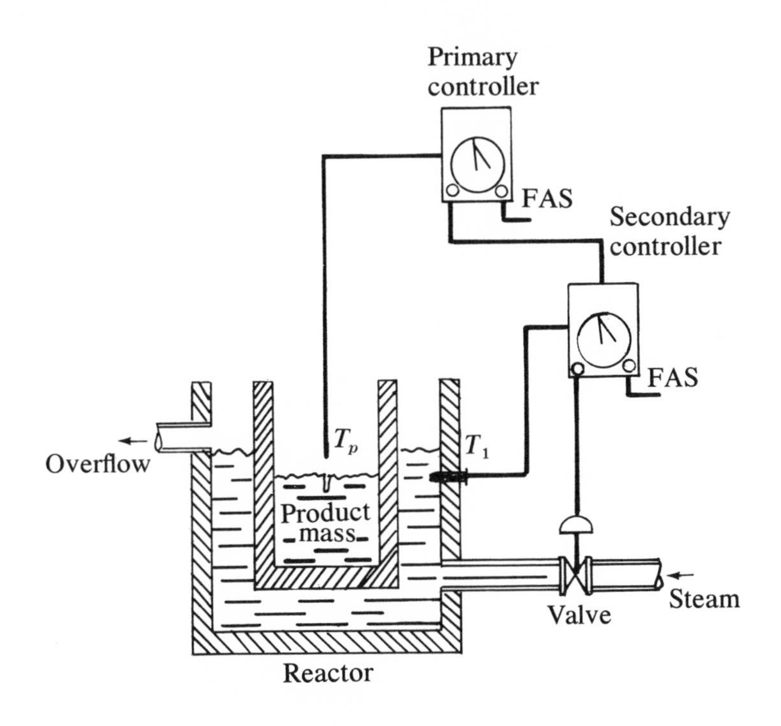

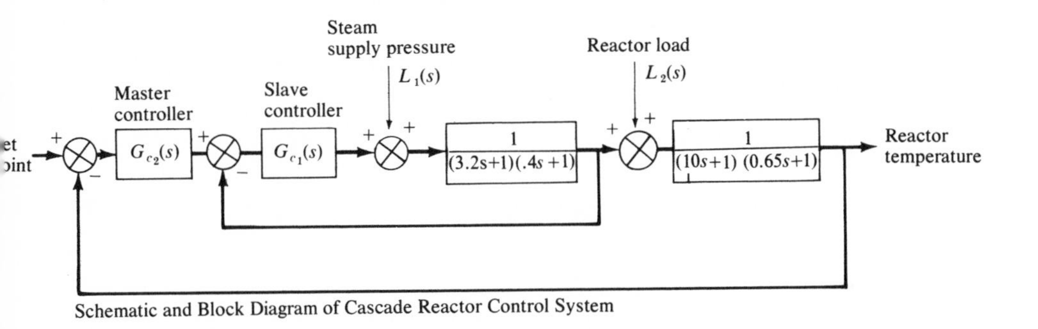

Schematic and Block Diagram of Cascade Reactor Control System

2. The block diagram of a second-order control system is shown below. Answer the questions asked in problem 1 with respect to this control system.

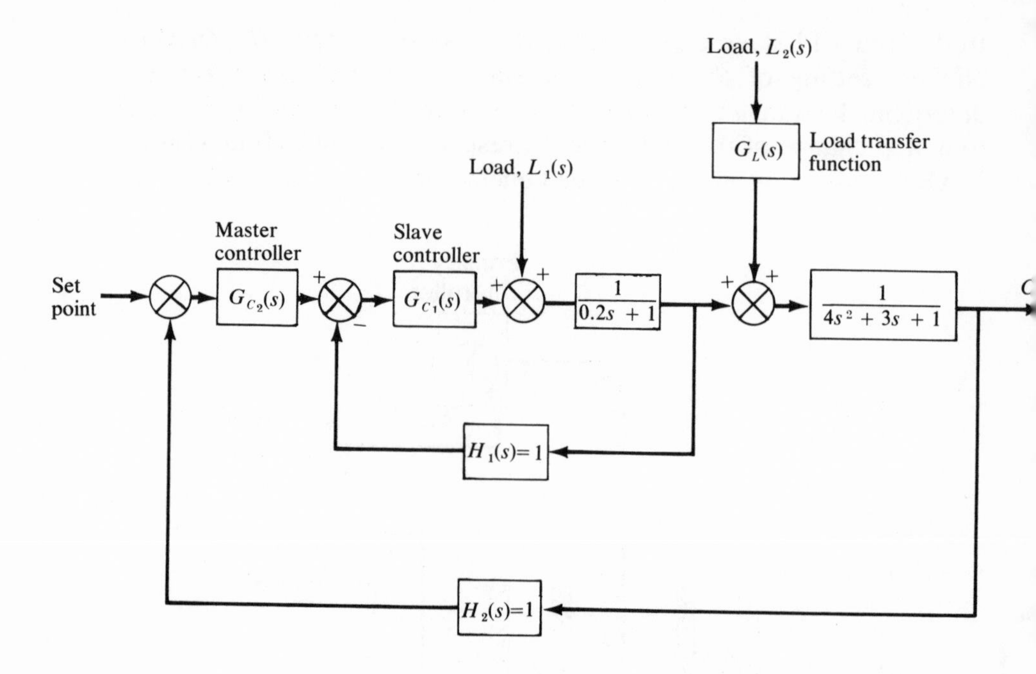

Load, $L_2(s)$
$G_L(s)$
Load transfer function
Load, $L_1(s)$
Master controller
Slave controller
Set point
$G_{C_2}(s)$
$G_{C_1}(s)$
$\frac{1}{0.2s + 1}$
$\frac{1}{4s^2 + 3s + 1}$
$H_1(s) = 1$
$H_2(s) = 1$
C

Appendix C1

Computer Program Listing for Z-Transform Inversion

This appendix presents a Fortran computer program for Z-transform inversion.* The program performs long division on the equation

$$C(Z) = \frac{P_0 + P_1Z^{-1} + P_2Z^{-2} + P_3Z^{-3} + \ldots . P_nZ^{-n}}{Q_0 + Q_1Z^{-1} + Q_2Z^{-2} + Q_3Z^{-3} + \ldots . Q_nZ^{-n}}$$

and outputs $C(nT)$ according to the equation

$$C(Z) = C(O) + C(T)Z^{-1} + C(2T)Z^{-2} + \ldots . C(mT)Z^{-m}$$

To use the program read in the following data:

1. $P_0, P_1, P_2, \ldots . P_{10}$
2. $Q_0, Q_1, Q_2, \ldots . Q_{10}$
3. Set MM equal to the number of output points desired.

* Crosby, H. A., Peterson, D. M. Fortran Subroutine Solves Z-Transform Inversion, *Control Engineering*, August 1967, pp. 92–93. Cited with Permission of the Publisher.

Given below is the program listing and a sample printout of results.

```
File: LONDIV.FOR
Edit: DSKE:LONDIV.FOR[424,246]
*PØ:*
ØØ1ØØ            DIMENSION PØ(1Ø),QØ(1Ø),CØ(1ØØ)
ØØ2ØØ            DO 7 I=1 , 1Ø
ØØ3ØØ            PØ(I)=Ø.
ØØ4ØØ            QØ(I)=Ø.
ØØ5ØØ       7    CONTINUE
ØØ6ØØ            TYPE 3ØØ
ØØ7ØØ     3ØØ    FORMAT (' ENTER # OF TERMS IN NUMERATOR & DENOMINATOR.')
ØØ8ØØ            ACCEPT 4ØØ,NNR,NDR
ØØ9ØØ     4ØØ    FORMAT (I3)
Ø1ØØØ            TYPE 5ØØ
Ø11ØØ     5ØØ    FORMAT(' ENTER COEFF. OF NUMERATOR TERM BY TERM.')
Ø12ØØ            DO 8 I=1,NNR
Ø13ØØ            ACCEPT 1ØØ,PØ(I)
Ø14ØØ       8    CONTINUE
Ø15ØØ            TYPE 6ØØ
Ø16ØØ     6ØØ    FORMAT(' ENTER COEFF. OF DENOMINATOR TERM BY TERM.')
Ø17ØØ            DO 9 I=1, NDR
Ø18ØØ            ACCEPT 1ØØ,QØ(I)
Ø19ØØ       9    CONTINUE
Ø2ØØØ     1ØØ    FORMAT (1ØF8.3)
Ø21ØØ            MM=25
Ø22ØØ            CALL ZTRANS(CØ,PØ,QØ,MM)
Ø23ØØ            TYPE 7ØØ
Ø24ØØ     7ØØ    FORMAT('          SAMP.INST.              OUTPUT')
Ø25ØØ            DO 1 K=1,MM
Ø26ØØ            N=K-1
Ø27ØØ            TYPE 2ØØ , N,CØ(K)
Ø28ØØ       1    CONTINUE
Ø29ØØ     2ØØ    FORMAT(1ØX,I3,2ØX,F5.3)
Ø3ØØØ            END
Ø31ØØ            SUBROUTINE ZTRANS (C,P,Q,MM)
Ø32ØØ            DIMENSION P(1Ø),Q(1Ø),C(1ØØ)
Ø33ØØ            DO 1 I=2,1Ø
Ø34ØØ            P(I)=P(I)/Q(1)
Ø35ØØ            Q(I)=Q(I)/Q(1)
Ø36ØØ       1    CONTINUE
Ø37ØØ            P(1)=P(1)/Q(1)
Ø38ØØ            Q(1)=1
Ø39ØØ            C(1)=P(1)
Ø4ØØØ            DO 6 N=2,MM
Ø41ØØ            SIGMA =Ø.
Ø42ØØ            IF (N-1Ø) 2,2,3
Ø43ØØ       2    ABC=P(N)
Ø44ØØ            M=N
Ø45ØØ            GO TO 4
Ø46ØØ       3    M=1Ø
Ø47ØØ            ABC=Ø.
Ø48ØØ       4    DO 5 I=2,M
Ø49ØØ            K=(N+1)-I
Ø5ØØØ            SIGMA=SIGMA+C(K)*Q(I)
Ø51ØØ       5    CONTINUE
Ø52ØØ            C(N)=ABC-SIGMA
Ø53ØØ       6    CONTINUE
Ø54ØØ            RETURN
Ø55ØØ            END
*
```

```
 EX LONDIV.FOR
FORTRAN: LONDIV
MAIN.
ZTRANS
LINK:   Loading
[LNKXCT LONDIV Execution]

ENTER # OF TERMS IN NUMERATOR & DENOMINATOR.
3
3

ENTER COEFF. OF NUMERATOR TERM BY TERM.
Ø.Ø
Ø.368
Ø.264

ENTER COEFF. OF DENOMINATOR TERM BY TERM.
1.Ø
-1.368
Ø.368

          SAMP.INST.              OUTPUT
            Ø                      .ØØØ
            1                      .368
            2                      .767
            3                      .914
            4                      .969
            5                      .988
            6                      .996
            7                      .998
            8                      .999
            9                     1.ØØØ
           1Ø                     1.ØØØ
           11                     1.ØØØ
           12                     1.ØØØ
           13                     1.ØØØ
           14                     1.ØØØ
           15                     1.ØØØ
           16                     1.ØØØ
           17                     1.ØØØ
           18                     1.ØØØ
           19                     1.ØØØ
           2Ø                     1.ØØØ
           21                     1.ØØØ
           22                     1.ØØØ
           23                     1.ØØØ
           24                     1.ØØØ

END OF EXECUTION
CPU TIME: Ø.16  ELAPSED TIME: 1:12.83
EXIT
```

Appendix C2

Closed-Loop Simulation Program for Sampled-Data Systems

This appendix presents a digital computer program, written in time-sharing Fortran for simulation of a sampled-data control system. The program can be used to obtain the closed-loop response of a first- or second-order system with dead time to a unit step change in the load variable or any specified change in the set point. The block diagram of the control system is shown in Figure C2.

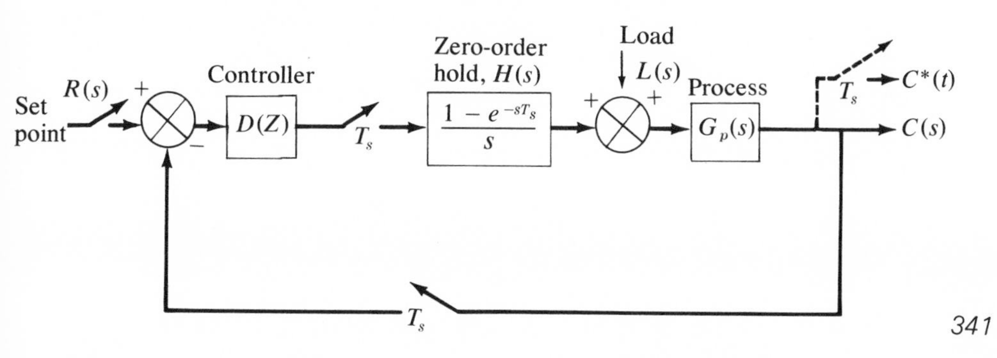

The following inputs are required for the program:

1. Order of process.

If process is first-order, enter 1; if second order, enter 2

2. Parameter of the process transfer function.

If process is first order, enter gain K_p, time constant τ_p, sampling period T_s, and dead time θ_d. If process is second order, enter gain K_p, the two process time constants τ_{p_1} and τ_{p_2}, sampling period T, and dead time θ_d.

3. Enter PID controller constants K_c, τ_I, and τ_D. (User may set τ_I equal to a high value and τ_D to zero if these modes are not required).

4. Is response required for a change in load or set point? Enter 1 if load response is desired or zero if set point response is desired.

5. If zero has been entered as answer to question number 4, enter 1 if response to a unit step change is desired or enter zero if response to some other kind of set point change is required.

6. If the answer to question number 5 is zero, enter the number of terms in the numerator and denominator of $R(Z)$ where

$$R(Z) = (a + bZ^{-1} + cZ^{-2})/(D + EZ^{-1} + FZ^{-2})$$

7. Enter the coefficients of the numerator polynomical from left to right.

8. Enter the coefficients of the denominator polynomical.

With these steps the computer will execute the program and output the response of the controlled variable at the various samples instants. The user can terminate execution of the program when the controlled variable reaches the final steady-state by typing Control C (↑C).

The program solves the following equations by multiplication and long division

For set point changes: $$C(Z) = \frac{D(Z)\,HG(Z)\,R(Z)}{1 + D(Z)\,HG(Z)}$$

For load changes $$C(Z) = \frac{LG(Z)}{1 + D(Z)\,HG(Z)}$$

For a PID controller

$$D(Z) = \frac{K_c\left\{\left(1 + \frac{T_s}{\tau_I} + \frac{\tau_D}{T_s}\right) - \left(1 + 2\,\frac{\tau_D}{T_s}\right) Z^{-1} + \left(\frac{\tau_D}{T_s}\right) Z^{-2}\right\}}{1 - Z^{-1}}$$

The listing of the program and a sample printout of the results are shown on the following pages:

```
File: ZTRANS.FOR
Edit: DSKE:ZTRANS.FOR[424,246]
*P0:*


00100   C       THIS PROGRAM EVALUATES CLOSED LOOP RESPONSE OF A FIRST
00200   C    ORDER OR SECOND ORDER SYSTEM WITH DEAD TIME TO A LOAD CHANGE
00300   C    OR SET-POINT CHANGE USING Z-TRANSFORM METHOD. THE CONTROLLER
00400   C    IS IN P-I-D MODE AND THE PROGRAM TAKES CONTROLLER CONSTANTS,
00500   C    PROCESS CONSTANTS AND SAMPLING TIME AS INPUT.
00600   C    THE LOOP ASSUMES A ZERO ORDER HOLD AFTER THE CONTROLLER.
00700   C    ...........WRITTEN & EDITED BY P.C.GOPALRATNAM...........
00800   C
00900           DIMENSION GCN(20),GPN(20),GCD(20),GPD(20),RD(20),RN(20)
01000           DIMENSION NUMER(100,20),PROD1(20),PROD2(20),PROD3(20)
01100           DIMENSION PROD4(20),PROD5(20),PROD6(20),DENOM(20)
01200           DIMENSION LGN(20),LGD(20),DUM(20)
01300           REAL KC,KP,M
01400           TYPE 105
01500     105   FORMAT(' ENTER THE ORDER OF THE PROCESS.')
01600           ACCEPT 210,IORDER
01700           GO TO (150,155) ,IORDER
01800     150   TYPE 100
01900     100   FORMAT(' ENTER VALUES OF KP,TAUP1,TSAMP,THETA AS REAL#')
02000           ACCEPT 200,KP,TAUP1,TSAMP,THETA
02100     200   FORMAT(F8.4)
02200           TAUP2=0.0
02300           GO TO 180
02400     155   TYPE 101
02500     101   FORMAT(' ENTER KP,TAUP1,TAUP2,TSAMP,THETA AS REAL#:')
02600           ACCEPT 200,KP,TAUP1,TAUP2,TSAMP,THETA
02700     180   TYPE 110
02800     110   FORMAT(' ENTER KC & TAUI,TAUD OF CONTROLLER AS REAL #.')
02900           ACCEPT 200,KC,TAUI,TAUD
03000           QUO=THETA/TSAMP
03100           N=INT(QUO)
03200           M=1.-(THETA/TSAMP-N)
03300           EX1=EXP(-TSAMP/TAUP1)
03400           L=N+6
03500           EXM1=EXP(-M*TSAMP/TAUP1)
03600           IF(TAUP2.EQ.0) GO TO 21
03700           EX2=EXP(-TSAMP/TAUP2)
03800           EXM2=EXP(-M*TSAMP/TAUP2)
03900           GO TO 22
04000      21   EX2=0.0
04100           EXM2=0.0
04200      22   TYPE 181
04300     181   FORMAT(' IS THIS LOAD CHANGE OR SET-POINT CHANGE ?')
04400           TYPE 185
04500     185   FORMAT(' ENTER 1 FOR LOAD CHANGE & 0 FOR SET-POINT CHANG
E')
04600           ACCEPT 210,KONST1
04700   C
04800   C    INITIALIZE THE POLYNOMIALS.
04900   C
05000           DO 300 I=1,L
05100           DUM(I)=0.0
05200           LGN(I)=0.0
05300           LGD(I)=0.0
05400           GCN(I)=0.
05500           GPN(I)=0.
05600           GCD(I)=0.
05700           GPD(I)=0.
05800           RD(I)=0.0
05900           RN(I)=0.
06000     300   CONTINUE
06100   C
06200   C    ESTABLISH THE POLYNOMIAL FOR SET-POINT.
06300   C
06400           IF(KONST1.EQ.1) GO TO 40
06500           TYPE 120
```

```
06600    120     FORMAT(' DO YOU WANT THE RESPONSE FOR A UNIT STEP ?')
06700            TYPE 125
06800    125     FORMAT(' ENTER 1 FOR YES & 0 FOR NO.    :')
06900            ACCEPT 210,KONST
07000    210     FORMAT(I2)
07100            IF(KONST) 10,10,20
07200     10     TYPE 130
07300    130     FORMAT(' ENTER # OF TERMS IN NUMER. & DENOM. OF R(Z)')
07400            ACCEPT 220,IRN,IRD
07500    220     FORMAT(I2)
07600            IR=MAX0(IRN,IRD)
07700            L=L-1+IR
07800            TYPE 140
07900    140     FORMAT(' ENTER COEFF.OF NUMERATOR TERM BY TERM.  :')
08000            ACCEPT 230,(RN(I),I=1,IRN)
08100    230     FORMAT(F12.6)
08200            TYPE 151
08300    151     FORMAT(' ENTER COEFF.OF DENOMINATOR TERM BY TERM.    :')
08400            ACCEPT 230,(RD(I),I=1,IRD)
08500            GO TO 40
08600     20     RN(1)=1.0
08700            RD(1)=1.0
08800            RD(2)=-1.0
08900    C
09000    C   ESTABLISH THE POLYNOMIAL FOR D(Z).
09100    C
09200     40     GCN(1)=KC*(1.0+TSAMP/TAUI+TAUD/TSAMP)
09300            GCN(2)=-KC*(1.0+2.0*TAUD/TSAMP)
09400            GCN(3)=KC*TAUD/TSAMP
09500            GCD(1)=1.0
09600            GCD(2)=-1.0
09700    C
09800    C   ESTABLISH THE POLYNOMIAL FOR THE PROCESS HG(Z) AND LG(Z).
09900    C
10000            GPN(N+2)=KP*(1.+(TAUP1*EXM1-TAUP2*EXM2)/(TAUP2-TAUP1))
10100            TEMP=(TAUP1*EXM1*(1.+EX2)-TAUP2*EXM2*(1.+EX1))
10200            GPN(N+3)=-KP*(EX1+EX2+TEMP/(TAUP2-TAUP1))
10300            TEMP1=TAUP1*EXM1*EX2-TAUP2*EXM2*EX1
10400            GPN(N+4)=KP*(EX1*EX2+TEMP1/(TAUP2-TAUP1))
10500            GPD(1)=1.0
10600            GPD(2)=-1.*(EX1+EX2)
10700            GPD(3)=EX1*EX2
10800            IF(KONST1.EQ.0) GO TO 80
10900            DUM(1)=1.
11000            DUM(2)=-1.
11100            CALL MPLY(DUM,GPD,LGD,L)
11200    C
11300    C   PERFORM MULTIPLICATIONS
11400    C
11500    80      CALL MPLY(GCN,GPN,PROD1,L)
11600            CALL MPLY(GPD,GCD,PROD3,L)
11700            DO 350 I=1,L
11800            PROD4(I)=PROD1(I)+PROD3(I)
11900    350     CONTINUE
12000            IF (KONST1.EQ.0) GO TO 410
12100            CALL MPLY(GPN,PROD3,PROD2,L)
12200            CALL MPLY(PROD4,LGD,DENOM,L)
12300            GO TO 420
12400    410     CALL MPLY(PROD1,RN,PROD2,L)
12500            CALL MPLY(PROD4,RD,DENOM,L)
12600    420     DO 360 I=2,L
12700            PROD2(I)=PROD2(I)/DENOM(1)
12800            DENOM(I)=DENOM(I)/DENOM(1)
12900    360     CONTINUE
13000            PROD2(1)=PROD2(1)/DENOM(1)
13100            DENOM(1)=1.0
13200            TYPE 11
13300      11    FORMAT(' SAMP.INST.                     OUTPUT VALUES.')
13400            I=0
13500            PROD2(L+1)=0.0
```

```
13600    500    RSPONS=PROD2(1)
13700           TYPE 170,I,RSPONS
13800    170    FORMAT(I5,25X,F6.3)
13900           DO 370 J=1,L
14000           PROD2(J)=PROD2(J+1)-DENOM(J+1)*RSPONS
14100    370    CONTINUE
14200           I=I+1
14300           IF (I.LE.100) GO TO 500
14400           END
14500           SUBROUTINE MPLY(A,B,C,N)
14600           DIMENSION A(20),B(20),C(20)
14700           DO 330 I=1,N
14800           C(I)=0.0
14900           DO 340 J=1,I
15000           C(I)=C(I)+A(J)*B(I+1-J)
15100    340    CONTINUE
15200    330    CONTINUE
15300           RETURN
15400           END
*
```

```
.EX ZTRANS.FOR
FORTRAN: ZTRANS
MAIN.
MPLY
LINK:   Loading
[LNKXCT ZTRANS Execution]

ENTER THE ORDER OF THE PROCESS.
1

ENTER VALUES OF KP,TAUP1,TSAMP,THETA AS REAL#
1.0
0.4
0.5
0.76

ENTER KC & TAUI,TAUD OF CONTROLLER AS REAL #.
0.3
0.5
0.0

IS THIS LOAD CHANGE OR SET-POINT CHANGE ?
ENTER 1 FOR LOAD CHANGE & 0 FOR SET-POINT CHANGE
1

SAMP.INST.                 OUTPUT VALUES.
   0                           0.000
   1                           0.000
   2                           0.451
   3                           0.843
   4                           0.833
   5                           0.592
   6                           0.314
   7                           0.123
   8                           0.035
   9                           0.016
  10                           0.024
  11                           0.030
  12                           0.028
  13                           0.020
  14                           0.011
  15                           0.005
  16                           0.002
  17                           0.001
  18                           0.001
  19                           0.001
  20                           0.001
  21                           0.001
  22                           0.000
```

```
  23                                      Ø.ØØØ
  24                                      Ø.ØØØ
  25                                      Ø.ØØØ
  26                                      Ø.ØØØ
  27                                      Ø.ØØØ
  28                                      Ø.ØØØ
  29                                      Ø.ØØØ
  3Ø                                      Ø.ØØØ
 EX ZTRANS.FOR
LINK:   Loading
[LNKXCT ZTRANS Execution]

ENTER THE ORDER OF THE PROCESS.
1

ENTER VALUES OF KP,TAUP1,TSAMP,THETA AS REAL#
1.Ø
Ø.4
Ø.5
Ø.76

ENTER KC & TAUI,TAUD OF CONTROLLER AS REAL #.
Ø.3
Ø.5
Ø.Ø

IS THIS LOAD CHANGE OR SET-POINT CHANGE ?
ENTER 1 FOR LOAD CHANGE & Ø FOR SET-POINT CHANGE
Ø

DO YOU WANT THE RESPONSE FOR A UNIT STEP ?
ENTER 1 FOR YES & Ø FOR NO.     :
1

SAMP.INST.                 OUTPUT VALUES.
    Ø                           Ø.ØØØ
    1                           Ø.ØØØ
    2                           Ø.271
    3                           Ø.641
    4                           Ø.888
    5                           Ø.993
    6                           1.ØØ4
    7                           Ø.983
    8                           Ø.968
    9                           Ø.967
   1Ø                           Ø.976
   11                           Ø.987
   12                           Ø.995
   13                           Ø.999
   14                           Ø.999
   15                           Ø.999
   16                           Ø.999
   17                           Ø.999
   18                           Ø.999
   19                           1.ØØØ
   2Ø                           1.ØØØ
   21                           1.ØØØ
   22                           1.ØØØ
   23                           1.ØØØ
   24                           1.ØØØ
   25                           1.ØØØ
   26                           1.ØØØ
   27                           1.ØØØ
   28                           1.ØØØ
   29                           1.ØØØ
   3Ø                           1.ØØØ
   31                           1.ØØØ
   32                           1.ØØØ
   33                           1.ØØØ
   34                           1.ØØØ
```

```
  35                              1.000
  36                              1.000
  37                              1.000
  38                              1. ↑C


.EX ZTRANS.FOR
LINK:    Loading
[LNKXCT ZTRANS Execution]

ENTER THE ORDER OF THE PROCESS.
1

ENTER VALUES OF KP,TAUP1,TSAMP,THETA AS REAL#
1.0
0.4
0.5
0.76

ENTER KC & TAUI,TAUD OF CONTROLLER AS REAL #.
0.3
0.5
0.0

IS THIS LOAD CHANGE OR SET-POINT CHANGE ?
ENTER 1 FOR LOAD CHANGE & 0 FOR SET-POINT CHANGE
0

DO YOU WANT THE RESPONSE FOR A UNIT STEP ?
ENTER 1 FOR YES & 0 FOR NO.     :
0

ENTER # OF TERMS IN NUMER. & DENOM. OF R(Z)
1
2

ENTER COEFF.OF NUMERATOR TERM BY TERM.   :
1.0


ENTER COEFF.OF DENOMINATOR TERM BY TERM.     :
1.0
-1.0

SAMP.INST.                  OUTPUT VALUES.
   0                              0.000
   1                              0.000
   2                              0.271
   3                              0.641
   4                              0.888
   5                              0.993
   6                              1.004
   7                              0.983
   8                              0.968
   9                              0.967
  10                              0.976
  11                              0.987
  12                              0.995
  13                              0.999
  14                              0.999
  15                              0.999
  16                              0.999
  17                              0.999
  18                              0.999
  19                              1.000
  20                              1.000
  21                              1.000
  22                              1.000
  23        ↑C
```

```
 EX ZTRANS.FOR
LINK:   Loading
[LNKXCT ZTRANS Execution]

ENTER THE ORDER OF THE PROCESS.
2

ENTER KP,TAUP1,TAUP2,TSAMP,THETA AS REAL#:
1.0
0.4
0.0
0.5
0.76

ENTER KC & TAUI,TAUD OF CONTROLLER AS REAL #.
0.3
0.5
0.0

IS THIS LOAD CHANGE OR SET-POINT CHANGE ?
ENTER 1 FOR LOAD CHANGE & 0 FOR SET-POINT CHANGE
1

SAMP.INST.                 OUTPUT VALUES.
    0                          0.000
    1                          0.000
    2                          0.451
    3                          0.843
    4                          0.833
    5                          0.592
    6                          0.314
    7                          0.123
    8                          0.035
    9                          0.016
   10                          0.024
   11                          0.030
   12                          0.028
   13                          0.020
   14                          0.011
   15                          0.005
   16                          0.002
   17                          0.001
   18                          0.001
   19                          0.001
   20                          0.001
   21                          0.001
   22                          0.000
   23                          0.000
   24                          0.000
   25                          0.000
   26                          0.000
   27                          0.000
   28                          0.000
   29                          0.000
   30                           ↑C
```

Appendix D

Pulse Analysis Program

This appendix (by permission of Instrument Society of America, Research Triangle Park, North Carolina) describes a Fortran program for computing the frequency response data from the pulse test data. A listing of the program is also included. When the input and output pulse data are entered as a function of time, the program prints a table of system gain (amplitude ratio) and phase angles as a function of frequency.

Program Input Instructions

Enter the problem data on input data sheets as described in the following paragraphs (see Tables I and IV):

Related input items are grouped into the lists described below.

List 1: Problem-Set Identification—Enter this list to initialize the program or to nullify data from a preceding set of related problems. The contents of columns 11 to 70 will be printed in the heading of each output page.

List 2: Case Identification—Enter this list to print a 60 character case title below the problem-set title on each output page. The list is optional.

List 3: Response Control Data—Enter this list to specify which of the pulses defined by list 6 is the input and which is the output. Omit this list to print only the Fourier transforms of the pulses.

Lists 4 and 5: Frequency Specifications—These lists provide two alternate methods of specifying the frequencies at which the program will compute the pulse amplitude and phase angle. Do not enter both lists. If neither is entered, the program provides ten frequencies per decade from 0.001 to 100 radians per unit time.

List 6: Pulse data—Enter this list for each pulse curve to be considered in the problem.

A curve may be split into as many as ten sections with equal time intervals in each section. The section starting at zero time is the first section.

The ordinates of the curve are numbered sequentially. Ordinate number one is at zero time after excitation. The maximum number of ordinates per curve is 1000. Ordinates may be entered with any units desired. If a chart factor or other multiplier is required to change the ordinate values to actual physical units, it can be entered for each curve.

The initial steady-state value (value prior to excitation) must be given, although it is usually the same as the first ordinate.

A pulse closure effect exists when the end of the curve fails to return to the drift corrected initial steady-state value. This may be handled in one of two ways by the pulse closure code:

(a) Enter 0 if the pulse should be closed. A step change between the last ordinate and the drift corrected initial steady state value is assumed.

(b) Enter 2 if it is believed that the curve need not (or should not) return to the initial steady-state value and that the final steady-state value is indeed different from the value prior to excitation. (An example of this type appears later in this appendix.)

List 99: End of Case Control—Enter this list to mark the end of the input data for a case.

Keypunching Rules

1. Punch data items in the order in which they appear on the input data sheets. Start at the top and read from left to right.

2. Do not punch a list number if it is immediately followed by a blank entry.

3. Punch numeric items into ten-column fields with eight fields per card.

4. Right-justify each integer within its ten-column field and omit the decimal point.

5. Punch each item preceded by a double dagger (‡) into colums one to ten of a new card. Do not punch the double dagger.

Tables II and V illustrate the punched card input prepared from the Input Data Sheets.

Output

The output corresponding to typical input data are given in Tables III and VI. For the Fourier transform the real part, imaginary part, amplitude and phase angle are given for each frequency. Any pulse closure corrections made are noted. When the frequency response option is requested, the amplitude ratio, $\log_{10}$ (amplitude ratio) and phase difference are given in addition to the Fourier transform items.

A warning message is printed in the frequency table after the first frequency for which the input pulse amplitude is less than 1% of its zero frequency amplitude. If either curve persists to infinity, the area under the curve is calculated from time 0 to time of the last point.

Program Execution

Add the control cards needed to load the program into the computer to the program deck which should be prepared in accordance with the listing shown in Table VII and the data cards prepared according to the instructions just described. The control cards and the job submission procedure will depend on the specific computer installation you use.

Table I

	INPUT DATA SHEET 1 OF 3 FREQUENCY RESPONSE FROM PULSE TEST DATA	BY ______ DATE ______

LIST 1: PROBLEM-SET IDENTIFICATION

‡ 1 ISA S26 FREQUENCY RESPONSE FROM PULSE TEST DATA

LIST 2: CASE IDENTIFICATION (Optional)

‡ 2 SAMPLE PROBLEM TO TEST COMPUTER PROGRAM PROPERLY EXCUTED

LIST 3: DYNAMIC RESPONSE CONTROL DATA (Optional)

Description	Entry
List Number	‡ 3
Input Pulse Number (integer 1 through 9)	‡ 1
Output Pulse Number (integer 1 through 9)	2
Response Normalization Code (Enter 0 for normalization, otherwise enter 1)	0
Transform Print Code (Enter 0 to print transforms of both pulses, Enter 1 to print Input Pulse Transform only, Enter 2 to print Output Pulse Transform only, Enter 3 to suppress printing of both transforms)	0

LIST 4: FREQUENCY RANGE (Optional *. Use List 5 if more appropriate.)

Description	Entry
List Number	‡ 4
Frequency Units (Enter 0 for radians/time, Enter 1 for cycles/time) (0 normal)	‡ 1
Number of Decades (5 normal, 5 max)	3
Number of Frequencies per Decade (10 normal, 10 max)	10
Initial Frequency (0.001 normal)	0.01

LIST 5; FREQUENCY POINTS (Alternate to List 4) *

Description	Entry
List Number	‡ 5
Frequency Units (Enter 0 for radians/time, Enter 1 for cycles/time)	‡
Number of Frequencies (50 max)	

Frequencies

‡

Note: The normal values in List 4 are used if both List 4 and List 5 are omitted.

Start new card. Right justify numeric entries in 10 column fields.

Table I (*continued*)

	INPUT DATA SHEET 2 OF 3 FREQUENCY RESPONSE FROM PULSE TEST DATA	BY ________ DATE ______

LIST 6. PULSE DATA (Repeat this list for each pulse using a unique pulse number, k)

Description	Entry
List Number and Pulse Identification	‡ 6 RUN 5.1.2I.
Pulse Number, k $(1 \leq k \leq 9)$	‡ 1
Pulse Height Prior to Excitation at Time 0⁻	0
Base Line Drift from Time 0 to T	0
Multiplier to Convert Pulse Height Units. Enter 1 if conversion not required.	1
Pulse Height Data Summary Code (0 normal. Enter 1 if printout desired.)	1
Pulse Closure Code. Enter 0 for closed pulses. Enter 2 for open pulses.	0
Number of Sections in Curve, M $(1 \leq M \leq 10)$	1

Section Number, m	1*	2*	3*	4*
Time Interval Between Points	‡ .05			
Number of Last Point in Section	18			

Number of Pulse Height Data Points, N $(2 \leq N \leq 1000)$ ‡ 18

Point No.	Pulse Height Data Starting at Time 0^+ and Ending at Time T^-				
1-5	‡ 0	4	11	43	56
6-10	69	76	75	70	63
11-15	48	35	22	14	5
16-20	3	1	0		
21-25					
26-30					
31-35					
36-40					
41-45					
46-50					
51-55					
56-60					
61-65					
66-70					
71-75					
76-80					
81-85					
86-90					
91-95					
96-100					
101-105					
106-110					
111-115					
116-120					
121-125					

LIST 99: END OF CASE CONTROL	‡ 99

* Punch entries below column-wise.

‡ Start new card. Right justify numeric entries in 10 column fields.

Table I (*concluded*)

	INPUT DATA SHEET 3 OF 3 FREQUENCY RESPONSE FROM PULSE TEST DATA	BY ______ DATE ______

LIST 6: PULSE DATA (Repeat this list for each pulse using a unique pulse number, k)

Description	Entry
List Number and Pulse Identification	‡ 6RUM S120 (columns 10, 15, 18)
Pulse Number, k $(1 \leq k \leq 9)$	‡ 2
Pulse Height Prior to Excitation at Time 0^-	0
Base Line Drift from Time 0 to T	0
Multiplier to Convert Pulse Height Units. Enter 1 if conversion not required.	1
Pulse Height Data Summary Code (0 normal. Enter 1 if printout desired.)	1
Pulse Closure Code. Enter 0 for closed pulses. Enter 2 for open pulses.	0
Number of Sections in Curve, M $(1 \leq M \leq 10)$	2

Section Number, m	1*	2*	3*	4*
Time Interval Between Points	‡ .05	0.1		
Number of Last Point in Section	15	31		

Number of Pulse Height Data Points, N $(2 \leq N \leq 1000)$	‡ 31

Pulse Height Data Starting at Time 0^+ and Ending at Time T^-

Point No.					
1-5	‡ 0	0	2	4	7
6-10	12	18	24	30	37
11-15	40	44	42	41	38
16-20	36	27	21	16	13
21-25	10	8	6	4.5	3.5
26-30	2.5	1.7	1.1	0.6	0.2
31-35	0				
36-40					
41-45					
46-50					
51-55					
56-60					
61-65					
66-70					
71-75					
76-80					
81-85					
86-90					
91-95					
96-100					
101-105					
106-110					
111-115					
116-120					
121-125					

LIST 99 END OF CASE CONTROL	‡ 99

* Punch entries below column-wise.

‡ Start new card. Right justify numeric entries in 10 column fields.

Table II
Printout Showing Input Data

```
1ISA - FREQUENCY RESPONSE FROM PULSE TEST DATA

  ISA S26 FREQUENCY RESPONSE FROM PULSE TEST DATA
  SAMPLE PROBLEM TO TEST COMPUTER PROGRAM PROPERLY EXCITED


 LIST 3 DATA          1.         2.         0.         0.


 LIST 4 DATA          1.         3.        10.     0.01000


 LIST 6 DATA     **********    PULSE NO. = 1  RUN S12I

   HEIGHT AT 0-      0.0000E+00  MULTIPLIER      1.000       CLOSURE CODE  0.
   BASE LINE DRIFT 0.0000E+00  SUMMARY CODE    1.          NO. SECTIONS  1.

   SECTION NUMBER          1
   TIME INTERVAL       0.500E-01
   LAST POINT NO.         18.

   PT. VALUE        PT. VALUE         PT. VALUE         PT. VALUE         PT. VALUE
     1  .000E+00     2  4.00            3  11.0            4  43.0            5  56.0
     6  69.0         7  76.0            8  75.0            9  70.0           10  63.0
    11  48.0        12  35.0           13  22.0           14  14.0           15  5.00
    16  3.00        17  1.00           18  .000E+00

 LIST 6 DATA     **********    PULSE NO. = 2  RUN S120

   HEIGHT AT 0-      0.0000E+00  MULTIPLIER      1.000       CLOSURE CODE  0.
   BASE LINE DRIFT 0.0000E+00  SUMMARY CODE    1.          NO. SECTIONS  2.

   SECTION NUMBER          1          2
   TIME INTERVAL       0.500E-01 0.100E+00
   LAST POINT NO.         15.        31.

   PT. VALUE        PT. VALUE         PT. VALUE         PT. VALUE         PT. VALUE
     1  .000E+00     2  .000E+00        3  2.00            4  4.00            5  7.00
     6  12.0         7  18.0            8  24.0            9  30.0           10  37.0
    11  40.0        12  44.0           13  42.0           14  41.0           15  38.0
    16  36.0        17  27.0           18  21.0           19  16.0           20  13.0
    21  10.0        22  8.00           23  6.00           24  4.50           25  3.50
    26  2.50        27  1.70           28  1.10           29  .600           30  .200
    31  .000E+00
```

Table III Results

```
1ISA - FREQUENCY RESPONSE FROM PULSE TEST DATA

  ISA S26 FREQUENCY RESPONSE FROM PULSE TEST DATA
  SAMPLE PROBLEM TO TEST COMPUTER PROGRAM PROPERLY EXCITED
 PULSE NO. = 1  RUN S12I     INPUT
          TOTAL TIME SPAN    =     0.8500
          PULSE AREA         =      29.75

 PULSE NO. = 2  RUN S12O   OUTPUT
          TOTAL TIME SPAN    =      2.300
          PULSE AREA         =      33.01

          STEADY STATE GAIN =       1.110
```

FREQUENCY RESPONSE (OUTPUT/INPUT)				FOURIER TRANSFORM OF THE INPUT CURVE				FOURIER TRANSFORM OF THE OUTPUT CURVE			
FREQUENCY CYC./TIME	AMPLITUDE RATIO	LOG AMP. RAT	PH. DIF DEGREES	REAL PART	IMAG. PART	AMPLITUDE	PH.ANG DEG.	REAL PART	IMAG. PART	AMPLITUDE	PH.ANG DEG.
1.000E-02	1.00	-0.00010	-1.52	29.7	-0.671	29.7	358.71	33.0	-1.62	33.0	357.19
1.300E-02	1.00	-0.00017	-1.97	29.7	-0.873	29.7	358.32	32.9	-2.10	33.0	356.35
1.600E-02	0.999	-0.00025	-2.42	29.7	-1.07	29.7	357.93	32.9	-2.58	33.0	355.51
2.000E-02	0.999	-0.00039	-3.03	29.7	-1.34	29.7	357.41	32.8	-3.23	33.0	354.38
2.500E-02	0.999	-0.00062	-3.79	29.7	-1.68	29.7	356.77	32.7	-4.03	33.0	352.98
3.200E-02	0.998	-0.00101	-4.85	29.7	-2.15	29.7	355.86	32.5	-5.14	32.9	351.02
4.000E-02	0.996	-0.00158	-6.06	29.6	-2.68	29.7	354.83	32.2	-6.40	32.9	348.77
5.000E-02	0.994	-0.00247	-7.57	29.5	-3.35	29.7	353.54	31.8	-7.95	32.8	345.97
6.300E-02	0.991	-0.00391	-9.52	29.4	-4.21	29.7	351.85	31.1	-9.91	32.7	342.33
8.000E-02	0.986	-0.00630	-12.07	29.2	-5.33	29.7	349.66	30.0	-12.4	32.4	337.58
0.100	0.978	-0.00984	-15.05	28.9	-6.63	29.6	347.07	28.4	-15.1	32.1	332.02
0.130	0.963	-0.01658	-19.47	28.3	-8.54	29.5	343.20	25.4	-18.7	31.6	323.72
0.160	0.944	-0.02503	-23.82	27.5	-10.4	29.4	339.32	22.0	-21.6	30.8	315.50
0.200	0.914	-0.03886	-29.46	26.3	-12.8	29.3	334.16	16.9	-24.4	29.7	304.69
0.250	0.871	-0.06010	-36.22	24.5	-15.5	29.0	327.71	10.3	-26.1	28.0	291.49
0.320	0.801	-0.09646	-44.97	21.4	-18.8	28.5	318.69	1.65	-25.3	25.3	273.73
0.400	0.715	-0.14551	-53.68	17.3	-21.8	27.9	308.41	-5.82	-21.3	22.1	254.73
0.500	0.613	-0.21263	-62.34	11.6	-24.2	26.8	295.61	-10.9	-14.6	18.3	233.27
0.630	0.506	-0.29587	-70.20	3.98	-24.9	25.3	279.07	-12.4	-6.84	14.2	208.87
0.800	0.414	-0.38306	-78.10	-4.87	-22.3	22.8	257.67	-10.5	7.883E-02	10.5	179.57
1.00	0.330	-0.48103	-87.87	-11.8	-15.6	19.5	232.96	-5.87	4.10	7.16	145.09
1.30	0.229	-0.64019	-97.73	-13.8	-4.28	14.4	197.29	-0.608	3.61	3.66	99.56
1.60	0.158	-0.80017	-105.15	-9.28	2.59	9.63	164.41	0.866	1.46	1.69	59.26
2.00	0.136	-0.86605	-99.40	-3.00	3.67	4.74	129.31	0.621	0.357	0.716	29.91
2.50	0.225	-0.64726	184.69	-0.870	1.58	1.81	118.80	0.249	-0.376	0.451	303.49
3.20	0.256	-0.59138	16.91	-0.510	1.21	1.31	112.93	-0.239	0.286	0.372	129.84
4.00	0.224	-0.64880	213.68	0.177	0.822	0.840	77.84	7.680E-02	-0.195	0.209	291.52
5.00	0.224	-0.64908	-11.89	0.365	0.243	0.438	33.69	0.101	4.053E-02	0.109	21.80
6.30	0.113	-0.94595	37.01	0.517	3.852E-02	0.518	4.26	4.893E-02	4.294E-02	6.510E-02	41.27
8.00	0.195	-0.70965	-56.80	-4.425E-02	-0.383	0.385	263.41	-7.461E-02	-3.737E-02	8.345E-02	206.61

Table IV

	INPUT DATA SHEET 1 OF 3 FREQUENCY RESPONSE FROM PULSE TEST DATA	BY ______ DATE ______

LIST 1: PROBLEM-SET IDENTIFICATION

‡1 ISA S26 FREQUENCY RESPONSE FROM PULSE TEST DATA

10 15 20 25 30 35 40 45 50 55 60 65 70

LIST 2: CASE IDENTIFICATION (Optional)

‡2 SAMPLE PROBLEM - NON CLOSING OUTPUT PULSE

10 15 20 25 30 35 40 45 50 55 60 65 70

LIST 3: DYNAMIC RESPONSE CONTROL DATA (Optional)

	Description	Entry
	List Number	‡ 3
	Input Pulse Number (integer 1 through 9)	‡ 1
	Output Pulse Number (integer 1 through 9)	2
	Response Normalization Code (Enter 0 for normalization, otherwise enter 1)	0
	Transform Print Code (Enter 0 to print transforms of both pulses, Enter 1 to print Input Pulse Transform only, Enter 2 to print Output Pulse Transform only, Enter 3 to suppress printing of both transforms)	0

LIST 4: FREQUENCY RANGE (Optional * Use List 5 if more appropriate.)

	Description	Entry
	List Number	‡ 4
	Frequency Units (Enter 0 for radians/time, Enter 1 for cycles/time) (0 normal)	‡ 1
	Number of Decades (5 normal, 5 max)	3
	Number of Frequencies per Decade (10 normal, 10 max)	10
	Initial Frequency (0.001 normal)	.01

LIST 5; FREQUENCY POINTS (Alternate to List 4) *

	Description	Entry
	List Number	‡ 5
	Frequency Units (Enter 0 for radians/time, Enter 1 for cycles/time)	‡
	Number of Frequencies (50 max)	

Frequencies

‡

* Note: The normal values in List 4 are used if both List 4 and List 5 are omitted.

‡ Start new card. Right justify numeric entries in 10 column fields.

Table IV (*continued*)

	INPUT DATA SHEET 2 OF 3 FREQUENCY RESPONSE FROM PULSE TEST DATA	BY ______ DATE ______

LIST 6: PULSE DATA (Repeat this list for each pulse using a unique pulse number, k)

Description	Entry
List Number and Pulse Identification	‡ 6 RUN 5,1,2,I
Pulse Number, k $(1 \le k \le 9)$	‡ 1
Pulse Height Prior to Excitation at Time 0⁻	0
Base Line Drift from Time 0 to T	0
Multiplier to Convert Pulse Height Units. Enter 1 if conversion not required.	1
Pulse Height Data Summary Code (0 normal. Enter 1 if printout desired.)	1
Pulse Closure Code. Enter 0 for closed pulses. Enter 2 for open pulses.	0
Number of Sections in Curve, M $(1 \le M \le 10)$	2

Section Number, m	1*	2*	3*	4*
Time Interval Between Points	‡ .125	.25		
Number of Last Point in Section	14	19		

Number of Pulse Height Data Points, N $(2 \le N \le 1000)$	‡ 19

Pulse Height Data Starting at Time 0⁺ and Ending at Time T⁻

Point No.					
1-5	‡ 0	.2	.9	2.3	4.5
6-10	6.7	8.3	9.7	10.7	11.4
11-15	11.8	10.2	7.9	5.4	2.9
16-20	1.5	.7	.2	.0	
21-25					
26-30					
31-35					
36-40					
41-45					
46-50					
51-55					
56-60					
61-65					
66-70					
71-75					
76-80					
81-85					
86-90					
91-95					
96-100					
101-105					
106-110					
111-115					
116-120					
121-125					

LIST 99: END OF CASE CONTROL	‡ 99

* Punch entries below column-wise.

‡ Start new card. Right justify numeric entries in 10 column fields.

Table IV (*concluded*)

	INPUT DATA SHEET 3 OF 3 FREQUENCY RESPONSE FROM PULSE TEST DATA	BY ________ DATE ________

LIST 6: PULSE DATA (Repeat this list for each pulse using a unique pulse number, k)

Description	Entry
List Number and Pulse Identification	‡ 6 RUN 3,1,2,0
Pulse Number, k $(1 \le k \le 9)$	‡ 2
Pulse Height Prior to Excitation at Time 0⁻	0
Base Line Drift from Time 0 to T	1.3
Multiplier to Convert Pulse Height Units. Enter 1 if conversion not required.	1
Pulse Height Data Summary Code (0 normal. Enter 1 if printout desired.)	1
Pulse Closure Code. Enter 0 for closed pulses. Enter 2 for open pulses.	2
Number of Sections in Curve, M $(1 \le M \le 10)$	2

Section Number, m	1*	2*	3*	4*
Time Interval Between Points	‡ .125	.25		
Number of Last Point in Section	14	22		

Number of Pulse Height Data Points, N $(2 \le N \le 1000)$ ‡ 22

Point No.	Pulse Height Data Starting at Time 0^+ and Ending at Time T^-				
1-5	‡ 0	.1	.5	1.2	2.1
6-10	3.2	4.3	5.7	6.8	7.8
11-15	8.4	8.8	8.3	7.7	6.1
16-20	4.8	3.6	2.7	2.0	1.8
21-25	1.4	1.3			
26-30					
31-35					
36-40					
41-45					
46-50					
51-55					
56-60					
61-65					
66-70					
71-75					
76-80					
81-85					
86-90					
91-95					
96-100					
101-105					
106-110					
111-115					
116-120					
121-125					

LIST 99: END OF CASE CONTROL	‡ 99

* Punch entries below column-wise.

‡ Start new card. Right justify numeric entries in 10 column fields.

Table V
Input Data—Non Closing Pulse Example

```
1ISA - FREQUENCY RESPONSE FROM PULSE TEST DATA

  ISA S26 FREQUENCY RESPONSE FROM PULSE TEST DATA
  SAMPLE PROBLEM - NON CLOSING OUTPUT PULSE


 LIST 3 DATA        1.        2.        0.        0.


 LIST 4 DATA        1.        3.       10.    0.01000

 LIST 6 DATA    **********    PULSE NO. = 1  RUN S12I

   HEIGHT AT 0-     0.0000E+00  MULTIPLIER     1.000       CLOSURE CODE  0.
   BASE LINE DRIFT 0.0000E+00  SUMMARY CODE   1.          NO. SECTIONS  2.

   SECTION NUMBER          1          2
   TIME INTERVAL       0.125      0.250
   LAST POINT NO.        14.        19.

   PT. VALUE       PT. VALUE       PT. VALUE       PT. VALUE       PT. VALUE
     1  .000E+00     2  .200         3  .900         4  2.30         5  4.50
     6  6.70         7  8.30         8  9.70         9  10.7        10  11.4
    11  11.8        12  10.2        13  7.90        14  5.40        15  2.90
    16  1.50        17  .700        18  .200        19  .000E+00

 LIST 6 DATA    **********    PULSE NO. = 2  RUN S120

   HEIGHT AT 0-     0.0000E+00  MULTIPLIER     1.000       CLOSURE CODE  2.
   BASE LINE DRIFT  1.300       SUMMARY CODE   1.          NO. SECTIONS  2.

   SECTION NUMBER          1          2
   TIME INTERVAL       0.125      0.250
   LAST POINT NO.        14.        22.

   PT. VALUE       PT. VALUE       PT. VALUE       PT. VALUE       PT. VALUE
     1  .000E+00     2  .100E+00     3  .500         4  1.20         5  2.10
     6  3.20         7  4.30         8  5.70         9  6.80        10  7.80
    11  8.40        12  8.80        13  8.30        14  7.70        15  6.10
    16  4.80        17  3.60        18  2.70        19  2.00        20  1.80
    21  1.40        22  1.30
```

ISA S26 FREQUENCY RESPONSE FROM PULSE TEST DATA
SAMPLE PROBLEM - NON CLOSING OUTPUT PULSE

PULSE NO. = 1 RUN S12I INPUT
TOTAL TIME SPAN = 2.875
PULSE AREA = 12.91

PULSE NO. = 2 RUN S12O OUTPUT
TOTAL TIME SPAN = 3.625
PULSE AREA = 12.00
LINEAR CORRECTION WAS MADE FOR BASE LINE DRIFT
FINAL VALUE ASSUMED TO PERSIST TO INFINITY

STEADY STATE GAIN = 0.9293

FREQUENCY RESPONSE (OUTPUT/INPUT)				FOURIER TRANSFORM OF THE INPUT CURVE				FOURIER TRANSFORM OF THE OUTPUT CURVE			
FREQUENCY CYC./TIME	AMPLITUDE RATIO	LOG AMP. RAT	PH. DIF DEGREES	REAL PART	IMAG. PART	AMPLITUDE	PH.ANG DEG.	REAL PART	IMAG. PART	AMPLITUDE	PH.ANG DEG.
1.000E-02	0.929	-0.03198	-1.30	12.9	-0.953	12.9	355.76	11.9	-1.16	12.0	354.47
1.300E-02	0.929	-0.03208	-1.69	12.8	-1.24	12.9	354.49	11.9	-1.50	12.0	352.81
1.600E-02	0.929	-0.03221	-2.07	12.8	-1.52	12.9	353.22	11.8	-1.84	12.0	351.15
2.000E-02	0.928	-0.03243	-2.59	12.8	-1.90	12.9	351.53	11.7	-2.30	12.0	348.94
2.500E-02	0.927	-0.03276	-3.24	12.7	-2.37	12.9	349.41	11.6	-2.85	11.9	346.17
3.200E-02	0.926	-0.03336	-4.14	12.5	-3.01	12.9	346.45	11.3	-3.62	11.9	342.31
4.000E-02	0.924	-0.03422	-5.18	12.3	-3.74	12.8	343.06	11.0	-4.46	11.9	337.89
5.000E-02	0.921	-0.03557	-6.46	11.9	-4.62	12.8	338.83	10.4	-5.46	11.8	332.37
6.300E-02	0.917	-0.03778	-8.13	11.4	-5.70	12.7	333.34	9.57	-6.65	11.6	325.21
8.000E-02	0.909	-0.04142	-10.29	10.4	-7.00	12.6	326.16	8.21	-7.96	11.4	315.88
0.100	0.898	-0.04684	-12.80	9.18	-8.34	12.4	317.74	6.37	-9.12	11.1	304.94
0.130	0.877	-0.05724	-16.48	6.94	-9.86	12.1	305.15	3.38	-10.0	10.6	288.67
0.160	0.850	-0.07039	-20.02	4.48	-10.7	11.6	292.62	0.448	-9.88	9.89	272.60
0.200	0.809	-0.09219	-24.47	1.16	-10.9	11.0	276.05	-2.80	-8.42	8.88	251.59
0.250	0.748	-0.12607	-29.38	-2.49	-9.71	10.0	255.61	-5.19	-5.41	7.50	226.23
0.320	0.654	-0.18410	-34.54	-5.76	-6.31	8.54	227.65	-5.45	-1.27	5.59	193.11
0.400	0.554	-0.25662	-37.00	-6.52	-1.97	6.82	196.84	-3.54	1.30	3.77	159.84
0.500	0.472	-0.32622	-35.56	-4.57	1.64	4.85	160.30	-1.30	1.88	2.29	124.74
0.630	0.421	-0.37538	-36.07	-1.23	2.61	2.88	115.26	0.228	1.19	1.22	79.19
0.800	0.311	-0.50712	-35.66	0.735	0.974	1.22	52.96	0.362	0.113	0.380	17.31
1.00	0.505	-0.29705	-14.09	2.272E-02	-0.266	0.267	274.88	-2.154E-02	-0.133	0.135	260.79
1.30	0.195	-0.70992	-36.74	-6.740E-02	0.477	0.481	98.05	4.508E-02	8.236E-02	9.390E-02	61.31
1.60	0.111	-0.95536	-23.21	0.300	-7.499E-02	0.310	345.98	2.731E-02	-2.076E-02	3.430E-02	322.76
INPUT AMPLITUDE IS LESS THAN ONE PER CENT OF PULSE AREA. RESULTS AT HIGHER FREQUENCIES ARE QUESTIONABLE.											
2.00	0.259	-0.58615	-123.53	-6.079E-02	4.053E-02	7.306E-02	146.31	1.747E-02	7.337E-03	1.895E-02	22.78
2.50	0.304	-0.51694	-38.22	-2.588E-02	-4.272E-02	4.995E-02	238.79	-1.422E-02	-5.338E-03	1.519E-02	200.57
3.20	9.276E-02	-1.03265	163.94	1.645E-02	2.613E-02	3.087E-02	57.81	-2.136E-03	-1.907E-03	2.864E-03	221.75
4.00	0.253	-0.59710	-180.00	4.559E-02	-6.333E-08	4.559E-02	360.00	-1.153E-02	-1.378E-07	1.153E-02	180.00
5.00	0.501	-0.30018	-217.75	-2.530E-03	-7.197E-03	7.629E-03	250.63	3.210E-03	2.075E-03	3.822E-03	32.88
6.30	0.169	-0.77188	108.48	9.270E-03	1.377E-02	1.660E-02	56.06	-2.706E-03	7.484E-04	2.807E-03	164.54
8.00	1.00	0.00057	0.00	3.203E-14	1.864E-20	3.203E-14	0.00	3.207E-14	5.978E-20	3.207E-14	0.00
10.0	0.259	-0.58613	-123.53	-2.432E-03	1.621E-03	2.922E-03	146.31	6.988E-04	2.935E-04	7.579E-04	22.78

Table VII
Program Listing

```
//LVMSR444      JOB 3LV5020, EXECUTE PULSE ANALYSIS PROGRAM    R
RJE*PRIORITY R
//              CLASS=D,PRTY=6,MSGLEVEL=(1,1)
//DOPGM         EXEC XXXSCGE1
//C.SYSIN       DD *
C      TITLE         = FREQUENCY RESPONSE FROM PULSE TEST DATA (ISA)
C      AUTHOR        = A C PAULS
C      LOCATION      = CED, MONSANTO CO, ST LOUIS, MO
C      DATE WRITTEN = 7/7/67, 1/13/69, 12/1/69
C      COMPUTER      = USASI FORTRAN
C      KEYPUNCH      = IBM 029 (EBCDIC)
C      FILES         = FORTRAN 5 (INPUT), 6 (OUTPUT)
C      SUBPROGRAMS   = FTRAN
C
C ABSTRACT
C      THIS PROGRAM COMPUTES FOURIER TRANSFORMS FOR CLOSED OR OPEN PULSES
C      AND TRANSFER FUNCTIONS FOR PAIRS OF CURVES.
C
C DEFINITION SECTION
C      A      FOURIER TRANSFORM (FT) AMPLITUDE
C      AI     FT AMPLITUDE OF INPUT PULSE
C      AIN    ALPHANUMERIC WORD
C      AR     AMPLITUDE RATIO (OUTPUT/INPUT)
C      ARL    ALOG10(AR)
C      AZ     AREA UNDER PULSE (FT AMPLITUDE AT ZERO FREQUENCY)
C      AZI    AREA UNDER INPUT PULSE
C      BLANK ALPAMERIC WORD
C      BN     NUMBER OF LAST POINT IN J-TH SECTION OF K-TH PULSE
C      CF     CHART FACTOR
C      CFI    CHART FACTOR FOR INPUT PULSE
C      DB     INPUT DATA BASE ARRAY
C      F      FREQUENCY ARRAY (CYCLES/TIME IF KFU=1, ELSE RAD/TIME)
C      FI     INITIAL FREQUENCY
C      FM     FREQUENCY ARRAY FOR ONE DECADE
C      FR     FREQUENCY (RAD/TIME)
C      FU     ALPHAMERIC WORD FOR FREQUENCY UNITS
C      GAIN   STEADY STATE GAIN (OUTPUT PULSE AREA / INPUT PULSE AREA)
C      H      TIME BETWEEN POINTS IN J-TH SECTION OF K-TH PULSE
C      IA     NUMBER OF THE FIRST POINT IN A PULSE SECTION
C      IB     NUMBER OF THE LAST  POINT IN A PULSE SECTION
C      K      PULSE NUMBER
C      KEE    PULSE CLOSURE CODE.  0=CLOSED, 2=OPEN
C      KEEI   PULSE CLOSURE CODE FOR INPUT PULSE
C      KFD    FREQUENCY DATA TYPE CODE
C      KFL    FREQUENCY LIMIT PRINT SWITCH
C      KFU    FREQUENCY UNITS CODE
C      KRN    RESPONSE NORMALIZATION CODE
C      KTP    TRANSFORM DATA PRINT CODE
C      LN     LIST NUMBER
C      NB     NUMBER OF LAST POINT IN J-TH SECTION OF A PULSE
C      NCI    PULSE NUMBER OF INPUT CURVE
C      NCO    PULSE NUMBER OF OUTPUT CURVE
C      ND     NUMBER OF DECADES
C      NF     NUMBER OF FREQUENCIES
C      NFD    NUMBER OF FREQUENCIES PER DECADE
C      NFP1   NF + 1
C      NP     NUMBER OF DATA POINTS
C      NS     NUMBER OF PULSE CURVE SECTIONS
C      OUT    ALPHAMERIC WORD
C      P      FT PHASE ANGLE
C      PA     ALPHAMERIC ARRAY FOR PULSE ID
```

Table VII (*continued*)

```
C       PD      PULSE DATA ARRAY
C       PH      PULSE HEIGHT DATA ARRAY
C       PI      FT PHASE ANGLE FOR INPUT PULSE
C       PL      PHASE LAG (OUTPUT - INPUT)
C       R       FT REAL PART
C       RI      FT REAL PART FOR INPUT PULSE
C       T       TIME
C       TEMP    TEMPORARY STORAGE FOR PROBLEM AND CASE TITLES
C       TT      TOTAL TIME
C       TTI     TOTAL TIME FOR INPUT PULSE
C       TTL     PROBLEM AND CASE TITLES
C       Y       PULSE HEIGHT ARRAY
C       YF      BASE LINE DRIFT FROM TIME 0 TO TT
C       YFI     BASE LINE DRIFT FROM TIME 0 TO TT FOR INPUT PULSE
C       YZ      PULSE HEIGHT AT TIME 0-
C       Z       FT IMAGINARY PART
C       ZI      FT IMAGINARY PART FOR INPUT PULSE
C
C       NONCE VARIABLES - AK, AN, ANFD, I, ID, J
C
C DECLARATIVE SECTION
        DIMENSION       DB(10000), TTL(30), TEMP(15)
        DIMENSION       BN(10,9), F(52), FM(10), H(10,9), PA(2,9),
       1                PD(10,9), PH(1000,9), R(52), RI(52),
       2                Y(1000), Z(52), ZI(52), NB(10)
        EQUIVALENCE     (DB(11),PD),    (DB(211),H),    (DB(311),BN),
       1                (DB(1001),PH)
        DATA  BLANK     /4H    /
        DATA  AIN/4H  IN/, OUT/4H OUT/, CYC/4HCYC./, RAD/4HRAD./
C
C FILE SECTION
  170 FORMAT(/ 3X, 15HHEIGHT AT 0-    , G11.4, 2X,
     1              12HMULTIPLIER  , G11.4, 2X,
     2              12HCLOSURE CODE, F4.0,
     3         / 3X, 15HBASE LINE DRIFT, G11.4, 2X,
     4              12HSUMMARY CODE, F4.0, 9X,
     5              12HNO. SECTIONS, F4.0)
  172 FORMAT(/ 3X, 18HSECTION NUMBER     , 10(I6,4X))
  174 FORMAT(  3X, 18HTIME INTERVAL      , 10G10.3)
  176 FORMAT(  3X, 18HLAST POINT NO.     , 10(F6.0, 4X))
  177 FORMAT(/ 3X, 5(15HPT. VALUE      ))
  178 FORMAT(  1X, I5, 1X, G9.3, I5, 1X, G9.3, I5, 1X, G9.3,
     1              I5, 1X, G9.3, I5, 1X, G9.3)
  181 FORMAT(//1X, 11HLIST 6 DATA, 4X, 10(1H*), 4X, 12HPULSE NO. = ,
     1              I1, 2X, 2A4)
  182 FORMAT(/ 1X, 12HPULSE NO. = , I1, 2X, 2A4, 2X, A4, 3HPUT)
  183 FORMAT( 10X, 20HTOTAL TIME SPAN    = , G13.4
     1       /10X, 20HPULSE AREA         = , G13.4)
  184 FORMAT( 10X, 46HLINEAR CORRECTION WAS MADE FOR BASE LINE DRIFT)
  185 FORMAT( 10X,42HFINAL VALUE ASSUMED TO PERSIST TO INFINITY)
  186 FORMAT(/20X, 50HFREQUENCY        REAL PART        IMAG. PART        AMPLI
     1           , 19HTUDE      PHASE ANGLE/
     2          20X, A4,  5H/TIME, 47X, 15HDEGREES (0-360))
  187 FORMAT( 18X, 4(1PG11.4,4X), 0PF8.2)
  190 FORMAT(/10X, 20HSTEADY STATE GAIN = , G13.4)
  191 FORMAT(/ 2X, 40H    FREQUENCY RESPONSE (OUTPUT/INPUT)     ,
     1              40H FOURIER TRANSFORM OF THE  INPUT CURVE  ,
     2              38H FOURIER TRANSFORM OF THE OUTPUT CURVE/
     3          2X, 40HFREQUENCY  AMPLITUDE   LOG     PH. DIF  ,
     4              40HREAL PART IMAG. PART  AMPLITUDE PH.ANG  ,
     5              38HREAL PART IMAG. PART  AMPLITUDE PH.ANG/
     6          2X, A4, 36H/TIME    RATIO   AMP. RAT  DEGREES  ,
     7              33X, 4HDEG., 36X, 4HDEG.)
  192 FORMAT(/ 2X, 40H    FREQUENCY RESPONSE (OUTPUT/INPUT)     ,
     1              25H FOURIER TRANSFORM OF THE, A4, 11HPUT CURVE  /
     2          2X, 40HFREQUENCY  AMPLITUDE   LOG     PH. DIF  ,
```

Table VII (*continued*)

```
    3              40HREAL PART IMAG. PART  AMPLITUDE PH.ANG  /
    4          2X, A4, 36H/TIME     RATIO   AMP. RAT  DEGREES  ,
    5              33X, 4HDEG.)
  193 FORMAT(/ 2X, 40H    FREQUENCY RESPONSE (OUTPUT/INPUT)     /
    1          2X, 40HFREQUENCY  AMPLITUDE   LOG     PH. DIF  /
    2          2X, A4, 36H/TIME     RATIO   AMP. RAT  DEGREES  )
  194 FORMAT(  1X, 50HINPUT AMPLITUDE IS LESS THAN ONE PER CENT OF PULSE
    1            , 50H AREA. RESULTS AT HIGHER FREQUENCIES ARE QUESTIONA
    2            ,  4HBLE.)
  195 FORMAT(2(1X,1PG10.3),0PF9.5,F9.2,2(3(1X,1PG10.3),0PF7.2))
  900 FORMAT(  8X, I2, 15A4 )
  901 FORMAT( 8F10.0 )
  902 FORMAT( 6X, I4, 5F10.0, 9X, I1 )
  903 FORMAT(  46H1ISA - FREQUENCY RESPONSE FROM PULSE TEST DATA
    1        //1X, 15A4/1X, 15A4)
  904 FORMAT( //  12H LIST 3 DATA,      4F10.0    )
  905 FORMAT( //  12H LIST 4 DATA, 3F10.0, F10.5   )
  906 FORMAT(/)
  907 FORMAT( 38H1JOB TERMINATED - BAD DATA BEFORE CARD // 8X, I2, 15A4)
C
C INPUT SECTION
C                                 DETERMINE LIST TYPE
    5 READ  (5,900,END=6)  LN,  (TEMP(I), I = 1,15)
      IF    (LN .GT. 0 .AND. LN .LE. 6)  GO TO (11,12,13,14,15,16), LN
      IF    (LN .EQ. 99) GO TO 30
      WRITE (6,907)  LN,  (TEMP(I), I = 1,15)
    6 CALL  EXIT
C
C                                 READ PROBLEM-TITLE, INITIALIZE DATA BASE
   11 DO 21 I       = 1,15
            TTL(I) = TEMP(I)
   21       TTL(I+15) = BLANK
      DO 22 I       = 1,400
   22       DB(I)   = 0.0
C                                 SET 'DEFAULT' OPTIONS
            DB(7)   = 5.0
            DB(8)   = 10.0
            DB(9)   = 0.001
            KFD     = 4
      GO TO 5
C                                 READ CASE-TITLE
   12 DO 23 I       = 1,15
   23       TTL(I+15) = TEMP(I)
      GO TO 5
C                                 LIST 3 - FREQUENCY RESPONSE DATA
   13 READ  (5,901)  (DB(I), I = 1,4)
      GO TO 5
C                                 LIST 4 - FREQUENCY RANGE DATA
   14 READ  (5,901)  (DB(I), I = 6,9)
            KFD     = 4
      GO TO 5
C                                 LIST 5 - FREQUENCY POINTS
   15 READ  (5,901)  DB(6), DB(10)
            NF      = DB(10)
      READ  (5,901)  (DB(I+100), I=1,NF)
            KFD     = 5
      GO TO 5
C                                 LIST 6 - PULSE DATA
   16 READ  (5,902)  K, (PD(I,K), I = 2,6), NS
            PD(1,K) = K
            PD(7,K) = NS
      READ  (5,901)  (H(I,K), BN(I,K), I = 1,NS)
      READ  (5,902)  NP
            PD(8,K) = NP
      READ  (5,901)  (PH(I,K), I = 1,NP)
            PA(1.K) = TEMP(1)
```

Table VII (*continued*)

```
            PA(2,K) = TEMP(2)
      GO TO 5
C
C INPUT DATA SUMMARY SECTION
C                                 PRINT CONTROL AND FREQUENCY RANGE DATA
   30 WRITE (6,903)   TTL
      IF    (DB(1) .NE. 0.0) WRITE (6,904) (DB(I), I = 1,4)
      IF    (KFD .EQ. 4) WRITE (6,905) (DB(I), I = 6,9)
C                                 PRINT PULSE DATA
      DO 31 K       = 1,9
            AK      = K
      IF    (PD(1,K) .NE. AK) GO TO 31
      WRITE (6,181)  K, PA(1,K), PA(2,K)
      WRITE (6,170)  (PD(I,K), I = 2,6,2), (PD(I,K), I=3,7,2)
            NS      = PD(7,K)
      WRITE (6,172)  (I, I = 1,NS)
      WRITE (6,174)  (H(I,K), I = 1,NS)
      WRITE (6,176)  (BN(I,K), I = 1,NS)
C                                 PRINT PULSE HEIGHT ON INITIAL ENTRY ONLY
      IF    (PD(5,K) .EQ. 0.0) GO TO 31
            PD(5,K) = 0.0
            NP      = PD(8,K)
      WRITE (6,177)
      WRITE (6,178)  (I, PH(I,K), I = 1,NP)
   31 CONTINUE
      WRITE (6,906)
C
C INITIALIZATION SECTION
C                                 SET CONTROL AND FREQUENCY RANGE VARIABLES
            NCI     = DB(1)
            NCO     = DB(2)
            KRN     = DB(3)
            KTP     = DB(4) + 1.01
            KFU     = DB(6)
            ND      = DB(7)
            NFD     = DB(8)
            FI      = DB(9)
            NF      = DB(10)
            FU      = RAD
      IF    (KFU .EQ. 1) FU = CYC
C                                 SET FREQUENCY ARRAY FROM RANGE DATA
      IF    (NF .GT. 0 .AND. KFD .EQ. 5) GO TO 35
            NF      = NFD*ND + 1
            ANFD    = NFD
            AN      = 0.0
      DO 32 I       = 1,NFD
            FM(I)   = 0.1*AINT(10.0**(1.0+AN/ANFD) + 0.6)
   32       AN      = AN + 1.0
      DO 33 J       = 1,NF
            ID      = (J - 1)/NFD
            I       = J - ID*NFD
   33       F(J+1)  = FI*FM(I)*10.0**ID
      GO TO 38
C                                 SET FREQUENCY ARRAY FROM INPUT POINTS
   35 DO 36 J       = 1,NF
   36       F(J+1)  = DB(J+100)
C                                 SET ZERO FREQUENCY FOR AREA CALCULATION
   38       F(1)    = 0.0
            NFP1    = NF + 1
C
C COMPUTATION SECTION
C                                 COMPUTE TRANSFORMS ONLY
      IF    (NCI .GT. 0) GO TO 45
   41       K       = 0
   42 IF    (K .GE. 9) GO TO 5
            K       = K + 1
```

Table VII (*continued*)

```
      IF     (IFIX(PD(1,K)) .NE. K) GO TO 42
      GO TO 50
C                                   COMPUTE TRANSFORMS AND FREQUENCY RESPONSE
   45        K       = NCI
C
C                                   SET PULSE PARAMETERS FOR K-TH CURVE
   50        CF      = PD(4,K)
             YZ      = PD(2,K)*CF
             YF      = PD(3,K)*CF
             KEE     = PD(6,K)
             NS      = PD(7,K)
             NP      = PD(8,K)
             TT      = 0.0
             IA      = 1
      DO 51  I       = 1,NS
             NB(I)   = BN(I,K)
             TT      = TT + H(I,K)*FLOAT(NB(I) - IA)
   51        IA      = NB(I)
C                                   ADJUST PULSE FOR TIME 0- AND CHART FACTOR
      DO 53  I       = 1,NP
   53        Y(I)    = PH(I,K)*CF - YZ
C                                   ADJUST PULSE FOR LINEAR BASE LINE DRIFT
      IF     (YF .EQ. 0.0) GO TO 60
             T       = 0.0
             J       = 1
      DO 56  I       = 2,NP
      IF     (I .GT. NB(J))  J = J + 1
             T       = T + H(J,K)
   56        Y(I)    = Y(I) - YF*T/TT
C                                   COMPUTE FOURIER TRANSFORM
   60 DO 63  J       = 1,NFP1
             FR      = F(J)
      IF     (KFU .EQ. 1) FR = FR*6.283185
   63 CALL   FTRAN      (NS, NB, H(1,K), Y, KEE, FR, R(J), Z(J))
C                                   AREA UNDER PULSE
             AZ      = SIGN(SQRT(R(1)**2 + Z(1)**2),R(1))
C
C                                   TRANSFORMS ONLY - PRINT PULSE ID
      IF     (NCI .GT. 0) GO TO 80
      WRITE (6,903)   TTL
      WRITE (6,906)
      WRITE (6,182)   K, PA(1,K), PA(2,K)
      WRITE (6,183)   TT, AZ
      IF     (YF .NE. 0.0) WRITE (6,184)
      IF     (KEE .EQ. 2)  WRITE (6,185)
C                                   PRINT TABLE HEADING AND TRANSFORM DATA
      WRITE (6,186)   FU
      DO 77 J        = 2,NFP1
             A       = SQRT(R(J)**2 + Z(J)**2)
             P       = 57.29578*ATAN2(Z(J),R(J))
      IF     (P .LT. 0.0) P = P + 360.0
   77 WRITE (6,187)   F(J), R(J), Z(J), A, P
      GO TO 42
C
C                                   RESPONSE NEEDED - SAVE INPUT
   80 IF     (K .EQ. NCO) GO TO 85
             CFI     = CF
             YFI     = YF
             KEEI    = KEE
             TTI     = TT
             AZI     = AZ
      DO 81  J       = 1,NFP1
             RI(J)   = R(J)
   81        ZI(J)   = Z(J)
             K       = NCO
      GO TO 50
```

Table VII (*continued*)

```
C                                STEADY STATE GAIN
  85       GAIN    = AZ/AZI
C
C                                PRINT RESPONSE TABLE
           KFL     = 0
      DO 120 J = 2,NFP1
      IF    (J .NE. 2 .AND. J .NE. 38) GO TO 100
      WRITE (6,903)  TTL
C                                PRINT INPUT PULSE ID
      IF    (J .NE. 2) GO TO 95
      WRITE (6,182)  NCI, PA(1,NCI), PA(2,NCI), AIN
      WRITE (6,183)  TTI, AZI
      IF    (YFI .NE. 0.0) WRITE (6,184)
      IF    (KEEI .EQ. 2)  WRITE (6,185)
C                                PRINT OUTPUT PULSE ID
      WRITE (6,182)  K, PA(1,K), PA(2,K), OUT
      WRITE (6,183)  TT, AZ
      IF    (YF .NE. 0.0) WRITE (6,184)
      IF    (KEE .EQ. 2)  WRITE (6,185)
      WRITE (6,190)  GAIN
C                                PRINT RESPONSE TABLE HEADINGS
  95 GO TO (96,97,98,99),KTP
  96 WRITE (6,191)  FU
     GO TO 100
  97 WRITE (6,192)  AIN, FU
     GO TO 100
  98 WRITE (6,192)  OUT, FU
     GO TO 100
  99 WRITE (6,193)  FU
C
C                                AMPLITUDE, AMP. RATIO (NORMALIZE IF KRN=0)
 100       AI      = SQRT(RI(J)**2 + ZI(J)**2)
           A       = SQRT(R(J)**2 + Z(J)**2)
           AR      = A/AI
      IF    (KRN .EQ. 0 .AND. MAXO(KEEI,KEE) .LT. 2) AR = ABS(AR/GAIN)
           ARL     = ALOG10(AR)
C                                PHASE ANGLES, PHASE DIFFERENCE
           PI      = 57.29578*ATAN2(ZI(J),RI(J))
           P       = 57.29578*ATAN2(Z(J),R(J))
      IF    (PI .LT. 0.0) PI = PI + 360.0
      IF    (P .LT. 0.0) P = P + 360.0
           PL      = P - PI
C                                PRINT INPUT AMPLITUDE WARNING
      IF    (KFL .NE. 0 .OR. AI .GE. 0.01*ABS(AZI)) GO TO 110
      WRITE (6,194)
           KFL     = 1
C                                PRINT J-TH FREQUENCY RESPONSE
 110 GO TO (111,112,113,114), KTP
 111 WRITE (6,195)  F(J), AR, ARL, PL, RI(J), ZI(J), AI, PI,
    1               R(J), Z(J), A, P
     GO TO 120
 112 WRITE (6,195)  F(J), AR, ARL, PL, RI(J), ZI(J), AI, PI
     GO TO 120
 113 WRITE (6,195)  F(J), AR, ARL, PL, R(J), Z(J), A, P
     GO TO 120
 114 WRITE (6,195)  F(J), AR, ARL, PL
C
 120 CONTINUE
     GO TO 5
     END
     SUBROUTINE     FTRAN  (NS, NB, H, Y, KEE, F, R, Z)
C
C    TITLE          = FOURIER TRANSFORM REAL AND IMAGINARY PARTS
C    AUTHOR         = A C PAULS
C    LOCATION       = CED, MONSANTO CO, ST LOUIS, MO
C    DATE WRITTEN   = 7/7/67
```

Table VII (*continued*)

```
C       COMPUTER       = USASI FORTRAN
C
C DEFINITION SECTION
C       F       FREQUENCY (RAD/TIME)
C       H       TIME BETWEEN POINTS IN J-TH SECTION OF PULSE
C       IA      NUMBER OF FIRST POINT IN A SECTION
C       IB      NUMBER OF LAST  POINT IN A SECTION
C       KEE     PULSE CLOSURE CODE.  0=CLOSED, 2=OPEN
C       NB      NUMBER OF LAST POINT IN J-TH SECTION OF PULSE
C       NS      NUMBER OF PULSE CURVE SECTIONS
C       R       FOURIER TRANSFORM REAL PART
C       TA      TIME AT FIRST POINT IN A SECTION
C       TB      TIME AT LAST  POINT IN A SECTION
C       Y       PULSE HEIGHT ARRAY
C       YF      PULSE HEIGHT AT INFINITE TIME
C       Z       FOURIER TRANSFORM IMAGINARY PART
C
C       NONCE VARIABLES - CA, CB, D, DH, FTA, FTB, FTI, IAP1, IBM1,
C               Q, R1, SA, SB, SYI, U, Z1
C
C
C DECLARATIVE SECTION
      DIMENSION       NB(1), H(1), Y(1)
C
C COMPUTATION SECTION
              R       = 0.0
              Z       = 0.0
              TA      = 0.0
              IA      = 1
C
C                                     POSITIVE FREQUENCY
      IF      (F .EQ. 0.0) GO TO 60
      DO 50 N         = 1,NS
              IB      = NB(N)
C                                     TIME AT END OF N-TH SECTION
              TB      = TA + H(N)*FLOAT(IB-IA)
C                                     D AND Q PARAMETERS
              U       = F*H(N)
              D       = (SIN(U/2.0)/(U/2.0))**2
              DH      = D/2.0
      IF      (U .GT. 0.1) GO TO 32
              Q       = U*(1.0 - U**2/20.0)/6.0
      GO TO 35
   32         Q       = (1.0 - SIN(U)/U)/U
C
C                                     CONTRIBUTION OF END POINTS
   35         FTA     = F*TA
              FTB     = F*TB
              SA      = SIN(FTA)
              CA      = COS(FTA)
              SB      = SIN(FTB)
              CB      = COS(FTB)
              R       = R + H(N)*(Y(IA)*(DH*CA - Q*SA) + Y(IB)*(DH*CB + Q*
     1                  SB))
              Z       = Z - H(N)*(Y(IA)*(DH*SA + Q*CA) + Y(IB)*(DH*SB - Q*
     1                  CB))
C
C                                     CONTRIBUTION OF INTERIOR POINTS
              IAP1    = IA + 1
              IBM1    = IB - 1
      IF      (IAP1 .GT. IBM1) GO TO 45
              R1      = 0.0
              Z1      = 0.0
      DO 41 I         = IAP1,IBM1
              FTI     = F*(TA + H(N)*FLOAT(I-IA))
              R1      = R1 + Y(I)*COS(FTI)
```

Table VII (*concluded*)

```
   41       Z1      = Z1 + Y(I)*SIN(FTI)
            R       = R + H(N)*D*R1
            Z       = Z - H(N)*D*Z1
C                               INITIALIZE FOR NEXT SECTION
   45       TA      = TB
            IA      = IB
   50 CONTINUE
C                               ADJUST FOR NON-CLOSURE
      IF    (KEE .NE. 2) GO TO 55
            YF      = Y(IB)
            R       = R - YF*SIN(F*TB)/F
            Z       = Z - YF*COS(F*TB)/F
   55 RETURN
C
C                               ZERO FREQUENCY
   60 DO 70 N       = 1,NS
            IB      = NB(N)
            SYI     = 0.0
            IAP1    = IA + 1
            IBM1    = IB - 1
      IF    (IAP1 .GT. IBM1) GO TO 65
      DO 61 I       = IAP1,IBM1
   61       SYI     = SYI + Y(I)
   65       R       = R + H(N)*(0.5*(Y(IA) + Y(IB)) + SYI)
   70       IA      = IB
C                               ADJUST FOR NON-CLOSURE
      IF    (KEE .NE. 2 .OR. Y(IB) .EQ. 0.0) GO TO 80
            Z       = -R
            R       = SIGN(1.0E-37,R)
   80 RETURN
      END
/* PLACE THE  FORTRAN SOURCE DECK IN FRONT OF THIS CARD
//X.FT05F001    DD *
         1 ISA S26 FREQUENCY RESPONSE FROM PULSE TEST DATA
         2 SAMPLE PROBLEM TO TEST COMPUTER PROGRAM PROPERLY EXCIT
         3
         1         2         0         0
         4
         1         3        10       .01
         6RUN S12I
         1         0         0         1         1         0
       .05        18
        18
         0         4        11        43        56        69
        70        63        48        35        22        14
         1         0
         6RUN S120
         2         0         0         1         1         0
       .05        15       0.1        31
        31
         0         0         2         4         7        12
        30        37        40        44        42        41
        27        21        16        13        10         8
       3.5       2.5       1.7       1.1        .6        .2
        99
         2 SAMPLE PROBLEM TO TEST COMPUTER PROGRAM PROPERLY EXCIT
         3
         1         3         0         0
         6RUN S130
         3         0         0         1         1         0
       .05        45
        45
         0         0         2         4         7        12
        30        37        40        44        42        41
        32        27        24        21        18        16
        12        10         9         8         7         6
```

Appendix E

Time Domain Process Identification Program

```
C        ----------MODEL----------
C        R.ALAN SCHAEFER
C        MASTERS THESIS PROGRAM
C        THIS PROGRAM FITS A SECOND ORDER PLUS DEAD TIME MODEL
C          TO A SET OF INPUT-OUTPUT DATA OF A PROCESS. A NON-
C          LINEAR REGRESSION IS USED TO SOLVE THE PROBLEM. THE
C          FOLLOWING NOMENCLATURE IS USED:
C        K=PROCESS GAIN
C        T1=PROCESS TIME CONSTANT 1
C        T2=PROCESS TIME CONSTANT 2
C        N=PROCESS DEAD TIME
C        INUM=NUMBER OF INPUT-OUTPUT POINTS
C        U(I)=PROCESS INPUT
C        X(I)=PROCESS OUTPUT
C        T=SAMPLING PERIOD
C        THE PROGRAM CALLS A DATA FILE NAMED RASDAX.DAT,
C          WHERE X IS A NUMBER BETWEEN 1 AND 5. THE DATA
C          IN THE FILES ARE SET UP IN THE FOLLOWING ORDER:
C        ENTRY 1   K(INITIAL GUESS)
C        ENTRY 2   T1(INITIAL GUESS)
C        ENTRY 3   T2(INITIAL GUESS)
C        ENTRY 4   INUM
C        ENTRIES 5 TO INUM+4   U(I) FOR I=1 TO INUM
C        LAST ENTRIES          X(I) FOR I=1 TO INUM
C
C        DATA COLLECTION
```

Appendix E (*continued*)
TIME DOMAIN PROCESS IDENTIFICATION PROGRAM

```
      REAL K
      COMMON X(100),U(100),XP(100),F(100)
      COMMON INUM,T,N,J,HGHT
      COMMON XAOPT(3),XBOPT(3),XCOPT(3),XSOPT(3)
      COMMON FAOPT(3),FBOPT(3),FCOPT(3),FSOPT(3)
      COMMON XMM2,XMM1,XM,XMP1,FMM2,FMM1,FM
      COMMON FMP1,DK,K,T1,T2
      COMMON PREOPT,FREOPT
      OPEN(UNIT=1,FILE='RASDA2.DAT')
      READ(1,130) K,T1,T2
      READ(1,140) INUM
      DO 100 I=1,INUM+1
      READ(1,130) U(I+9)
100   CONTINUE
      DO 110 I=1,INUM+1
      READ(1,130) X(I+9)
110   CONTINUE
      DO 120 I=1,9
      U(I)=0.
      X(I)=0.
120   CONTINUE
130   FORMAT(7X,F12.5)
140   FORMAT(7X,I2)
C
C     IC=CYCLE NUMBER
      N=0
      T=1.0
      IC=1
      TYPE 160
160   FORMAT('0STARTING VALUES')
      TYPE 165
165   FORMAT(10X,'K',14X,'T1',13X,'T2')
C
C     IN THE NEXT 3 SECTIONS OF THE PROGRAM, K,T1,AND
C       T2 ARE OPTIMIZED SEPARATELY USING A DSC-POWELL
C       SEARCH.  A PART FO THE DSC SEARCH IS IN A
C       SUBROUTINE.  THE POWELL SEARCH IS AT THE END
C       OF THE MAIN PROGRAM.
171   FORMAT( 3(F15.6))
170   P1=K
      P2=T1
      P3=T2
      WRITE(5,171) K,T1,T2
      WRITE(5,175) IC
175   FORMAT(' CYCLE',I3)
C
C     OPTIMIZATION OF K
C     DSC SEARCH
      L=1
      DK=.001
      J=1
      PREOPT=K
      F(J)=ERROR(K,T1,T2)
      IW1=J+1
      F(J+1)=ERROR(K+DK,T1,T2)
      IF(F(J+1).GE.F(J)) DK=-DK
      IF(K+DK.LE.0.) GO TO 218
200   IW1=J+1
      F(J+1)=ERROR(K+DK,T1,T2)
      IF(F(J+1).GT.F(J)) GO TO 210
      OLDDK=DK
      K=K+DK
      PREOPT=K
      FREOPT=F(J+1)
      DK=2.*DK
```

Appendix E (*continued*)

TIME DOMAIN PROCESS IDENTIFICATION PROGRAM

```
      IF(DK.GT.Ø.5) DK=.5
      IF(DK.LT.-Ø.5) DK=-.5
      IF(K+DK.LT.Ø.) GO TO 2Ø5
      J=J+1
      GO TO 2ØØ
2Ø5   DK=OLDDK
      K=K-DK
21Ø   XM=K+DK
      XMM1=K
      XMM2=K-.5*DK
      XMP1=K+.5*DK
      FM=F(J+1)
      FMM1=F(J)
      FMM2=ERROR(XMM2,T1,T2)
      FMP1=ERROR(XMP1,T1,T2)
      CALL DSC(1)
      K=XSOPT(L)
      GO TO 25Ø
C
C     OPTIMIZATION OF T1
C     DSC SEARCH
218   L=2
      DK=.Ø1
      J=1
      PREOPT=T1
      F(J)=ERROR(K,T1,T2)
      IW1=J+1
      F(J+1)=ERROR(K,T1+DK,T2)
      IF(F(J+1).GE.F(J)) DK=-DK
      IF((T1+DK)/T.LT.Ø.Ø145) GO TO 235
225   IW1=J+1
      F(J+1)=ERROR(K,T1+DK,T2)
      IF(F(J+1).GT.F(J)) GO TO 23Ø
      OLDDK=DK
      T1=T1+DK
      PREOPT=T1
      FREOPT=F(J+1)
      DK=2.*DK
      IF(DK.LT.-Ø.5) DK=-.5
      IF(T1+DK.LT.Ø.) GO TO 227
      IF((T1+DK)/T.LT.Ø.Ø145) GO TO 227
      J=J+1
      GO TO 225
227   DK=OLDDK
      T1=T1-DK
23Ø   XM=T1+DK
      XMM1=T1
      XMM2=T1-.5*DK
      XMP1=T1+.5*DK
      FM=F(J+1)
      FMM1=F(J)
      FMM2=ERROR(K,XMM2,T2)
      FMP1=ERROR(K,XMP1,T2)
      CALL DSC(2)
      T1=XSOPT(L)
      GO TO 25Ø
C
C     OPTIMIZATION OF T2
C     DSC SEARCH
235   L=3
      DK=.Ø1
      J=1
      PREOPT=T2
      F(J)=ERROR(K,T1,T2)
      IW1=J+1
```

Appendix E (*continued*)
TIME DOMAIN PROCESS IDENTIFICATION PROGRAM

```
          F(J+1)=ERROR(K,T1,T2+DK)
          IF(F(J+1).GE.F(J)) DK=-DK
          IF((T2+DK)/T.LT.Ø.Ø145) GO TO 48Ø
 245      IW1=J+1
          F(J+1)=ERROR(K,T1,T2+DK)
          IF(F(J+1).GE.F(J)) GO TO 247
          OLDDK=DK
          T2=T2+DK
          PREOPT=T2
          DK=2.*DK
          FREOPT=F(J+1)
          IF(DK.GT.Ø.5) DK=.5
          IF(DK.LT.-Ø.5) DK=-.5
          IF(T2+DK.LT.Ø.Ø) GO TO 246
          IF((T2+DK)/T.LT.Ø.Ø145) GO TO 246
          J=J+1
          GO TO 245
 246      DK=OLDDK
          T2=T2-DK
 247      XM=T2+DK
          XMM1=T2
          XMM2=T2-.5*DK
          XMP1=T2+.5*DK
          FM=F(J+1)
          FMM1=F(J)
          FMM2=ERROR(K,T1,XMM2)
          FMP1=ERROR(K,T1,XMP1)
          CALL DSC(3)
          T2=XSOPT(L)
C
C         POWELL SEARCH
C         THIS SECTION OF THE PROGRAM CALCULATES "XSTAR" WHICH
C           IS THE OPTIMUM VALUE OF ONE OF THE VARIABLES.
 25Ø      IF(XSOPT(L).EQ.PREOPT) GO TO 47Ø
          PREOPT=XSOPT(L)
          FREOPT=FSOPT(L)
          XA=XAOPT(L)
          XB=XBOPT(L)
          XC=XCOPT(L)
          XSTAR=XSOPT(L)
          FA=FAOPT(L)
          FB=FBOPT(L)
          FC=FCOPT(L)
          FSTAR=FSOPT(L)
 255      XACCUR=.ØØ5
          I1=INT(FA*1.E6+.5)
          I2=INT(FB*1.E6+.5)
          I3=INT(FC*1.E6+.5)
          I4=INT(FSTAR*1.E6+.5)
          IF(I1.EQ.I2.AND.I2.EQ.I3.AND.I3.EQ.I4) GO TO 47Ø
 29Ø      IF(FA.GE.FB.AND.FA.GE.FC) GO TO 3ØØ
          IF(FB.GE.FA.AND.FB.GE.FC) GO TO 31Ø
          IF(FC.GE.FA.AND.FC.GE.FB) GO TO 32Ø
 3ØØ      FBIG=FA
          XBIG=XA
          IF(FB.LE.FC) GO TO 3Ø7
 3Ø5      FSMALL=FC
          XSMALL=XC
          GO TO 33Ø
 3Ø7      FSMALL=FB
          XSMALL=XB
          GO TO 33Ø
 31Ø      FBIG=FB
          XBIG=XB
          IF(FA.GT.FC) GO TO 3Ø5
 315      FSMALL=FA
```

Appendix E (*continued*)
TIME DOMAIN PROCESS IDENTIFICATION PROGRAM

```
      XSMALL=XA
      GO TO 330
320   FBIG=FC
      XBIG=XC
      IF(FA.LE.FB) GO TO 315
      GO TO 307
330   ABEROR=ABS(XSTAR-XSMALL)
      IF(ABEROR.LE.XACCUR) GO TO 470
      IF(FSTAR.LT.FBIG) GO TO 350
      XSTAR=XSMALL
      GO TO 470
350   IF(FBIG.NE.FB) GO TO 370
      GO TO 470
370   IF(FBIG.EQ.FC) GO TO 380
      AB1=ABS(XC-XB)-ABS(XC-XSTAR)
      IF(AB1) 390,400,400
380   AB1=ABS(XA-XB)-ABS(XA-XSTAR)
      IF(AB1) 400,390,390
390   XC=XB
      FC=FB
      XB=XSTAR
      FB=FSTAR
      GO TO 410
400   XA=XB
      FA=FB
      XB=XSTAR
      FB=FSTAR
410   ANUM=(XB**2-XC**2)*FA+(XC**2-XA**2)*FB+(XA**2-XB**2)*FC
      DENOM=(XB-XC)*FA+(XC-XA)*FB+(XA-XB)*FC
      XSTAR=(ANUM/DENOM)/2.
      IF(XSTAR.LT.0.) XSTAR=PREOPT
      IF(L-2.GE.0.AND.XSTAR/T.LT.0.0145) XSTAR=PREOPT
      IF(L-2) 430,440,450
430   FSTAR=ERROR(XSTAR,T1,T2)
      K=XSTAR
      IF(FSTAR.GE.FREOPT) K=PREOPT
      GO TO 460
440   FSTAR=ERROR(K,XSTAR,T2)
      T1=XSTAR
      IF(FSTAR.GE.FREOPT) T1=PREOPT
      GO TO 460
450   FSTAR=ERROR(K,T1,XSTAR)
      T2=XSTAR
      IF(FSTAR.GE.FREOPT) T2=PREOPT
460   IF(K.EQ.PREOPT) GO TO 470
      IF(T1.EQ.PREOPT) GO TO 470
      IF(T2.EQ.PREOPT) GO TO 470
      GO TO 255
470   IF(L-2) 218,235,480
C
C     THIS SECTION OF THE PROGRAM DETERMINES IF ALL 3 OF
C       THE VARIABLES ARE WITHIN THE PRESCRIBED ACCURACY
C       (YACCUR) OF THE PREVIOUS CYCLE VALUES. IF THEY
C       ARE NOT THEN THE OPTIMIZATION CYCLE IS REPEATED.
480   IC=IC+1
      YACCUR=.001
      IF(ABS(K-P1).GT.YACCUR) GO TO 170
      YACCUR=YACCUR*2.
      IF(ABS(T1-P2).GT.YACCUR) GO TO 170
      IF(ABS(T2-P3).GT.YACCUR) GO TO 170
      TYPE 490
490   FORMAT('0OPTIMUM VALUES')
      WRITE(5,500) K,T1,T2
500   FORMAT( 3(F15.6))
      STOP
      END
```

Appendix E (*continued*)
TIME DOMAIN PROCESS IDENTIFICATION PROGRAM

```
C
C
C
C       THIS SUBPROGRAM CALCULATES "ERROR" WHICH IS
C         A MEASURE OF HOW WELL THE VALUES OF K,T1,AND
C         T2 FIT THE DATA.
C       XP(I)=THE PREDICTED PROCESS OUTPUT
C
        FUNCTION ERROR(K,T1,T2)
        REAL K
        COMMON X(100),U(100),XP(100),F(100)
        COMMON INUM,T,N,J,HGHT
        ERROR=0.0
        XP(9)=0.0
        XP(10)=0.0
        A1=EXP(-T/T1)+EXP(-T/T2)
        A2=EXP(-T/T1)*EXP(-T/T2)
        B1=K*(1.-A1+(T1*EXP(-T/T2)-T2*EXP(-T/T1))/(T1-T2))
        B2=K*(A2-(T1*EXP(-T/T2)-T2*EXP(-T/T1))/(T1-T2))
        DO 900 I=11,INUM+10
890     XP(I)=A1*XP(I-1)-A2*XP(I-2)+B1*U(I-N-1)+B2*U(I-N-2)
        E=X(I)-XP(I)
        ERROR=ERROR+E*E
900     CONTINUE
        RETURN
        END
C
C
C
C       DSC SEARCH (CONTINUED)
C       THIS SUBROUTINE FINDS "XSOPT(L)" WHICH IS THE PSEUDO-
C         OPTIMUM VALUE OF THE PARTICULAR VARIABLE BEING
C         SEARCHED.
C       L=1   K(SEARCHED VARIABLE)
C       L=2   T1(SEARCHED VARIABLE)
C       L=3   T2(SEARCHED VARIABLE)
C
        SUBROUTINE DSC(L)
        REAL K
        COMMON X(100),U(100),XP(100),F(100)
        COMMON INUM,T,N,J,HGHT
        COMMON XAOPT(3),XBOPT(3),XCOPT(3),XSOPT(3)
        COMMON FAOPT(3),FBOPT(3),FCOPT(3),FSOPT(3)
        COMMON XMM2,XMM1,XM,XMP1,FMM2,FMM1,FM
        COMMON FMP1,DK,K,T1,T2
        COMMON PREOPT,FREOPT
        IF(FMM1.GE.FMP1) GO TO 600
        XAOPT(L)=XMM2
        XBOPT(L)=XMM1
        XCOPT(L)=XMP1
        FAOPT(L)=FMM2
        FBOPT(L)=FMM1
        FCOPT(L)=FMP1
        GO TO 610
600     XAOPT(L)=XMM1
        XBOPT(L)=XMP1
        XCOPT(L)=XM
        FAOPT(L)=FMM1
        FBOPT(L)=FMP1
        FCOPT(L)=FM
610     ANUM=(DK/2.)*(FAOPT(L)-FCOPT(L))
        DENOM=2.*(FAOPT(L)-2.*FBOPT(L)+FCOPT(L))
        XSOPT(L)=XBOPT(L)+(ANUM/DENOM)
        FSOPT(L)=1E+6
        IF(XSOPT(L).LE.0.) GO TO 670
        IF(L-2.GE.0.AND.XSOPT(L)/T.LT.0.0145) GO TO 670
```

Appendix E (*concluded*)
TIME DOMAIN PROCESS IDENTIFICATION PROGRAM

```
      IF(L-2) 630,640,650
630   FSOPT(L)=ERROR(XSOPT(L),T1,T2)
      GO TO 660
640   FSOPT(L)=ERROR(K,XSOPT(L),T2)
      GO TO 660
650   FSOPT(L)=ERROR(K,T1,XSOPT(L))
660   IF(FSOPT(L).LE.FREOPT) GO TO 690
670   XSOPT(L)=PREOPT
690   RETURN
      END
```

Appendix E1

Data Set Listings—RASDA1

```
RASDA1
.5
.5
3.Ø
15
1.5
1.5
1.5
1.5
1.5
Ø.
Ø.
Ø.
Ø.
Ø.
Ø.
Ø.
Ø.
Ø.
Ø.
Ø.
Ø.
.Ø5
.18
.22
.3Ø

RASDA1
.34
.35
.25
.2Ø
.13
.1Ø
.Ø8
.Ø5
.Ø3
.Ø2
Ø.Ø

RASDA2
.5
.5
3.
19
-1.5
-1.5
-1.5
-1.5
-1.5
Ø.
Ø.
Ø.
Ø.

RASDA2
Ø.
Ø.
Ø.
Ø.
Ø.
Ø.
Ø.
Ø.
Ø.
Ø.
Ø.
Ø.
-.Ø4
-.15
-.2
-.3
-.37
-.38
-.35
-.27
-.23
-.2Ø
-.18
-.17
-.12

RASDA2
-.1
-.1
-.Ø6
-.Ø5
-.Ø4
Ø.Ø

RASDA3
6.
5.
1.
3Ø
1.5
1.5
1.5
1.5
1.5
1.5
1.5
1.5
1.5
1.5
1.5
1.5
1.5
1.5
```

Appendix E1 (*continued*)
DATA SET LISTINGS—RASDA3 (*continued*)

```
RASDA3   1.5
         1.5
         1.5
         1.5
         1.5
         1.5
         1.5
         1.5
         1.5
         1.5
         1.5
         1.5
         1.5
         1.5
         1.5
         1.5
         1.5
         0.
         1.0
         2.0
         3.3
         4.25
         5.0
         5.5
         6.3
         7.0
         7.2
         7.5
         7.75
         8.2
         8.5
         8.7
         8.8
         8.9
         9.05
         9.15
         9.25
         9.35
         9.45
         9.5
         9.6
         9.7
         9.75
         9.8
         9.9
         9.9
         9.95
         10.0

RASDA 4  6.
         5.
         1.
         40
         1.5
         1.5
         1.5
         1.5
         1.5
         1.5
         1.5
         1.5
         1.5
         1.5
         1.5
         1.5

RASDA 4  1.5
         1.5
         1.5
         1.5
         1.5
         1.5
         1.5
         1.5
         1.5
         1.5
         1.5
         1.5
         1.5
         1.5
         1.5
         1.5
         1.5
         1.5
         1.5
         1.5
         1.5
         1.5
         1.5
         1.5
         1.5
         1.5
         1.5
         1.5
         1.5
         0.
         .3
         1.79
         2.65
         3.46
         4.20
         4.85
         5.41
         5.9
         6.31
         6.65
         6.94
         7.18
         7.38
         7.54
         7.68
         7.79
         7.88
         7.96
         8.02
         8.07
         8.12
         8.15
         8.17
         8.2
         8.22
         8.24
         8.25
         8.26
         8.27
         8.28
         8.29
         8.29
         8.30
         8.30
         8.30

RASDA4   8.30
         8.31
         8.31
         8.31
RASDA5
         6.0
         5.0
         1.0
         44
         1.5
         1.5
         1.5
         1.5
         1.5
         0.
         0.
         0.
         0.
         0.
         0.
         0.
         0.
         0.
         0.
         0.
         0.
         0.
         0.
         0.
         0.
         0.
         0.
         0.
         0.
         0.
         0.
         0.
         0.
         0.
         0.
         0.
         0.
         0.
         0.
         0.
         0.
         0.
         0.
         0.
         0.
         0.
         0.
         0.
         0.
         0.
         .297
         .969
         1.79
         2.65
         3.46
         3.90
         3.88
         3.62
         3.25
         2.85

RASDA5   2.45
         2.09
         1.76
         1.48
         1.23
         1.03
         .851
         .705
         .583
         .481
         .397
         .327
         .269
         .222
         .183
         .150
         .124
         .102
         .084
         .069
         .057
         .047
         .038
         .032
         .026
         .021
         .018
         .014
         .012
         .010
         .008
         .007
         .005
         0.
```

Index

STATE OF MICHIGAN
JUNG S.
PARK
ENGINEER
NO.
30616
LICENSED PROFESSIONAL ENGINEER